Previous books published in the Perspectives on Cognitive Science series:

Creativity, Cognition and Knowledge — An interaction
Terry Dartnall

Language Universals and Variation
Mengistu Amberber and Peter Collins

Perspectives on Cognitive Science, Vol 2 — Theories, Experiments, and Foundations
Janet Wiles and Terry Dartnall

Perspectives on Cognitive Science, Vol 1 — Theories, Experiments, and Foundations
Peter Slezak, Terry Caelli and Richard Clark

Representation in Mind: New Approaches to Mental Representation
Hugh Clapin, Phillip Staines and Peter Slezak

Perspectives on Cognitive Science
Series editor: Peter Slezak

COMPETITION AND VARIATION IN NATURAL LANGUAGES

THE CASE FOR CASE

EDITED BY

MENGISTU AMBERBER
Linguistics Department, The University of New South Wales, Sydney, Australia

HELEN DE HOOP
Linguistics Department, Radboud University Nijmegen, Nijmegen, The Netherlands

2005

ELSEVIER

Amsterdam – Boston – Heidelberg - London – New York – Oxford
Paris – San Diego – San Francisco – Singapore – Sydney – Tokyo

ELSEVIER B.V.
Radarweg 29
P.O. Box 211, 1000 AE
Amsterdam, The Netherlands

ELSEVIER Inc.
525 B Street, Suite 1900
San Diego, CA 92101-4495
USA

ELSEVIER Ltd
The Boulevard, Langford Lane
Kidlington, Oxford OX5 1GB
UK

ELSEVIER Ltd
84 Theobalds Road
London WC1X 8RR
UK

First edition 2005

British Library Cataloguing in Publication Data
A catalogue record is available from the British Library.

ISBN: 0-08-044651-5

∞ The paper used in this publication meets the requirements of ANSI/NISO Z39.48-1992 (Permanence of Paper).
Printed in The Netherlands.

Contents

Volume Editors

Mengistu Amberber (Ph.D., McGill) is a Senior Lecturer in Linguistics at the University of New South Wales, Sydney. His main research interest is the (lexical) semantics–syntax interface, with particular reference to transitivity and valence. He is the author of *Verb Classes and Transitivity in Amharic* (Lincom Studies in Afro-Asiatic Linguistics, 06, 2002), and (with P. Collins) editor of *Language Universals and Variation* (Praeger, 2002). His other recent publications include: 'The grammatical encoding of Thinking in Amharic', *Cognitive Linguistics* 14-2/3 (2003); and book chapters: 'Testing emotional universals in Amharic' (in *Emotions in Crosslinguistic Perspective*, Mouton de Gruyter, 2001), 'Valency-changing and valency-encoding devices in Amharic' (in *Changing Valency: Case Studies in Transitivity*, Cambridge University Press, 2000). He has also published in the *Proceedings of the Berkeley Linguistics Society and the West Coast Conference in Formal Linguistics.*

Helen de Hoop (Ph.D., University of Groningen) is an Associate Professor in Linguistics at the Radboud University Nijmegen, The Netherlands. Her research interests include case, interpretation, and Optimality Theory. Since 2002 she is the principal investigator of the PIONIER research project *Case cross-linguistically* which integrates theoretical, typological, and processing approaches to the syntax and semantics of case (financed by NWO, the Netherlands Organisation of Scientific Research). She is also a co-project leader of the research project *Conflicts in Interpretation*, financed by the Cognitive Science program of NWO. Her Ph.D. dissertation Case configuration and noun phrase interpretation appeared in the Garland series Outstanding Dissertations in Linguistics in 1996. She has recently published (co-authored) articles in the journals *Linguistics and Philosophy*, *Journal of Semantics*, and *Lingua*, and together with Reinhard Blutner and Petra Hendriks she co-authored a book entitled *Optimal Communication* that will appear with CSLI, Stanford, in 2005.

Contributors

Cynthia L. Allen	School of Language Studies, Australian National University, Canberra, Australia
Mengistu Amberber	Linguistics Department, School of Modern Language Studies, The University of New South Wales, Sydney, Australia
Alice Gaby	Max Planck Institute for Psycholinguistics, Nijmegen, The Netherlands
Rik van Gijn	Department of Linguistics, Radboud University Nijmegen, Nijmegen, The Netherlands
Helen de Hoop	Department of Linguistics, Radboud University Nijmegen, Nijmegen, The Netherlands
Monique Lamers	Department of Linguistics, Radboud University Nijmegen, Nijmegen, The Netherlands
Andrej L. Malchukov	Department of Linguistics, Radboud University Nijmegen, Nijmegen, The Netherlands
Bhuvana Narasimhan	Max Planck Institute for Psycholinguistics, Nijmegen, The Netherlands
Ryo Otoguro	Department of Language & Linguistics, University of Essex, Colchester, UK

Andrew Spencer — Department of Language & Linguistics, University of Essex, Colchester, UK

Peter de Swart — Department of Linguistics, Radboud University Nijmegen, Nijmegen, The Netherlands

Lisa deMena Travis — Department of Linguistics, McGill University, Montréal, Canada

Mario van de Visser — Utrecht Institute of Linguistics OTS, Utrecht University, Utrecht, The Netherlands

Preface

First of all, we wish to thank all our contributors for making this volume possible.

Our heartfelt thanks go to the anonymous reviewers who assisted us in reviewing the chapters. Their constructive and helpful criticisms have been invaluable in improving the volume. Remaining errors and shortcomings of course remain with the authors and us.

Many thanks go to Peter Slezak, the Series Editor of *Perspectives on Cognitive Science*, for his encouragement and support throughout the development of this project.

Thanks also to Fiona Barron, our acquisition editor at Elsevier, for her diligent and professional help.

Special thanks go to Sander Lestrade for his indispensable help in preparing the final manuscript for publication.

Helen de Hoop gratefully acknowledges the support of the Netherlands Organisation of Scientific Research (grants 220-70-003 for the PIONIER project *Case cross-linguistically* and 051-02-070 for the NWO Cognition project *Conflicts in interpretation*).

Mengistu Amberber
Sydney

Helen de Hoop
Nijmegen

February, 2005

Chapter 1

Competition and Variation in Natural Languages: The Case for Case

Mengistu Amberber and Helen de Hoop

> I am going to suggest below that there are many semantically relevant syntactic relationships involving nouns and the structures that contain them, that these relationships are in large part covert but are nevertheless empirically discoverable, that they form a specific finite set, and that observations made about them will turn out to have considerable cross-linguistic validity. I shall refer to these as 'case' relationships.
>
> Fillmore (1968)

Suppose a person witnessed the killing of Frank by Harry and wishes to report on that. When the speaker chooses to express this as *Harry killed Frank*, then it seems important that a hearer will understand the chosen expression correctly. In particular, the hearer should be able to distinguish between an interpretation such that Harry is the killer and Frank the victim and an interpretation according to which Frank would be the killer and Harry the victim. In English, the distinction is obviously made with the help of word order. Other languages display different strategies, one of them being the use of morphological case. For example, in some languages the object is accusatively case-marked and/or the subject is case-marked with ergative case. In German, morphological case overrules the effect of word order. Hence, *Den Harry hat der Frank getötet* 'The$_{\text{ACC}}$ Harry killed the$_{\text{NOM}}$ Frank' can only mean that Frank killed Harry.

Competition and Variation in Natural Languages: The Case for Case
Edited by M. Amberber and H. de Hoop

ISBN: 0-08-044651-5

Cases manifest and disguise themselves differently in the languages of the world. The present book deals with competition and variation among cases and case-marking devices. Why case? Case is a linguistic phenomenon that (i) plays a key role in many different languages, (ii) is very diverse in its manifestations, (iii) involves both form and meaning elements, and (iv) cannot be reduced to non-linguistic cognitive capacities. In this introductory chapter we provide a brief synopsis of each chapter by highlighting some of the more salient hypotheses and findings.

1. Variation

A simple division of case-realizing languages into nominative–accusative and ergative–absolutive is insufficient. One cross-linguistic generalization is that the richer the case system, the more (specific) information may be conveyed by case. We observe case-alternations that are much more fine-grained than the well-known alternations between subjects and objects or agents and patients. Examples of such alternations are the one between volitional and non-volitional agents, specific versus non-specific objects, or between pronouns and other nominal constituents. In general, morphological case reflects all kinds of semantic and pragmatic information, grammatical role being only one of them. This may seem obvious for all kinds of lexical or specific cases, but it holds for the core cases (nominative, accusative, ergative) as well. One clear piece of evidence in this respect is that elements that do not bear any grammatical function may get case under the same conditions as arguments of a predicate.

In Kuuk Thaayorre (a Paman language of Queensland, Australia), there is a split between the pronominal paradigm that exhibits a nominative–accusative marking pattern and other nominals that show an ergative–absolutive marking pattern. But things are more complicated than that. As **Alice Gaby** points out in her chapter on **case and the composition of arguments in Kuuk Thaayorre**, it is not very likely that each participant in an event will be imbued with the same degree of importance or emphasis, either with respect to the particular event described, or across a larger stretch of discourse. In Kuuk Thaayorre, the subtleties of the relationships between participants, and the different roles they play in a group action, are largely expressed through case-marking. Gaby discusses a range of Thaayorre constructions in which part and whole of an entity or of a group of participants are encoded by separate noun phrases. In all of these instances, case-marking marks a relationship of dependency between nominal constituents with overlapping reference. The chapter by Gaby also considers one usage of a particular case morpheme, the ergative suffix, that is argued to have a pragmatic function. The different uses of case markers are united by their role in identifying and tracking the relationships between multiple participants (or multiple parts of a single participant) involved in an event.

The central claim of **Rik van Gijn** is that not only argument structure but also topicality influence the split between **head marking and dependent marking of grammatical relations in Yurakaré** (an unclassified language spoken in central Bolivia). Subject and object relations are always head marked, while outer adverbials are always dependent marked. But for most categories in the middle we find both head marking and dependent marking. Highly topical arguments are head marked. The factor animacy plays an important role. Most of the head marked applied objects are animate (often human), while among the dependent marked participants only a few are animate. Accompanying objects are always animate, but comitative objects can be either animate or inanimate, while they can be head marked or dependent marked. However, the dependent marked comitatives display a lower degree of pragmatic salience, they are less topical. To sum up, wherever there is competition and variation between head marking and dependent marking, it can be argued that the degree of topicality determines which participants get to be marked on the head and which are dependent marked.

2. Universals

Despite the overwhelming variation in case patterns cross-linguistically, typological investigations have also shown remarkably similar interdependencies between case oppositions and interpretation in genetically unrelated languages.

Andrej Malchukov shows how a small set of functionally based constraints can account for **case pattern splits of particular verb types**. The purpose of the chapter is to construct a universally valid hierarchy that can explain and predict case-marking variation conditioned by the verb class of the predicate. The starting point of the investigation is a scale that stretches from more to less transitive verb types. For example, a verb that encodes an effective action is typically more transitive than a verb that encodes feeling. Some verb types in the hierarchy are subdivided into further subclasses. Thus, verbs of effective action come in two types, resultatives that include canonical transitives like *kill* as well as non-resultatives such as *hit*. There seem to be cross-linguistic preferences for particular verb types to select for particular case frames. Malchukov decomposes the verb type hierarchy into two different dimensions: a (sub-)hierarchy of decreased patienthood of the object (a path that leads from *break* to *go*) and another (sub-)hierarchy that additionally involves decreased agentivity on the part of the subject (a path that leads from *break* to *freeze*. Malchukov develops an Optimality Theoretic analysis of the cross-linguistic variation in case-marking of arguments for individual verb types in terms of a number of universal constraints on case-marking.

One of the puzzling aspects of case is that a language can have a morphological (inflectional) case system that is not completely congruent to its system of syntactic case relationships. **Andrew Spencer and Ryo Otoguro** tie together a number of observations in their **critical survey of the notion 'case'** in order to deepen our understanding of the nature of case. A language can have grammatical markers to which we conveniently give the label of 'case,' but in many instances no generalizations are lost if we would simply refer directly to form-based categories such as 'the *de*-suffixed form of the word.' For instance, in languages in which case does not enter into an agreement system, there seems to be no motivation for setting up a special syntactic feature of case. However, there are languages which show case agreement or other syntactic processes appealing to a syntactic property of case. A clear sign of the need to distinguish phrase-level syntactic case of this sort from word level morphological case is the phenomenon of 'non-autonomous case' where we can define a syntactic case that has no unique morphological correspondent. Within this metagrammatical perspective on case, it becomes less puzzling to see so many non-canonical types of case-marking, whether at the level of form or function. As a central testing ground for their approach, Spencer and Otoguro offer a detailed re-examination of 'case' in Japanese. They defend the view that there is no justification of adopting traditional case labels for any aspect of the Japanese language. They present a battery of empirical arguments, which collectively show that there is no need in Japanese to set up abstract case features such as nominative, accusative, genitive, dative, etc. These include showing that the nominative case particle *na* can be dropped or replaced by adverbial particles of various discourse functions (such as contrastive focus, topicalization).

One well-known characteristic of Western Malayo-Polynesian languages such as Malagasy and Tagalog is their complex voice system that allows for a range of noun phrases to become a subject (or topic), i.e., the unmarked argument. **Lisa deMena Travis** in her chapter on **case in predicate initial languages** investigates the role of abstract nominative case and proposes a language typology that is created through a feature-based system of movement as outlined in the Minimalist Program. Languages like English check nominal and verbal features of the functional domain through maximal constituent and head movement, respectively, whereas languages like Malagasy do not have maximal noun phrase movement. Therefore nominative case, if anything, would relate to head movement. However, Travis strongly suggests that in this type of language case is not a syntactic force for movement at all. The relevant arguments in these languages are neither subjects nor topics but are noun phrases base-generated in a clitic-left dislocated structure.

While Basque is often cited as a typical example of an ergative language, **Mario van de Visser** argues that underlyingly **Basque exhibits a nominative-accusative system**, just like English. The only difference is that transitive sentences in Basque are obligatorily realized as passive constructions, resulting in an ergative case system and an agreement system which is only seemingly ergative. What looks like agreement with the transitive subject is argued to be the result of a process of pronoun incorporation. Van de Visser argues in favor of a universal treatment of case and agreement, which allows arguments to be licensed by either case or agreement. In his view, a main advantage of his analysis over other accounts is the fact that it relates ergative patterns to a construction that occurs independently in non-ergative languages as well: the passive. In answer to the question why Basque should lack an active counterpart altogether, he suggests that the passive became widely used and eventually took over the function of the active construction in Basque.

3. Competition

Manifestations of case-marking in expressions can be motivated in terms of maximization of contrasts for ease of comprehension from the point of view of the hearer. If an object has a case separate from that of the subject, it is easier to determine what is what. On the other hand, minimization of effort from the point of view of the speaker may lead to a reduction of morphological variability, hence a loss of case. If there are no case endings, a speaker need not produce them. Obviously, languages differ in the balance they have found between these different maximization (avoid ambiguity) and minimization (economy) strategies. This can also be dependent on the nature of the semantic or grammatical information that is reflected by the case morphology. That is, it has been noted in the literature that those cases that are most relevant from a semantic point of view in that they carry non-redundant (very specific) semantic information tend to be retained in morphology (i.e., they resist neutralization or deletion). There is an obvious cognitive motivation for such a tendency, as it aids communication and helps to avoid ambiguity.

What role, if any, does the avoidance of ambiguity play in the assignment of case to the arguments of a transitive predicate? **The correlation between case and ambiguity** is investigated from a broad cross-linguistic perspective by **Peter de Swart**. Languages that do not mark grammatical functions explicitly on their noun phrases have a greater chance of ambiguity of grammatical relations. De Swart argues that due to this danger of potential ambiguity, these languages place restrictions on constructions with noun phrases in order to increase the success of communication. In languages that have morphological case, the danger of ambiguity is less likely to occur. Case itself is a means of ambiguity avoidance. But of course, noun phrase ambiguity is not restricted to languages without case. When case markers disappear in daily speech or because of case syncretism, languages tend to place a strong restriction on the possible word order.

Cynthia Allen shows that **English underwent a major typological shift** from a case-marking language to a 'fixed word order' language in Early Middle English. This chapter offers a case study into the implementation of this change, focusing particularly on the disappearance of one type of genitive construction. Allen argues against one particular analysis of this fact, namely that the disappearance is the result of the loss of the genitive as a case, itself caused by the loss of the accusative/dative distinction. Allen argues that a closer look at the facts reveals that this analysis is not only incorrect but also fails to predict some basic facts. She then argues that the decline of the post-head genitives originated in the increasing tendency to rely on adpositions and a probabilistic processing strategy to decode grammatical relations. This tendency is apparent in Old English also with sentence-level constituents, although at that stage, morphology was decisive when there was any conflict with processing strategy. The processing strategy account has the advantage of allowing for the variation that we find during the long period of typological shift.

Across languages and even across certain configurations in languages, similar semantic or grammatical features may be expressed in syntactically diverse ways. What is expressed by case morphology in one language, may be reflected by word order, agreement features, voice alternations or cross-reference in other languages. In order to accomplish the task of understanding the meaning of a sentence, there are several sources of information that readers and listeners use: pragmatic, semantic, and syntactic information. This is not at all an easy task, but people seem to do it very rapidly and without much effort. A question might be what is the difference in processing between a language without case and a language with case. And in case of a competition between different types of information, which type is most likely to win? The chapter by **Monique Lamers** on **the on-line resolution of subject–object ambiguities with and without case marking in Dutch** presents psycholinguistic evidence, which can be brought to bear on the investigation of how potential ambiguity between subjects and objects is resolved. Lamers reports research that makes effective use of Event Related brain Potentials to examine the resolution of ambiguity with and without case-marking with particular reference to Dutch. She concludes that not only case information, but also animacy information is used in resolving subject-object ambiguities in Dutch.

The relation between morphological case and ambiguity is probably most clear from languages with differential case-marking. Two chapters in particular — those by Amberber and De Hoop and Narasimhan — provide a detailed discussion of case marking with particular reference to the phenomenon of differential case-marking. **Mengistu Amberber** examines **differential subject marking in Amharic**. He focuses on the semantic factors that determine differential subject marking, the interaction between differential subject marking and agreement on the verb and the interaction between differential subject marking and topicality. Amberber shows that differential subject marking in Amharic is found with a semantically homogeneous class of verbs — broadly construed as experiencer predicates. He argues that intransitive experiencer predicates such as *worry* are conceptually more complex, with an abstract causer argument that agrees with the subject pronominal suffix, while the experiencer argument agrees with the object pronominal suffix on the verb. The latter fact is accounted for with a movement analysis where the experiencer argument is left-dislocated into an adjunct position. The pronominal suffix on the verb is then analyzed as an anaphoric pronoun rather than a mere agreement marker. Amberber's study shows the complex interaction between the three coding strategies of grammatical relations — case, verbal agreement, and word order — in the manifestation of the so-called non-canonical subjects.

On a related theme, **Helen de Hoop and Bhuvana Narasimhan** attempt to account for **differential case-marking in Hindi**. In Hindi, a certain class of transitive predicates allows for case alternations on the subject as well as case alternations on the object. De Hoop and Narasimhan show that differential object marking in Hindi can be explained in terms of a general principle of ambiguity avoidance with respect to the subject and the object. But they argue against the idea that differential case-marking on subjects can also be motivated by a principle of ambiguity avoidance. If it were, one would expect object-like subjects to be case-marked. By contrast, ergative case is assigned to 'prototypical' subjects. Situating their analysis within the theoretical framework of Optimality Theory, they show that two different functions of case, an economy constraint that penalizes morphological case, and the strength of the argument under consideration, together characterize the complex case system in transitive clauses in Hindi.

4. Conclusion

The aim of this volume is to combine different perspectives on case-marking: (1) typological and descriptive approaches of various types and instances of case-marking in the languages of the world as well as comparison with languages that express similar types of relations without morphological case-marking; (2) formal analyses in different theoretical frameworks of the syntactic, semantic, and morphological properties of case-marking; (3) a historical approach of case-marking; (4) a psycholinguistic approach of case-marking: how is case being processed?

Although there are a number of publications on case-related issues, to the best of our knowledge, there is no volume like the present one, which exclusively looks at case-marking, competition, and variation from a cross-linguistic perspective and within the context of different contemporary theoretical approaches to the study of language. We hope the present volume will be of interest to researchers and advanced students in cognitive science, general linguistics, typology, historical linguistics, formal linguistics, and psycholinguistics. The book will interest scholars working within the context of formal syntactic and semantic theories as it provides insight into the properties of case from a cross-linguistic perspective. We believe the book also will be of interest to cognitive scientists interested in the relationship between meaning and grammar, in particular, and the human mind's capacity in the mapping of meaning onto grammar, in general.

References

Fillmore, C. J. (1968). The case for case. In: E.Bach & R.T. Harms (Eds), Universals of linguistic theory. New York: Holt, Rinehart & Winston.

Chapter 2

Some Participants are More Equal than Others: Case and the Composition of Arguments in Kuuk Thaayorre

Alice Gaby

1. Introduction

1.1. Similarities and Differences between Coparticipants

In any event involving a plural number of participants, it is unlikely that these participants are engaged in exactly the same way, throughout each moment of the event. Nor is it likely that each of the participants will be imbued with the same degree of importance or emphasis, either with respect to the particular event described, or across a larger stretch of discourse. This chapter surveys a range of strategies available to the Thaayorre speaker for emphasising individuals within a larger group. After giving some background as to the morphosyntax of case-marking and the structure of the noun phrase in Section 2, Section 3 presents two constructions that involve the apposition of multiple noun phrases, the overlapping reference of which is signalled by referential case. It draws parallels between the apposition of noun phrases referring to part and whole of a single entity, and those referring to part and whole of a participant group. Section 3 explores constructions in which noun phrases referring to the participant group and subsets thereof are marked by different cases. Finally, I conclude that it is largely through case marking that the subtleties of the relationships between participants, and the different

Competition and Variation in Natural Languages: The Case for Case
Edited by M. Amberber and H. de Hoop

ISBN: 0-08-044651-5

roles they play in a group action, find expression in alternative Thaayorre argument structures.

1.2. Kuuk Thaayorre and its Speakers

Kuuk Thaayorre is a Paman language of Queensland, Australia. Its more precise genetic affiliation has not yet been conclusively established, although it has been classified as belonging to the Western Paman branch by Grimes and Grimes (2000) and Dixon (2002; p.xxxii) offers the areal classification Ea1 (upper southwest Pama group of the Western Cape York Peninsula areal group). Whatever its heritage, it is clear that there has been enduring and extensive contact between the Thaayorre people and the Yir Yoront (to the south), and the Kugu Nganhcara patrilects (to the north). This has left its mark on the language in the form of mutual loan words and calques.

Kuuk Thaayorre is the traditional language of more than 300 people, most of whom live in or around the community of Pormpuraaw (located on the west coast of Cape York Peninsula). Originally an Anglican mission, Pormpuraaw lies in the north west of the Thaayorre traditional territory, and many families divide their time between it and outstations on their own land. Many Thaayorre now use English as their primary means of communication, and few children are fully fluent in the language. The majority of Kuuk Thaayorre speakers are also fluent, to a greater or lesser degree, in a number of other languages indigenous to the region (notably the Kugu Nganhcara patriletcs, Yir Yoront, Pakanh, Wik Mungkan and Olkolo).

Most of the data reported here were collected by the author over three field trips between 2002 and 2004. Remaining data were obtained from published and unpublished materials prepared by Tom Foote and Allen Hall, and made available to me by the Pormpuraaw Community Council. Examples drawn from the work of Foote and Hall have been modified to fit the practical orthography used throughout this chapter, and in many cases have been glossed and translated by the present author. Sources other than my own field notes are acknowledged following the relevant examples.

2. Relevant Linguistic Features of Kuuk Thaayorre

2.1. Case and Case-Marking

Kuuk Thaayorre is a predominantly dependent-marking language, in which grammatical function is signalled solely by nominal/pronominal case forms. The Thaayorre case system distinguishes the grammatical functions of intransitive subject, transitive subject and transitive object. In terms of morphological form, however, there is a split between the pronominal paradigm (in which the syncretism of nominative and ergative case forms results in a nominative-accusative marking pattern) and other nominals (in which nominative/accusative syncretism gives rise to a so-called ergative-absolutive marking pattern). This can be seen in Table 1.

Formal identity between case suffixes exists across many non-core cases as well. For example, whilst inflection of the noun *pam* 'man' distinguishes genitive, dative and locative cases, the same suffix (*-thak*) marks all three of these cases for *peep* 'net'. Similarly, singular pronouns have different forms for accusative and genitive cases, where non-singular (dual and plural) pronouns have a single form. Does this mean that the genitive case exists for singular but not plural pronouns (for which the accusative case marks possession)? Or that *pam* 'man' may be in locative case but *peep* 'net' may not? Or that a predicate will assign ergative case to a noun but nominative case to a coreferential pronoun? The importance of separating case categories from both grammatical function and morphological form has been stressed by Mel'čuk (1979), Goddard (1982), Blake (1994) and Spencer (2003). Following these, I maintain that any distinction overtly marked by some subpart of the case system, should be analytically maintained throughout. Hence, the accusative/genitive distinction is maintained across both singular and plural pronouns (as well as nouns), just as is the genitive/dative/locative distinction and the distinction between ergative, nominative and accusative cases.

Table 1: Comparative case forms of a pronoun and a noun.

	Nouns		**Free pronouns**		**Clitic Pronouns**
Case	***pam* 'man'**	***peep* 'net'***	***nhul* 'he'**	***peln* 'they'**	**=*ul* 'he'**
Ergative	*pam-al*	*peep-nthrr*	*nhul*	*peln*	=*ul*
Nominative	*pam*	*peep*	*nhul*	*peln*	=*ul*
Accusative	*pam*	*peep*	*nhunh*	*pelnan*	=*unh*
Genitive	*pam-ak*	*peep-thak*	*nhangan*	*pelnan*	–
Dative	*pam-a*	*peep-thak*	*nhangun*	*pelnungun*	–
Locative	*pam-an*	*peep-thak*	*nhangun*	*pelnungun*	–
Ablative	*pam-am*	*peep-mam*	*nhanganma*	*pelnantam*	–
Instrumental	*pam-al*	*peep-an*	–	–	–

*Some of the case forms of *peep* 'net' were obtained from Hall (1972, p. 241).

Only one case suffix may appear in any single noun phrase, and it is always attached to the final eligible element of the noun phrase. Accordingly, in (1) it is the quantifier *mong* 'many' that inflects for ergative case, since it follows the head noun *kuta* 'dog', with which it forms a phrase:

(1) *nganh* *kuta* *mong-thurr* *patharr*
1sgACC dog many-ERG bit
'many dogs have bitten me'

The fact that Thaayorre case marking appears at the periphery of the noun phrase, suggests that the case morphs are postpositional enclitics. Yet their formal irregularity is more characteristic of inflectional affixes. Compare, for example, the forms of the ergative morphs given in Table 2.

As such lexically determined allomorphy cannot be reconciled with an analysis of the ergative morpheme as an independent grammatical unit, I will continue to refer to the ergative markers as (inflectional) case suffixes. This analysis is given weight by Zwicky's (1987) analysis of phrasal affixes in English. It does seem likely, however, that the Thaayorre ergative morpheme is in the process of becoming an enclitic.[1] A single form *-(n)thurr* ERG is used productively with almost all

Table 2: Comparison of nominative and ergative forms of 10 Thaayorre nouns.

	Nominative	**Ergative**
'meat'	*minh*	*minh-al*
'good'	*min*	*min-thurr*
'tooth'	*kiin*	*kiin-a*
'dog'	*kuta*	*kuta-ku*
'woman'	*paanth*	*paanth-u*
'eye'	*meer*	*meer-e*
'cat'	*thok*	*thok-un*
'saltwater croc.'	*pinj*	*pinj-i*
'child'	*parr_r* *	*parr-an*
'boomerang'	*werngr*	*werng-arr*

*Note that an underscore is used in the Thaayorre practical orthography in order to disambiguate a sequence of rhotics (i.e. in *parr_r* 'child' the alveolar tap/trip II represented by *rr* is immediately followed by the retroflex *r*).

[1] The term *grammaticalisation* does not seem appropriate to describe this process, since it is usually assumed that affixes are more grammatical than clitics. The shift from affix to enclitic would therefore seem to represent reverse grammaticalisation.

loan nouns, and most younger speakers are using it in place of the original (irregular) ergative suffixes of an increasing number of Thaayorre nouns. Such regularisation would bring ergative marking more in line with other case markers, the forms of which are sufficiently regular to be considered enclitics.

2.2. The Noun Phrase

There are several clear indicators of the existence of a noun phrase in Kuuk Thaayorre (unlike some other Australian languages). These include: (a) case marking; (b) internal word order; and (c) prosody. As mentioned above (and exemplified by (1)), case suffixes are attached only to the final eligible constituent of the noun phrase. The same is visible in (2):

(2)	*ngan*	*pumun*	*ngathan-thurr*	*kuta*	*theerngarr*
	RELATIVE	yB	1sgGEN-ERG	dog.ACC	hit

‘my younger brother hit a dog’

Although clausal constituent order is pragmatically, and not syntactically, determined, word order within the noun phrase generally adheres to the following template:

(3) NP ➔ (GENeric)(N)(AdjP[2])*(Genitive.PRO)(QUANTifier)(DEMonstrative)

It is possible, however, to permute this order for pragmatic or stylistic effect. Smith and Johnson (2000, p. 388) propose a similar invariant NP constituent order for Kugu Nganhcara, and Wilkins (1989) also describes fixed ordering of NP constituents in Mparntwe Arrernte. As the brackets in (3) indicate, each element of the Thaayorre noun phrase may be omitted, including the head noun.[3] It is extremely rare (and distinctly odd-sounding) for all of the elements to co-occur in a single noun phrase, but (2) above illustrates the ordering GEN– N – G.PRO and (4) below illustrates GEN – N – AdjP – AdjP:

(4)	*minh*	*kothon*	*pam*	*ngamal*
	MEAT	wallaby	man	large.NOM

‘an enormous male wallaby’
(Hall 1972, p. 67, gloss the present author’s)

---

[2] An adjectival phrase is composed of a (head) adjective and optional degree adverb(s).
[3] Note that Wilkins (2000, p. 150) also finds that “an overt head noun is not obligatory in a noun phrase”.

Semantically, the generic is always the most general of the noun phrase constituents, determining the 'type' of referent described by the phrase as a whole, with all other elements further specifying its reference. For this reason, it is tempting to analyse the generic (e.g. *minh* 'MEAT' in (4)) as the NP head, with the following 'specific' noun (*kothon* 'wallaby' in (4)) relegated to nominal modifier. However, McGregor (1990) and Wilkins (2000) argue persuasively for an analysis of the specific noun as head, with the generic noun fulfilling a classifying function.[4] For this reason, I label the specific nouns 'N' rather than 'specific'.

Adnominal demonstratives typically occur at the end of a noun phrase, but may not inflect for case.[5] Where a case-marked noun phrase has a demonstrative as its final constituent, then it is the penultimate constituent that bears case inflection (whether this be noun, adjective, quantifier, ignorative, etc.). Thus in (5), the ergative suffix *-al* appears in the middle of the noun phrase *pam ith* 'that man', since *pam* 'man' is the last element that may inflect for case:

(5) *pam-al* *ith* *kuta* *thanp-arr=ul*
man-ERG dem:dist dog.ACC kick-PERF=3sg.ERG
'that man kicked a dog'

In order to explain its following the case affixes, it is tempting to analyse the demonstrative as a determiner that heads a determiner phrase with a (case-marked) NP complement. There are several factors that complicate such an analysis, however. To begin with, pronouns serve a similar, determiner-like function. The presence of either a demonstrative or pronoun entails that the reference be definite (nb. Wilkins, 2000, p. 157 finds the same for Arrernte). However, their presence is not required for a NP to be a referring expression, and a bare noun (or noun phrase) with no demonstrative or pronoun may also have definite reference. Alpher (1973, pp. 281–282) finds the same for Yir Yoront, although he continues to classify pronouns and demonstratives together as 'determiners'. Blake (2001, p. 417) similarly establishes for Pitta-Pitta a 'pronoun-determiner' class that heads the phrase. However, if the Thaayorre demonstratives and pronouns formed a single word class, we would expect to find them occupying the same syntactic slots. In contrast, demonstratives are tightly bound to the NP with respect to word order, whilst pronouns are entirely free and inflect for case independently (even where contiguous with the N-headed NP). Additionally, pronouns are often found in combination with a (coreferential) NP that already includes a demonstrative.

[4] There is insufficient space to detail their arguments here.

[5] Pronominal demonstratives may inflect for case, but are not involved in the constructions described in this chapter, and so will not be discussed further.

Demonstratives can even modify pronouns themselves. This is seen in example (6), in which the demonstrative *inh* 'this' directly follows the pronoun *nhul* 'she', in order to specify that the speaker is referring to the woman sitting next to him:

(6) *nhul* *inh* *kanpa-tam* *inh*
3sg.NOM dem.prox first-ABL dem.prox
'she's the first (born)'

It is difficult, then, to retain an analysis under which both demonstratives and pronouns head DPs. Instead, I propose to analyse the demonstrative as a specifier. This allows us to posit the following phrase structure rules for Thaayorre noun phrases:

(7) **NP** ➔ N′ (DEM)

(8) **N′** ➔ (GEN) (N) (AdjP)*(G.PRO) (QUANT)

Moving the demonstrative out of N′ and into the specifier position, helps explain why case markers are never suffixed to the demonstrative: case is marked at the end of the N′.[6]

An analysis of pronouns as determiners, would result in the following phrase structure rule:

(9) **DP** ➔ (NP) D (DEM)

In clauses where the pronoun immediately follows a NP (e.g. (10)), the two would form a DP together as follows:

(10) *pam* *-al* *ith* *nhul* *may carrots*
[[[man]$_{N'}$ -ERG dem.dist]$_{NP}$ 3sg.ERG]$_{DP}$ [[VEG carrots]$_{N'}$ACC]$_{NP}$
yakakerr
cut.up
'the man cut up the carrots'

In clauses where the NP and pronoun are separated by other clausal constituents, the pronoun would form a DP in isolation, which then forms a (discontinuous) DP at some higher level[7] with the NP as complement.

Arguments for a Thaayorre DP are mostly theory-driven. Noun phrases do not require pronouns in order to refer, and often have definite reference in their absence. Indeed, there is no obvious reason to class the Thaayorre pronouns as

---

[6] A minor complication for this analysis is the fact that numeral quantifiers (with a discourse function of referent tracking) can occasionally follow the demonstrative.

[7] Exactly what this higher level represents (e.g. 'f-structure' (in LFG) or 'deep structure' (in GB)) will depend on the theory employed.

determiners other than the fact that they make the reference of a noun phrase definite. The same semantic effect could be achieved, however, by an appositional analysis in which independent (but coreferential) noun phrases each contribute information about a particular referent. Evans (2003, p. 350) explores in detail several ways in which "unification from verbal and nominal material is used to construct referring expressions" in Bininj Gun-Wok. Many of his observations also apply to Kuuk Thaayorre, except that information about referents is generally gleaned entirely from nominal sources.

If the determiner phrase is abandoned, and pronouns are analysed simply as pronouns, we may modify the phrase structure rule in (8) to read as in Fig.1.

(11)

$$\mathbf{N'} \rightarrow \left\{ \begin{array}{l} \text{(GEN) (N) (AdjP)* (G.PRO) (QUANT)} \\ \text{Pro} \end{array} \right\}$$

This allows us to make the generalisation that case may only be marked on the rightmost constituent of N′ (whether this is a head noun and modifiers, or a pronoun), and that this N′ may form an NP in isolation, or with a demonstrative as specifier. The clause given in example (10), then, would instead be analysed as in (12):

(12)	*pam*	*-al*	*ith*	*nhul*
	$[[\text{man}]_{N'}$	-ERG	$\text{dem.dist}]_{NP}$	$[[\text{3sg}]_{N'}\text{ERG}]_{NP}$
	may carrots			*yakakerr*
	$[[\text{VEG carrots}]_{N'}\text{ACC}]_{NP}$			cut.up
	'that man cut up the carrots'			

2.3. Composition of Arguments

There is no necessary one-to-one relationship between Thaayorre arguments and noun phrases. As mentioned in Section 2.2, frequent argument ellipsis means many arguments are not realised as NPs at all. Alternatively, an argument may be encoded by multiple noun phrases apposed in the same case. In (12) above, for example the two noun phrases *pamal ith* 'that man' and *nhul* 'he' do not form a syntactic phrase, yet both contribute information about a single referent. Similarly, it is common for free pronouns to be repeated several times in a single clause (as in (15) below), or for (partially or wholly) coreferential noun phrases to be apposed (e.g. (29)). The syntactic structure of such constructions is discussed in Section 3.

As well as noun phrases (composed either of head noun and modifiers or pronouns), an argument may be encoded by a pronominal enclitic. These are basically reduced, phonologically dependent forms of the free pronouns, apparently in the early stages of grammaticalising into verbal cross-referencing. Synchronically, though, they remain formally, and functionally very close to the free pronouns, and hence are still analysed as having case (though only nominative, ergative and accusative cases are attested so far). Whilst they tend to be encliticised to the verb, these pronouns may instead attach to most other clausal constituents (including noun phrases). There is no restriction on the position of the host element with respect to other clausal constituents. They are freely omissible, and (13) is equally grammatical with or without the free pronoun *ngay* 'I' and/or the encliticised=*ay* 'I':

(13)	*ngay*	*iirrkuw*	*Darwin-ak*	*yat=ay*
	1sg.NOM	to:west	Darwin-DAT	went=1sg.NOM
	'I went to Darwin'			

Because they are still an emergent syntactic class, enclitic pronouns are particularly difficult to elicit. When questioned on the subject, or asked to repeat clauses slowly and carefully, consultants are likely to replace any enclitic pronouns with the corresponding free forms.

3. Referential Case

3.1. Meronymic Apposition

It was pointed out in Section 2.2 that a single argument may be composed of coreferential noun phrases apposed in the same case. Further to this, it is possible in Kuuk Thaayorre to appose noun phrases that are not wholly coreferential, but rather refer to Part and Whole of a single entity. This is exemplified by (14), a clause headed by an intransitive verb, *rokr* 'enters', that subcategorises for a single argument in nominative case:

(14)	*kuta*	*nhul*	*paant*	*glass-ak*	*rokr*
	dog.NOM	3sg.NOM	head.NOM	glass-DAT	enters
	'the dog puts his head into the jar'				

The three nominative noun phrases — *kuta* 'dog', *nhul* 'he', and *paant* 'head' — together comprise the subject argument, each contributing different kinds of information. *Kuta* 'dog' identifies the participant involved in the event, while *paant* 'head' informs the addressee of the particular Part of the dog that enters the jar. *nhul* 'he', then, contributes deictic information and aids in the pragmatic tracking of referents across discourse. But what exactly is the relationship between these three noun phrases?

Meronymic apposition of noun phrases referring to Part and Whole is a common strategy for representing inalienable possession, particularly across the Australian continent. Analogous constructions in other Australian languages have been labelled 'external possession' (e.g. McGregor, 1999 for Nyulnyulan languages), the 'favorite construction' (e.g. Hale, 1981 for Warlpiri), and 'inalienable possession' (e.g. Evans, 1996 for Mayali; Goddard, 1982 for Yidiny; Austin, 1981 for Diyari). Each of these constructions function to link both possessor and possessum to a single argument of a predicate. Investigators differ, however, as to the relative status of Whole and Part. Evans (1996, p. 87) asserts that the Mayali "part and whole are syntactically in apposition, with 'head' like properties shared between the part and the whole". Hale (1981) also argues for an appositional analysis of Warlpiri Part and Whole nominals, with both being linked to the same argument position of the verb. Unlike Evans, though, Hale (1981, p. 338) views the Whole NP as head, proposing that the Part NP may function as a secondary predicate with scope over the Whole NP. Harvey (1996, p. 134) provisionally adopts an analysis of the Warray construction under which the Whole is head and the Part modifier.

What, then, is the relationship between Whole and Part NPs in Kuuk Thaayorre? To begin with, the two noun phrases do not occupy a fixed position relative to each other. Whilst they are often contiguous (as in example (14) above), they need not be. This can be seen in (15), in which the pronoun representing the possessor (*nhinh* 'you') appears clause-initially, while the possessum *name nhangkan* 'your name' does not appear until after the predicate. It should be noted that many indexes of a person (e.g. their name, voice, shadow and footprints) are treated as 'Part' of them by this construction:

(15) *nhinh* *ngay* *wuump* *walmeerem* *name*
2sg.ACC 1sg.NOM CNTR remember name
nhangkan *ngay*
2sgGEN.ACC 1sg.NOM
'I remember your name'

The reverse ordering of Part and Whole is seen in example (16), in which the Part NP occurs clause-initially, and the Whole NP follows at the end of the clause:

(16) *koow* *rathirr=eln=unh*
nose.ACC chopped=3pl.ERG=3sgACC
'they chopped his nose off' (Hall, 1972, p. 77)

It seems clear, then, that Part and Whole are referred to by separate NPs. How, then, are these phrases related? Both Part and Whole NPs exhibit some characteristics of a head. McGregor (1985), Evans (1996) and Harvey (1996) each give examples (from Warrwa, Mayali and Warray, respectively) in which the predication holds for the Part but not the Whole. In Kuuk Thaayorre, too, what is predicated of the Part need not be entailed of the Whole, suggesting that the Part NP has head status. This can be seen in (14), of which it is not true that the Whole dog goes into the glass, only its head:

(14′)	*kuta*	*nhul*	*paant*	*glass-ak*	*rokr*
	dog.NOM	3sg.NOM	head.NOM	glass-DAT	enters
	'the dog puts his head into the jar'				

But it is true to say that the dog enters the glass inasmuch as its head does, which is no doubt a motivation for these kinds of constructions. In example (17), however, the difference between walking on a man's footprints and walking on the man himself is abundantly clear:

(17)	*kamp-an*	*yarr*	*pam-an*
	track-LOC	go:IMPER	man-LOC
	'we must walk in his tracks' (Foote & Hall; Reader 9)		

The following is also an example of the predicate having scope over the Part but not the Whole. (18) was uttered by a man who knew the referent of the Whole expression, *pam nhangkan* 'your husband', very well, but had temporarily forgotten his *nhamp* 'name':

(18)	*ngay*	*pam*	*nhangkan*	*nhamp*	*ngay*	*pamngoongkom*
	1sg.NOM	man	2sgGEN.ACC	name.ACC	1sg.NOM	be.ignorant[8]
	'I've forgotten your husband's name'					

It is possible even to omit the Whole NP, as in the following statement about an recently deceased pet dog:

(19)	*nhamp*	*ulp*	*Buddy*	*yanjim*
	name.NOM	dem.fam	Buddy	go:IMPERF[9]
	'(his) name was Buddy'			

There are, however, several factors that favour an analysis of the Whole NP as head. To begin with, it is the Whole that determines the selection of predicate in copular clauses. The predicate in a copular clause is drawn from small set of (mostly postural) verbs, selected according to semantic features of the subject[10]. In example (19) above, the motion verb *yan* 'go' is the 'default' copular where the

---

[8] *pamngoongkom* 'be ignorant' is a bivalent predicate nominal.

[9] In this context, the motion verb *yan* 'go' is being used as a copular verb.

[10] A detailed discussion of how copular verb selection operates is given in Gaby (in preparation).

subject is animate. This is clearly conditioned by the (omitted) Whole referent — Buddy the dog — rather than his inanimate name. Similarly in (20), the copular verb *nhiinin* 'sit' can be used in the description of an ongoing state associated with a human subject, but may not be used with an inanimate subject:

(20) *meer* *purng* *nhiinim* *ngul* *nhul* *meer*
eye.NOM closed sit:IMPERF then 3sg.NOM eyes.NOM
parrkarr
shone
'(she had) closed eyes then she opened them'

The fact that *nhiin* 'sit' is used, as opposed to the default verb for inanimates, *wun* 'lie', again suggests that the (elided) Whole NP is the head.

Furthermore, where a pronominal enclitic is attached to the verb this always agrees with the Whole and not the Part:

(21) *yuur* *stitchm* *rirkr=anh*
hand.ACC stitch DO=1sgACC
'they put stitches in my hand'

If the Part were the head, we would expect the pronominal enclitic to agree with it in (third) person, but instead we find the enclitic=*anh* '1sgACC' attached to the verb. Similarly, wherever the meronymic construction contains a free pronoun, it must have the person and number of the Whole, rather than the Part.

Whilst the properties of head seem split between the Part and Whole, those attributed to the Whole probably outweigh those associated with the Part. The scope of predication is the key feature that favours an analysis of the Part as head. However, even the claim that the predicate does not have scope over the Whole argument in clauses such as (18), repeated below, may be a misapplication of English rules of entailment to Kuuk Thaayorre:

(18′) *ngay* *pam* *nhangkan* *nhamp* *ngay* *pamngoongkom*
1sg.NOM man 2sgGEN.ACC name.ACC 1sg.NOM be.ignorant
'I've forgotten your husband's name'

To say in English *I don't remember your husband* entails that the speaker does not remember the person as a Whole, not just one small aspect of them. The same is not necessarily true of Kuuk Thaayorre, though. Example (22) was uttered by one of my consultants shortly after I'd arrived for a visit:

(22) *ngay* *nhinh* *pamngoongkom*
1sg.NOM 2sgACC be.ignorant
'I didn't know you (were here)'

As she herself explained, she meant that she did not know I was coming to Pormpuraaw, not that she didn't know who I was. In Kuuk Thaayorre, then, *I don't know you* can mean *I don't know [something associated with] you*. Even where a Part is not specified, then, something predicated of an argument that refers to a Whole, may in fact apply only to one Part or aspect of that Whole. If this is true, then it seems even more plausible that the Whole is head, and the Part complement at some level of representation.

Alternatively, it is possible to analyse the meronymic construction as double-headed, with Part and Whole both functioning as heads, with a single grammatical function' but not operating as a single syntactic unit. Such an analysis fits Dench and Evans's (1988, p. 13) characterisation of referential case:

> The referential case function involves the marking of some NP or adverb in agreement with some other (usually core) NP in the same clause...the identically marked words are separate constituents.

Finally, it should be noted that speakers may alternatively encode Part and Whole within a single noun phrase, as in the following:

(23)	*nhunt*	*[minh*	*kothon*	*thamr*	*kamp]-ak*[11]	*thanan*	*ulp*
	2sg.NOM	[MEAT	wallaby	foot	track]-LOC	standing	dem.fam

'you're standing on wallaby tracks'

Here, the Whole *minh kothon* 'wallaby' and Part *thamr kamp* 'tracks' form a single phrase (marked by square brackets), as evidenced by their contiguity and — most importantly — the presence of only a single suffix marking locative case.

As mentioned above, the terms *possessor raising* and *external possession* have been applied to analogous constructions in many other languages (see, e.g., the chapters in Chappell & McGregor 1996). The syntactic transformation of 'possessor raising' (as described by, e.g., Baker, 1988; O'Connor, 1996) derives Part–Whole constructions such as the one described here from an underlying structure in which the possessor is a dependent of the possessum. There is no evidence, however, that the Thaayorre possessive construction (in which the genitive-marked possessor is a dependent of the unmarked possessum) is more basic than the Part–Whole appositional construction described in this section. As argued by Hale (1981) for Warlpiri, and Evans (1996) for Mayali, the same-case Part–Whole construction appears to be less marked and more frequently used

[11] Note the two allomorphs of the dative suffix (*-an* and *-ak*) associated with the noun *kamp* 'track'. Their usage is determined by the age and dialect of the speaker, rather than any semantic or functional contrast.

than a possessive construction in which the possessor is encoded as a modifier. Further, Baker (1988, p. 274) predicts that possessor raising is only possible if the relevant noun phrase is an internal argument:

> 'Possessor Raising' should only be allowed if the raised NP is the possessor of a transitive verb's direct object, or of an unaccusative verb's surface subject.

In Kuuk Thaayorre, however, apposed noun phrases referring to Part and Whole may have a wide range of functions in the clause. The functions of intransitive subject and transitive object (both of which are predicted by Baker) are illustrated by (14) and (15), respectively. Additionally, however, they may function as transitive subject, as illustrated by (24):

(24)	*nhul*	*ngalnan*	*kiin-a*	*paatharr*
	3sg.ERG	1du.incl:ACC	teeth-ERG	bit

'he bit us with his teeth'

For most nouns, a single nominal suffix form expresses both ergative and instrumental cases. For this reason, it might be argued that the noun phrase *kiina* 'teeth' represents an instrument, rather than a part of the transitive subject. It is possible to disambiguate these two scenarios, however. Possessive pronouns inflect for case (when in phrase-final position), and have unambiguous ergative and instrumental forms. This can be seen in a comparison of (25) with (26):

(25)	*kuta*	*ngalin-man*	*yak*	*theerngarr*
	dog	1du:excl:GEN-ERG	snake.ACC	killed

'our dog killed the snake'

(26)	*Tom-thurr*	*ngali*	*kay*	*ngalin-antamn*	*yak*	*theerngarr*
	Tom-ERG	1du:excl.ERG	gun	1du:excl:GEN-INSTR	snake:ACC	killed

'Tom and I killed a snake with our gun'

The transitive subject possessum of (25), *kuta ngalinman* 'our dog', is marked by the ergative suffix-*man*, whilst the instrumental possessum of (26), *kay ngalinantamn* 'with our gun' is marked by the instrumental suffix-*antamn*. Crucially, in a Part–Whole apposition construction, the possessum receives the unambiguous ergative marking:

(27)	*ngul*	*William-n*	*pam*	*thono*	*nhaawr*	*meer*	*nhangan-man*
	then	William-ERG	man	one.ACC	saw	eye	3sgGEN-ERG

'and so William saw one man with his own eyes'

If (27) described William as using his eyes as an instrument, the noun phrase *meer nhangan* 'his eyes' should be marked by the instrumental suffix *-antamn*, rather than the ergative suffix *-man*. As it is, the two apposed ergative noun phrases *Williamn* 'William' and *meer nhanganman* 'his eyes' together function as transitive subject.

In addition to the three core arguments, apposed Part–Whole noun phrases function as optional clausal adjuncts. In (17), repeated below, the noun phrases *kampan* 'on the tracks' and *paman* 'on the man' are apposed in locative case to refer to a path (the verb *yan* 'go' subcategorises only for a single nominative-case subject argument):

(17′)	*kamp-an*	*yarr*	*pam-an*
	track-LOC	go:IMPER	man-LOC

'we must walk in his tracks' (Foote & Hall; Reader 9)

3.2. Simple Inclusory Construction

Inclusory constructions are employed to make explicit reference to a particular individual (or a subset of individuals) within a larger participant group. This is illustrated by (28), in which the referent of *wangath* 'doctor' is included within the participant group referred to by *ngali* 'we two':

(28)	*ngali*	*wangath-an*	*hour-ak*	*ngat*	*pit-im*
	1du:excl.ERG	doctor-ERG	hour-DAT	fish.ACC	hold-IMPERF

'The doctor and I were fishing for an hour' (Foote & Hall, Primer 10)

Kuuk Thaayorre is unusual in having several alternative inclusory constructions (Singer, 2001). All of them consist of two or more noun phrases, at least one of which refers to the superset of participants, the other(s) referring to the subset(s) in focus. The key difference between the different inclusory constructions is the case-marking of the subset noun phrase. In the simple inclusory construction, superset and subset are apposed in the same case. This is illustrated by (29), which contains three coreferential noun phrases representing the superset (*ngali* 'we two', twice instantiated, and *kuthirr* 'the two (of us)') and one NP representing the subset *I.C.* (a personal name). All four noun phrases are in nominative case, assigned to the subject argument of the intransitive verb *yan* 'go':

(29)	*ngali*	*I. C.*	*ngali*	*yat*	*kuthirr*
	1du:excl.NOM	I. C..NOM	1du:excl.NOM	went	two.NOM

'I. C. and I went, the two of us'

Inclusory constructions are most frequently associated with the grammatical function of subject. This is not always the case, however, as the following example shows:

(30)	*ngalinan*	*pam*	*thuump-ak*	*inhul*
	1du:exc:GEN	man	old-GEN	dem.pro:prox

'this one is my husband and mine'

Whilst the subset NP most commonly appears immediately following the superset pronoun, this need not be so. In (31), for instance, the subset *ngan waanharr* 'elder brother' and superset *ngali* 'we two' are separated by the verb:

(31)	*wey,*	*ngali*	*yanjim*	*ngan*	*waanharr*	*iipal*
	hey	1du:excl.NOM	go:IMPERF	relative	eB.NOM	from.there

'hey, my brother and I have come here'
(Foote, 1977 — gloss and translation the present author's)

Superset and subset do not, then, appear to form a single phrase. Like the Part–Whole construction, it seems that the linkage of the two phrases to a single argument slot and grammatical function is achieved via referential case. That the number and person features of the superset pronoun are those of the argument overall, suggests that it is the superset that is head and the subset complement. This is also true of many other Australian languages (Singer, 2001), but not all. In Mparntwe Arrernte, for example, the superset pronoun and subset NP must be contiguous, and case markers — which are suffixed to the final element of a NP — only appear following the entire construction (Wilkins, 1989, p. 409). This suggests that the Arrernte inclusory construction forms a single phrase.

The simple Thaayorre inclusory construction is labelled such because of its relative lack of morphological marking (compared with the comitative and dative inclusory constructions). It is not intended to signify cognitive primacy or diachronic priority.

The superset noun phrase in a simple inclusory construction is most commonly a first or second person dual pronoun. In such cases, one referent is specified by the subset NP, and the second participant is fully recoverable as either the speaker or addressee respectively. The simple inclusory construction is therefore used frequently in discourse to introduce new participants, or to clarify the membership of a plural participant group. In (32), for example, an inclusory construction formed from a second person dual pronoun is used specifically in order to request the identity of the second, non-addressee participant:

(32)	*nhip*	*wanh*	*wun*	*iikaw*
	2du.NOM	who.NOM	reside	in.east

'who do you live with?'

Analogous constructions have been documented in many different language families, for example by Lichtenberk (2000), Schwartz (1988) and Singer (2001). Schwartz's typology takes a slightly different approach to the other two, characterising the construction as 'verb-coded coordination'. She focuses on the relationship between free noun phrases representing subset(s) and verbal agreement or cross-referencing that reflects the plural number of the superset. Her prediction that "a language will have VCCs only if its verb independently has syntactic number coding of arguments" (1988, p. 57) does not apply to Kuuk Thaayorre, except marginally to the optional enclitic pronouns (rarely present in inclusory constructions). Nevertheless the constructions she considers are very similar to the inclusory constructions of Kuuk Thaayorre and many other Australian languages. Take, for example, the following Yapese clause:

(33) *ka* *ra* *bow* *Tamag*
T/A 3DU/PL come+DU Tamag
'he and Tamag came'
(Jenson, 1977; quoted in Schwartz, 1988, p. 54)

As in Kuuk Thaayorre, a non-singular superset pronoun is apposed to a subset noun phrase (here referring to an individual by name). The key difference between the two constructions, then, is the coding of dual number on the verb in Yapese.

Singer (2001, pp. 22–25), in her typology of inclusory constructions in Australian languages, offers the following defining features of inclusory constructions: (1) the relationship between superset and subset is one of proper inclusion; (2) entailments of the predicate always hold for the superset and subset; (3) the superset is encoded by a pronoun; (4) the inclusory construction is equivalent to a single argument of a single predicate. These features straightforwardly apply to each of the Thaayorre inclusory constructions (although not the inclusory pronouns, in which both superset and subset are encoded by a single pronoun).

4. Adnominal Case

4.1. Comitative Inclusory Construction

Section 3.2 described the simple inclusory construction, in which the noun phrases referring to both superset and subset are apposed in the same case. Alternatively, the noun phrase referring to the subset may be marked by the comitative enclitic *-kak*, as seen in example (34). (Note that where an example consists of multiple sequential clauses, this is indicated by a clause number following the example number (i.e. '34.1' labels the first clause of the 34th example)):

(34.1) *pam-kak* *ii* *yan* *pul*
man-COM there go 3du.NOM
'that husband and wife are going along'

(34.2) *pul* *ii* *yan* *kuthirr* *pam-kak*
3du.NOM there go two.NOM man-COM
'they two are going along there, two of them, husband and wife'

In (34.1), the participants are identified by means of both the nominative pronoun *pul* 'they two' (referring to the superset) and the comitative noun phrase *pamkak* 'with the man' (the 'man' being a subset of the subject participant group). The comitative enclitic functions to relate *pamkak* 'with the man' to the subject pronoun *pul* 'they two'; a function labelled 'adnominal case' by Dench and Evans (1988). Crucially, a single participant (the man) is referred to by both the subject pronoun *pul* 'they two' and the NP *pam kak* 'with the man'. Contrast this with a literal English translation of (34.1); *they two are going along with the man*, which entails three rather than two participants, since the reference of the adjunct PP *with the man* must not overlap with the reference of the subject NP *they two*. An analogous construction is also available in Kuuk Thaayorre, with the comitative adjunct referring to a participant wholly distinct from that referred to by the subject NP. This is illustrated by (35), in which the reference of the subject pronoun *ngay* 'I' is wholly distinct from that of the comitative adjunct *pam puukath kak* 'with an initiated man':

(35) *pam* *puukath-kak* *wak-m* *ngay*
MAN initiated-COM hunt-IMPERF 1sg.ERG
'I was hunting with an initiated man'
(Foote & Hall, Primer 10, p. 6)

Why, then, would a speaker choose to utter the comitative inclusory construction over the simple inclusory construction, or a clause in which subject and adjunct are not partially coreferential? The Thaayorre comitative inclusory construction is advantageous in the following respects. Firstly, the fact that the subject pronoun *pul* 'they two' in (34) incorporates both participants in its reference captures the fact that both the man and woman participate in the event in the same way (i.e. they are both walking), and thus a single relationship holds between them and the predicate (the nominative case assigned to the subject pronoun hence characterised as 'relational case' (Dench & Evans 1988), marking a dependency relationship between the noun phrase and predicate). Meanwhile, the extraposed comitative NP allows the speaker to simultaneously comment on the relationship that holds between the respective subject participants (adnominal case). Specifically, in example (34), the comitative adjunct is used to imply that the subject participants are married. The noun *pam* means both 'man' and 'husband', and the clauses in (34) were uttered both to point out to the addressee a couple walking past, and (more importantly) to communicate the relationship that holds between them. This is an implicature, though, and not an entailment. Where the superset pronoun is plural (as opposed to dual), the implicature of marriage is no longer present[12]:

(36) *pam-kak* *nganjin* *yat*
man-COM 1pl.NOM went
'we went with the man'

This use of the comitative inclusory construction to assert a particular relationship between participants is common. In such cases, it is possible for each member of the superset to be explicitly mentioned (although it is more prototypical of inclusory constructions for some participant(s) not to be enumerated, according to Singer, 2001). This is illustrated by the following:

(37) *kanangkarr* *pul* *pam* *ngumpurr-kak* *pul* *may-i* *yat*
long.ago 3du.NOM man.NOM old.lady-COM 3du.NOM VEG-DAT went
'long ago an old man and his wife went out to gather food'

In this example, one subset noun phrase (*pam* 'man') takes the same (referential) nominative case as the pronoun denoting the superset (*pul* 'they two'), as in a simple inclusory construction. The second subset NP (*ngumpurr-kak* 'with the old lady'), however, is marked comitative. This overtly marks the subset NP as subject complement, modifying the head pronoun which receives its relational case from the predicate.

Example (38) expresses a different kind of relationship between participants. Here again, both the participants represented by the dual pronoun *nhip* 'you two' are exhorted to engage in the same activity. The extraposition of one of these in the comitative *parr_r-kak* 'with the child', suggests that an unequal relationship holds between the two. Specifically, we may infer from (38) that the child is less centrally involved in the event, which is to be initiated by the addressee (the unspecified member of the superset pronoun *nhip* 'you two'). The choice of construction thus places responsibility on the addressee for the actualisation of the event, and the involvement of the subset 'child'.

[12] Note also that comitative inclusory constructions containing plural superset pronouns are often indistinguishable from simple comitative constructions in which the referent of the comitative phrase is not included within the reference of the plural pronoun.

(38) *parr_r-kak* *nhip* *yarr!*
child-COM 2du.NOM go:IMPER
'you go with that child'

This construction is not exotic, and both Schwartz (1988) and Singer (2001) find many languages in their samples employ an adnominal case (often comitative, dative or instrumental cases, or some other marker of nominal conjunction) to relate subset NPs to the superset. For example, the following Hungarian example is taken from Schwartz (1988, p. 54):

(39) *elindultunk* *a* *vezetömmel*
went+1pl DET guide+1GENS+&
'my guide and I left'

Here the person and number of the superset is encoded on the verb, while the subset is encoded by a free NP *a vezetömmel* 'with my guide'. Russian possesses a construction in which a subset NP (in instrumental case, signifying accompaniment) complements a free pronoun representing the superset:

(40) *my* *s* *vami* *gde-to* *vstrechal-i-s'*
1plNOM with 2pl[13]:INSTR somewhere met –PL-RFL
'you and I have met somewhere' (film title)

4.2. Dative Inclusory Construction

The dative inclusory construction is structurally very similar to the comitative inclusory construction; the two differ only in the eponymous case assigned to the subset NP. The dative inclusory construction is exemplified by (41), in which the head NP *pul* 'they two' is followed by its dative-marked complement *pam kunyangkar ngathanmun* 'at my brother':

(41) *ngali* *pam* *kunyangkar* *ngathan-mun* *nhaanhath-rr*
1du:excl.NOM MAN sibling 1sgGEN-DAT look.at-RCP
'my brother and I are looking at each other'

The dative inclusory construction is often found in reciprocal clauses derived from base verbs that subcategorise for an indirect object.[14] In (41), for example,

[13] The Russian second person plural pronoun is used as a respectful second person singular form.

[14] The base verb *nhaa-* has a nominative-dative case frame, which alternates with an ergative-accusative case frame. See Gaby (in preparation) for further discussion.

both participants are associated with two semantic roles (agent and goal) and also two cases (nominative and dative). The ascription of two semantic roles to a single argument is licensed by the reciprocal construction (marked by the verbal suffix *-rr*). The fact that the referent of the dative-marked subset NP is included within the reference of the subject pronoun, however, makes it difficult to analyse (41) as a reciprocal clause that has retained both a (nominative) subject and a (dative) indirect object. Instead, I argue, the two participants are encoded as a single subject argument, encoded by a pronominal head and dative complement.

It is also common for dative inclusory constructions to contain multiple subset NPs specifying each member of the superset (as we saw in example (37) above). In all the examples collected to date, one of these subset NPs is marked by adnominal dative case, whilst the other receives the referential case-marking appropriate to the grammatical function of the argument as a whole. This can be seen in (42), in which the first subset NP *Jimmynthurr* 'Jimmy' agrees with the superset pronoun *pul* 'they two' (twice instantiated) in ergative case, whereas the second subset NP *Johnnyn* 'at Johnny' is dative-marked:

(42)	*Jimmy-nthurr*	*Johnny-n*	*pul*	*ngerngkan*	*thanp-rr-r*	*pul*
	Jimmy-ERG	Johnny-DAT	2du.ERG	yesterday	kick-RECP-PST	2du.ERG

'Jimmy and Johnny kicked each other yesterday'
(Hall, 1971, p. 244)

Such clauses thus contain two subject complements in different cases.

The fact that dative case-marking is adnominal in function, rather than being assigned by the predicate (i.e. relational), is evidenced by the construction's compatibility with head verbs that do not subcategorise for an indirect object (e.g. (42) above). Instead, the dative-marking of subset(s) appears to be semantically, rather than syntactically, motivated. It was pointed out earlier that this construction is frequently employed in reciprocal clauses. A more accurate generalisation that holds for all the examples collected to date, however, is that the dative inclusory construction is used in the description of semantically symmetric (i.e. reciprocated) events, in which either: (a) the participants do not have the same role at the same time (as in (42) and (43)); or (b) the relationship that holds between the two participants is one marked by the dative case in clauses describing non-symmetric events (e.g. (41), (43) and (44)). To clarify what exactly is meant by this, consider the following two examples:

(43)	*pul*	*pam*	*kunyangkar*	*nhangan-mun*	*kuuk*	*yiik*
	3du.NOM	MAN	sibling	3sgGEN-DAT	WORD	saying

'he and his brother are talking'

(44)	*pam*	*ith*	*pul*	*paanth-ak*	*nhiinat*	*pul*
	man.NOM	dem:dist	3du.NOM	woman-DAT	sat	3du.NOM

'the man and the woman sat down next to each other'

Both describe semantically symmetric events (expressed by the reciprocal construction in their English translations), but are are not marked by the reciprocal suffix. Nevertheless, the symmetry of the event is indicated by the fact that both participants are encoded as a single argument. The fact that two semantic roles are associated with this argument is meanwhile expressed by the dative marking of the subset NP. So in (43), it is clear that the two brothers are directing their speech to one another, rather than merely broadcasting simultaneously. This fits both the conditions given above, in that (a) while one brother is agent, the other is recipient and vice versa; and (b) in the corresponding non-symmetric clause (i.e. describing one brother talking to the other), the recipient would be encoded in dative case. Similarly, in (44), the relationship of adjacency (i.e. that the two participants sit *beside* one another) is usually marked by dative case in non-symmetric clauses, as in the following:

(45)	*yuk*	*thongkn*	*church-ak*	*thanan*
	TREE	tree	church-DAT	stands

'the tree is next to the church'

Thus, the dative suffix on the subset *paanthak* 'to/at the woman' captures the relationship of adjacency, whilst the fact that both participants are included within the reference of the superset subject pronoun *pul* 'they two' marks the event as symmetrical.

4.3. Inclusory Pronoun Construction

The inclusory pronoun construction, discussed here because of its functional similarity to the other inclusory constructions, is syntactically very different. Inclusory prounouns express, in a single lexeme, both the subject superset and an oblique subset. An example of this is (46), in which the inclusory pronoun *ngalathun* makes reference to both a dual subject superset (usually encoded as *ngal* 'we two (inclusive)') and a singular oblique subset (usually encoded by *ngathun* 'to me'):

(46)	*ngalathun*	*kuuk*	*yik*
	1du:incl>1sgINCLS	WORD	saying

'you and I are speaking together in the vernacular'
(Hall, 1972, p. 380)

The inclusory pronominal paradigm is, in most instances, transparently related to both the nominative and the dative case forms of the personal pronouns, as can be seen in Table 3.

Inclusory pronouns are attested so far only in the work of Hall (1972), and further investigation is required to provide a full explanation of their semantics and morphosyntax. Pending this, however, the following observations can be made. Firstly, the inclusory pronouns appear to be the product of compounding (of nominative and dative case pronouns) at some earlier stage of the language's history. Synchronically, though, the pronouns appear to be monomorphemic, rather than productively derived, since phonological reduction and (in some cases) modification have obscured the source morphemes. Nevertheless, it is useful for the discussion below to discuss the inclusory pronouns in terms of two component parts: an initial element drawn from the nominative paradigm; and a second element drawn from the dative paradigm. It is acknowledged, however, that this most likely does not reflect the synchronic morphological status of these forms.

The fact that the apparent case-marking of the component elements is not assigned by the predicate, is evident in the following example:

(47) *pelnathun* *kanpa* *yaat*
3pl>1sgINCLS before went
'they and I all went previously'
(Hall, 1972, p. 381)

The dative case instead expresses a comitative relationship that obtains between the participants.

There is a distinction between those inclusory pronouns in which the initial (nominative) element refers to the superset, and those in which it refers to the subset. The former is illustrated by rows 1–4 of Table 3 and the latter by rows

Table 3: Comparison of inclusory, nominative and dative pronouns.

	Inclusory pronoun	Nominative pronoun	Dative pronoun
1	*ngalathun* 'you and I'	*ngal* '1du:incl'	*ngathun* '1sg'
2	*ngalngun* 'you and I'	*ngal* '1du:incl'	*nhangkun* '2sg'
3	*ngalingathun* 'he and I'	*ngali* '1du:excl'	*ngathun* '1sg'
4	*ngampathun* 'we and I'	*ngamp* '1pl:incl'	*ngathun* '1sg'
5	*pelnathun* 'they and I'	*peln* '3pl'	*ngathun* '1sg'
6	*nhunturra* 'all of you'	*nhunt* '2sg'	*nhurrnungun* '2pl'
7	*nhipurra* 'the two of you'	*nhip* '2du'	*nhurrnungun* '2pl'

6–7. Discussion of the semantics of the latter case is deferred to below. Where the subset follows the superset, the inclusory pronoun appears to focus attention on the subset referent. This is evident in the comparison of (48) with (49):

(48)

ngamp	*ngerngkan*	*ngampathun*	*thangkarnam*	*raak*	*min-m*
1pl:incl.NOM	yesterday	1pl:incl>1sgINCLS	laugh:IMPERF	thing	good-ABL

'all of us had a roaring good time yesterday from those jokes with me'
(Hall, 1972, p. 380)

(49)

nhunt	*ngalngun*	*pam.thaaw*
2sg.NOM	1du:incl>2sgINCLS	friend.NOM

'you and I are good friends together' (Hall, 1972, p. 380)

In (48) (as well as (46) above), particular attention is drawn to the speaker (the individuated subset). In (49), however, it is the addressee that is in focus, and specified as subset. Note that examples (48) and (49) present what could be analysed as an inclusory pronoun construction embedded within a simple inclusory construction. The simple inclusory construction requires that a pronoun referring to the superset be apposed to one or more NPs referring to subset(s), but since the inclusory pronoun in itself refers to both superset and subset separately, it can be apposed to pronouns referring to either the superset (e.g. *ngamp* in (48)) or the subset (e.g. *nhunt* in (49)).

The semantics of the inclusory pronouns in which the subset precedes the superset, are less easily specified. The two examples for which this holds are as follows:

(52)

nhunturra	*pam;*	*paanth*	*nhunturra;*	*ongkorr*	*reepon*	*murk=th*
2sg>2plINCLS	man	woman	2sg>2plINCLS	PROHIB	steal:IMPER	others'.poss

'every one of you men — and women — don't steal what's another's'
(Hall, 1972, p. 380)

(53)

nhipurra	*yaarr*
2du>2plINCLS	go:IMPER

'just two of you all go'
(Hall, 1972, p. 381)

In both, the inclusory pronouns focus attention on particular members of a larger group. In (52), though, the individualisation of the plural subject group (by means of the inclusory pronoun) is actually intended to stress that the message applies to each individual member of the superset. By contrast, in (53) the exhortation is apparently directed towards only the subset of the larger group. Whether this difference in interpretation is due to contextual implicature, constructional frame or some feature of the (potentially polysemous) inclusory pronouns themselves, requires further investigation.[15]

It should be noted here that there is some formal irregularity in the composition of these two inclusory pronouns; *nhunturra* “2sg>2plINCLS” and *nhipurra* “2du>2plINCLS”. In Table 3, a comparison was made between the second element of the inclusory pronoun, and the dative personal pronoun corresponding in person and number. In these two cases, however, the second element *-urra* seems less like the second person plural dative pronoun *nhurrnungun* and more like the unmarked *nhurr* “2pl.NOM”. It is worth noting also that the suffix *-a* marks the dative case for some declensions of free noun (e.g. *pama* “to the man”). Further investigation may reveal that *nhurra* was a dative second person plural pronominal form at some stage of the language’s history.

Finally, the lexeme *pelnathun* “they and I” in row 5 of Table 3 is labelled an ‘inclusory pronoun’ due to its formal resemblance of the other inclusory pronouns, despite the fact that the label is in this case somewhat of a misnomer. Used in clauses such as (47), repeated below, the dative referent *-athun* ‘to me’ is not actually a subset of the plural nominative form *peln* ‘they’, but rather specifies an additional participant who acts in conjunction with the participants referred to by the ‘superset’:

(47′) *pelnathun* *kanpa* *yaat*
3pl>1sgINCLS before went
‘they and I all went previously’ (Hall, 1972, p. 381)

We may assume that here again some additional focus is given the ‘subset’ referent (as opposed to a clause with a first person plural subject; “we all went previously”).

5. Pragmatic Case

There is one further circumstance under which a noun phrase representing a subset of subject participants receives different case-marking from that representing the superset. Contrary to its syntactic function of (always and only) marking the subject of a transitive clause, the ergative suffix is sometimes affixed to the subject NP of an intransitive (or semi-transitive) clause. So, in the second clause of example (54), the subset NP *Edwardnthurr* ‘Edward’ is ergative-marked:

[15] Further speculation here is impossible, since the context of utterance of (53) is not given by Hall (1972), and the present author has not recorded analogous uses of the inclusory pronouns.

(54) *G.Y.* *nhul* *driver.* *Ngali* *mit* *rirk* *Edward-nthurr*
G.Y. 3sg.NOM driver.NOM 2du:excl.NOM work DO[16] Edward-ERG
'G. Y. is the driver. We two work, [me and] Edward'

The presence of ergative-marking in cases such as this is motivated by pragmatic considerations, rather than being assigned by the predicate or intra-clausal morphosyntax. The Thaayorre ergative morpheme is in the process of developing new discourse functions, and may be employed to mark any 'unexpected subject,' whether in a transitive or intransitive clause. A subject is defined as 'unexpected' where it refers to a newly introduced (or re-introduced) participant in a section of discourse that has had a stable protagonist/topic chain that has been displaced by this new participant, as characterised by McGregor's (1998:p. 516) 'Expected Actor Principle':

> The episode protagonist is — once it has been established — the expected (and unmarked) Actor of each foregrounded narrative clause of the episode; any other Actor is unexpected.

This helps to explain why the intranstive subject subset *Edwardnthurr* 'Edward' (of example (54)) is ergative-marked. The first clause of (54) sets up G.Y. as topic. When the second clause goes on to state 'we two work', the addressee might be misled into thinking that the speaker is referring to himself and G.Y., rather than himself and Edward. For this reason, the speaker affixes the ergative marker to *Edward* to stress the fact that (perhaps contrary to the addressee's expectations), he is the subject of the second clause.

As well as introducing an entirely new subject referent, the ergative morpheme may be used to signal a change in membership of a plural subject group. Example (55) is excerpted from a story about the narrator's childhood, and the time he and his siblings first saw a plane. The enduring topic of the whole narrative is *nganjin* 'we', which is used most of the time in reference to the speaker and his siblings (as it is in (55.1)). In the second line of the example, however, the referential scope of *nganjin* 'we' expands to include the speaker's parents. It is this unexpected shift in reference that triggers the suffixation of the ergative morpheme to

[16] The verb *rirkr* (which, in its basic sense, means 'arise'), commonly combines with Thaayorre nouns or loan verbs to produce a new complex predicate. The complex predicate *mit rirk* 'work' is intransitive, taking an unmarked nominative subject, as illustrated by the following:

(i) *Sitika* *mit* *rirk-m*
Sitika.NOM work DO-IMPERF
'Sitika was working'

the two noun phrases in (55.2) that refer to the new subset participants; *nganipi* 'father' and *mami ngathanman* 'my mum':

(55.1)	*nganjin*	*kanangkarr*	*parr_r*	*mant*
	3pl:excl.NOM	long.ago	child	small.NOM

'when we used to be small kids'

(55.2)	*ngul*	*nganjin*	*nganip-i*	*nganjin,*	*mami*	*ngathan-man*	*nganjin…*
	then	1pl:excl.NOM	father-ERG	1pl:excl.NOM	mum	1sgGEN-ERG	1pl:excl.NOM

'we, including Dad and Mum…'

(55.3)	*nganjin*	*irrkaw*	*yat*
	1pl:excl.NOM	to:east	went

'we went out bush'

Pragmatic and discourse contexts similarly influence the distribution of ergative morphemes in several other Australian languages. In Jaminjung, for example, the ergative marker is optionally replaced by the ablative, depending on the ease with which the addressee is expected to retrieve the appropriate referent (Schultze-Berndt, 2000, p. 168). McGregor (1992), (1998) and (2004), discuss in detail the influence of discourse pragmatics on the inclusion or omission of the ergative morphemes in Gooniyandi and Warrwa. He argues that the Warrwa focal ergative marker combines the functions of marking syntactic ergative case and pragmatic focus. Pensalfini (1999) finds that the Jingulu ergative markers have grammaticalised into focus markers (which now operate in parallel to ergative case markers), due to the pressures of language contact and obsolescence. Meakins and O'Shannessy (2004) similarly attribute to language contact the fact that the ergative morphemes of Gurindji Kriol and Light Warlpiri have come to mark topics and contrastive focus, regardless of clausal transitivity. Gaby (submitted) discusses in detail exactly how pragmatic ergativity operates in Kuuk Thaayorre, including a comparison between the Thaayorre system and those documented in other Australian languages.

6. Conclusion

This chapter has discussed a range of Thaayorre constructions in which Part and Whole of an entity or of a group of participants are encoded by separate noun phrases. In each of the constructions, it appears to be the (usually pronominal) NP with the most inclusive reference that heads the argument. Section 3 presented two constructions in which referential case marking linked the subpart/subset NP to its head, the two being apposed in the same case. Section 4 discussed three inclusory constructions (further to the one described in Section 3.2) in which the suffixation of adnominal case to subset NPs marks them as complements to the superset. In all of these instances, case marking is not assigned by the predicate to the subset (since this is related to the predicate only indirectly via the superset head), but rather marks a relationship of dependency between NPs with overlapping reference. What is particularly interesting, is the fact that these NPs do not form a phrase together, but are independent with respect to word order and case marking.

Finally, this chapter considered one usage of a case morpheme that serves neither a referential nor adnominal function (nor is it assigned by the predicate). Instead, it was argued, the ergative suffix can have a pragmatic function. All of these uses of case markers are united by their role in indentifying and tracking the relationships between multiple participants (or multiple parts of a single participant) involved in an event.

Acknowledgements

The research reported in this chapter was supported by The University of Melbourne and the Max-Planck-Institut für Psycholinguistik. Comments from an anonymous reviewer were extremely helpful in developing my analysis, as were some long and engaging discussions with Nick Evans and Rachel Nordlinger, who helped in particular with identifying the structure of Thaayorre noun phrases. The analysis presented herein, though, does not necessarily represent their views. Finally, I would like to thank the Pormpuraaw community in general, and my language speak teachers in particular, for their support, encouragement and collaboration.

References

Alpher, B. (1973). *Son of Ergative: The Yir Yoront language of Northeast Australia.* Unpublished Ph.D. thesis, Ithaca: Cornell University.

Austin, P. (1981). *A grammar of Diyari.* Cambridge: Cambridge University Press.

Baker, M. (1988). *Incorporation.* Chicago: University of Chicago Press.

Blake, B. (1994). *Case.* Cambridge: Cambridge University Press.

Blake, B. (2001). The noun phrase in Australian languages. In: J. Simpson, D. Nash, M. Laughren, P. Austin & B. Alpher (Eds), *Forty years on: Ken Hale and Australian languages.* pp. 415–426, Canberra: Pacific Linguistics.

Chappell, H., & McGregor, W. (Eds) (1996). *The grammar of inalienability*. Berlin: Mouton de Gruyter.

Dench, A., & Evans, N. (1988). Multiple case-marking in Australian languages. *Australian Journal of Linguistics, 8*, 1–47.

Dixon, R. M. W. (2002). *Australian languages: their nature and development*. Cambridge: Cambridge University Press.

Evans, N. (1996). The syntax and semantics of body part incorporation in Mayali. In: H. Chappell & W. McGregor (Eds), *The grammar of inalienability* (pp. 65–110). Berlin: Mouton de Gruyter.

Evans, N. (2003). *Bininj Gun-Wok: a pan-dialectal grammar of Mayali, Kunwinjku and Kune*. Canberra: Pacific Linguistics.

Foote, T. (1977). *Kuta Woochorrm: Dog of the dreamtime*. Brisbane: Crusade Press.

Foote, T., & Hall, A. (unpublished). *Readers and primers for Kuuk Thaayorre*. Pormpuraaw: Unpublished manuscript.

Gaby, A. R. (submitted). Pragmatically case-marked: non-syntactic functions of the Thaayorre ergative suffix. In: I. Mushin & B. Baker (Eds), *Discourse and grammer in Australian languages*. Amsterdam: John Benjamins.

Gaby, A. R. (in preparation). *A grammar of Kuuk Thaayorre*. Unpublished Ph.D. thesis, Melbourne: University of Melbourne.

Goddard, C. (1982). Case systems and case marking. *Australian Journal of Linguistics, 2*, 167 196.

Grimes, B., & Grimes, J. (Eds) (2000). *Ethnologue: languages of the world*. (14th ed.) Dallas: Summer Institute of Linguistics.

Hale, K. (1981). Preliminary remarks on the grammar of part–whole relations in Warlpiri. In: J. Hollyman & A. Pawley (Eds), *Studies in Pacific languages and cultures in honour of Bruce Biggs*. (pp. 333–344). Auckland: Linguistic Society of New Zealand.

Hall, A. H. (1972). *A study of the Thaayorre language of the Edward River tribe, Cape York Peninsula, Queensland: being a description of the grammar*. Unpublished Ph.D. thesis, Brisbane: University of Queensland.

Harvey, M. (1996). Body parts in Warray. In: H. Chappell & W. McGregor (Eds), *The grammar of inalienability* (pp. 111–154). Berlin: Mouton de Gruyter.

Jenson, J. (1977). *Yapese reference grammar*. Honolulu: University of Hawai'i Press.

Lichtenberk, F. (2000). Inclusory pronominals. *Oceanic Linguistics*, *39*(1): 1–32.

McGregor, W. (1985). Body parts in Kuniyanti clause grammar. *Australian Journal of Linguistics, 5*, 209–232.

McGregor, W. (1990). *A functional grammar of Gooniyandi*. Amsterdam: John Benjamins.

McGregor, W. (1992). The semantics of ergative marking in Gooniyandi. *Linguistics, 30*, 275–318.

McGregor, W. (1998). Optional ergative marking in Gooniyandi revisited: Implications to the theory of marking. *Leuvens contributions in linguistics and philology, 87*(3–4); 491–571.

McGregor, W. (1999). External possession constructions in Nyulnyulan languages. In: D. L. Payne & I. Barshi (Eds), *External possession*. Amsterdam, John Benjamins. 429–448.

McGregor, W. (2002) Ergative and accusative patterning in Warrwa. In: K. Davidse & B. Lamiroy (Eds), *Case and grammatical relations across languages. Volume 4: the nominative/accusative*. Amsterdam: John Benjamins.

Meakins, F., & O'Shannessy, C. (2004). Shifting functions of ergative case-marking in Light Warlpiri and Gurindji Kriol. Paper presented at *ALS 2004*. University of Sydney.

Mel'čuk, I. (1979). *Studies in dependency syntax*. Ann Arbor: Karoma.

O'Connor, M. C. (1996). The situated interpretation of possessor-raising. In: M. Shibatani & S. Thompson (Eds), *Grammatical constructions: their form and meaning*. (pp. 125–156). Oxford, Clarendon Press.

Pensalfini, R. (1999). The rise of case suffixes as discourse markers in Jingulu – A case study of innovation in an obsolescent language. *Australian Journal of Lingusitics, 19*, 225–240.

Schultze-Berndt, E. (2000). *Simple and complex verbs in Jaminjung: a study of event categorisation in an Australian language.* Unpublished Ph.D. thesis. Nijmegen: Katholieke Universiteit Nijmegen.

Schwartz, L. (1988). Conditions for verb-coded coordinations. In: M. Hammond, E. Moravcsik & J. Wirth (Eds), *Studies in syntactic typology* (53–73). Amsterdam: John Benjamins.

Singer, R. (2001). *The inclusory construction in Australian languages*. Unpublished BA(Hons) thesis. Melbourne: University of Melbourne.

Smith, I., & Johnson, S. (2000). Kugu Nganhcara. In: R. M. W. Dixon & B. J. Blake (Eds), *Handbook of Australian languages* (Vol. 5, pp. 357–489). Melbourne: Oxford University Press.

Spencer, A. (2003). Syntactic vs. morphological case: implications for morphosyntax. Paper presented at the *Pionier Workshop: Case, Valency and Transitivity*, 17th June. Nijmegen: Katholieke Universiteit Nijmegen.

Wilkins, D. (2000). Ants, ancestors and medicine: a semantic and pragmatic account of classifier constructions in Arrernte (Central Australia). In: G. Senft (ed.), *Systems of nominal classification* (pp. 147–216). Cambridge: Cambridge University Press.

Wilkins, D. (1989). *Mparntwe Arrernte (Aranda): studies in the structure and semantics of grammar*. Unpublished Ph.D. thesis. Canberra: Australian National University.

Zwicky, A. (1987). Suppressing the Zs. *Journal of Linguistics*. *23*, 133–148.

Appendix: Glossary of terms

&	connector (Schwartz, 1988, p. 54)
1	first person
2	second person
3	third person
ABL	ablative case
dem	demonstrative
dist	distal

du	dual number
eB	elder brother
ERG	ergative case
excl	exclusive of addressee
GEN	genitive case
IMPER	imperative mood
IMPERF	imperfect tense/aspect
incl	inclusive of addressee
INCLS	inclusory pronoun
LOC	locative case
others'.poss	the possession of somebody other than speaker and addressee
pl	plural number
PRO	pronoun
PROHIB	prohibitive mood
PROP	proprietive
prox	proximal
sg	singular number
T/A	tense/aspect marker
VEG	generic noun for plant foods
yB	younger brother

Chapter 3

Head Marking and Dependent Marking of Grammatical Relations in Yurakaré

Rik van Gijn

1. Introduction

Yurakaré, a language spoken by approximately 2500 people in central Bolivia, has a mixed system of head and dependent marking of grammatical relations.[1] Subject and direct object are obligatorily marked on the verb,[2] and there is a complex system of applicative markers with which a number of other grammatical relations can be marked on the verb. Other arguments are dependent marked by means of postpositional enclitics.

In this chapter I will discuss what determines the split between head and dependent marking of grammatical relations in Yurakaré. I will first give a descriptive account of the head marking (Section 2) and dependent marking (Section 3) strategies for encoding grammatical relations before returning to the question of what determines the split (Section 4).

2. Head Marking

Head-marked participants are marked at two different positions in the verbal template of Yurakaré. Subjects are suffixed, objects are prefixed. Furthermore, there

---

[1] With grammatical relations I mean syntactic relations between a verb and dependent arguments and adjuncts. This definition subsumes both case and verb agreement (cf. Blake, 2001, p. 3).
[2] When a verb is marked for progressive, the direct object may be left out.

Competition and Variation in Natural Languages: The Case for Case
Edited by M. Amberber and H. de Hoop

ISBN: 0-08-044651-5

is a position in the verbal template between the object prefixes and the verb root where applicative markers may appear.

Dixon and Aikhenvald (2000, p. 3) distinguish between transitivity and valency: 'there are two main transitivity types — intransitive (with a core argument S) and transitive (with A and O)- and plain and extended subtypes of each. Valency relates to the number of core arguments.'

In Yurakaré, formal encoding of participants on the verb is based on transitivity rather than valency in the sense that roots are either intransitive or transitive. Additional arguments, whether obligatorily present or not, are encoded as applicative objects formally. Subjects are suffixed to the verb stem and objects are prefixed. Direct objects are encoded simply by prefixing object cross-reference markers to a transitive stem. Other types of objects are encoded by the same or at least a related paradigm of object cross-referencing prefixes plus some additional formal encoding indicating the type of object. I interpret this additional formal encoding to be a marker for extending basic transitivity. This means that there are only two types of roots in Yurakaré: intransitive and transitive roots, which can both be extended, transitives with maximally one additional argument, and intransitives with either one or two. On this basis I distinguish four types of clauses in Yurakaré with regard to transitivity (Table 1).

Note that, as far as valency goes, there are only three types of predicates: one place,[3] two place and three place.

Even though some verbs, like *-kaya-* 'give' require three participants, the indirect object is morphologically encoded as an applicative object (i.e. by means of

Table 1: Types of clauses in terms of transitivity and their valency.

Transitivity	**Valency**
intransitive	one
extended intransitive	two
transitive	two
extended transitive	three
doubly extended intransitive	three

[3] Even whether verbs seem to have a subject, since they can trigger different or same subject morphology:

a. *kwarenta dia kwarenta noche mashi-ø=**ja** tom wita-ø=ya ashaa=chi*
forty day forty night rain-3=SS IDEO arrive.SG-3=REP above=DIR
When it had rained forty days and forty nights it (the water) reached the sky'

b. *nij mashi-ø=**ti** nij shuñe-jti-ø=w arush*
NEG rain-3=DS NEG grow-HAB-3=PL rice
'If it doesn't rain, the rice won't grow'

a formal applicative marker). Therefore I treat this verb as an extended transitive, the recipient being the added argument.

I will first discuss this basic transitivity of roots (i.e. without derivational morphology): how are subjects and direct objects encoded,[4] and what semantic roles can the markers of these syntactic relations encode (Section 2.1)? Subsequently, I will discuss the five different applicatives of Yurakaré (Section 2.2).

2.1. Core Relations: Subject and Direct Object

Subjects are marked by suffixes that follow the root, derivational information and Tense, Mood and Aspect (TMA) markers. Three persons and two numbers are distinguished by the following morphemes:

(1) Cross-referencing paradigm subjects

1SG	-y	1PL	-tu
2SG	-m	2PL	-p
3SG	-ø[5]	3PL	-ø=w

These subject markers are obligatory. Participants that are cross-referenced on the verb may but need not be expressed overtly in addition to the cross-reference marker on the verb. Subject markers can encode a variety of semantic roles. Among them are the following:

(2) a. *së=ja*[6] *bobo-y* *ti-tiba* (agent)
I=S hit;kill-1SG.S 1SG-pet
'I hit my pet'

---

[4] I take as the first diagnostic for the different categories the formal differences in the way they are encoded. Frajzyngier and Shay (2003, p. 3) claim that 'meaning does not exist as some universe outside of language but exists only within a given language and only to the extent that the language codes it through some formal means.' Different means of encoding means different function. What this function is should be determined by looking at the different situations in which a certain coding means (e.g. morpheme, paradigm) appears. I will indicate for each subcategory of grammatical relations marking what the formal encoding is, and in what (semantic) circumstances it can appear.

[5] I assume a zero pronominal element here, since the third-person subject marker is pluralized with the nominal plural enclitic=*w* which, being an element that attaches to the last element of a noun phrase, needs a nominal element to have scope over. This is in accordance with the system of independent pronouns, in which the first and second person are suppletive for number, while the third person is referred to by means of a demonstrative pronoun which can be pluralized like other noun phrases: by means of the enclitic=*w*.

[6] The enclitic=*ja* is found on overtly expressed subjects, but not on all of them. It emphasizes the subject participant. Although it is only found on subjects, I do not primarily consider it to be a marker of grammatical relations. Instead, I consider it an emphasis marker, which is constrained by the subject relation. Therefore, I will not take this marker into consideration in this chapter.

b. *yujshë=ja* *bobo-ø* *yee* (force)
wind=S hit;kill-3 woman
'The tornado killed the woman'

c. *mitchi=w* *peta-ø=w* *elle=y* (theme)
cat=PL lie.PL-3=PL ground=LOC
'The cats lie on the ground'

d. *së=ja* *adyindye-y* (experiencer)
I=S sad-1SG.S
'I am sad'

Direct objects have less variety in the semantic roles they encode: canonically they encode participants that are directly affected by the action (patients):

(3) *ma-bobo-y*
3PL-hit;kill-1SG.S
'I hit/killed them'

The following paradigm is used to mark direct objects:

(4)
1SG	ti-	1PL	ta-
2SG	mi-	2PL	pa-
3SG	no marking[7]	3PL	ma-

Although most direct objects encode patients — they are directly affected by the action because they undergo a change in place or state — there are some cases in which they encode semantic roles which cannot be defined in this way:

(5) a. *ma-bëjta-y*
3PL-see-1SG.S
'I see them'

[7] There are a number of situations where the third person object of a transitive verb is cross-referenced with the prefix *ka-*. It is at this point unclear what exactly determines the presence or absence of *ka-* as a direct object marker, but it has to do with characteristics of the noun referred to rather than with voice, valency or predicate semantics. There are some nouns, such as animate nouns, that cannot trigger *ka-* as an object marker, there are some nouns that always trigger *ka-*, these nouns are always singular syntactically, and finally there is a small set of nouns, also always singular, which may or may not trigger the cross-reference marker *ka-*, for reasons not yet understood. The distribution and functions of *ka-* look like the description by Dixon (1999, p. 301) of this morpheme in Arawan languages. The morpheme *ka-* as far as it is connected to reference may even have a wider areal distribution, into Tucanoan and Arawak languages (cf. Metzger, 1998). For more information on the system of nominal number and classes of nouns in Yurakaré, cf. Van Gijn (2004).

b. *mi-willa-y*
2SG-love-1SG
'I love you'

The perceived entity in the perception verb in (5), as well as the stimulus in the emotion verb in (5) are encoded in the same way as canonical patients in many languages of the world (cf. Palmer, 1994, pp. 26–27). Palmer mentions that this has to do with the fact that perceivers and experiencers are animate and are therefore more like agents than the perceived or emotion stimulus which may or may not be animate. This may explain why the perceiver and the experiencer rather than the perceived and the stimulus are encoded as subjects, but it does not explain why the perceived and the stimulus are encoded as direct objects rather than another type of objects. This seems to be a lexically determined matter: some stems simply are transitive, i.e. take direct objects, while others are not, i.e. they take other types of objects.

2.2. *Applicatives*

Yurakaré has five applicatives, each with a different meaning. Table 2 shows the five applicatives with their meaning, formal representation and the kind of predicates they occur with.

The applicative markers are prefixed to the root of the predicate, and combined with the object prefixes of (4) above, with the exception of the third person singular, which is consequently cross-referenced with the prefix *ka-* when we are dealing with an applied object. In the verbal template, this looks as follows:

(6) object marker - applicative marker - root - subject marker

2.2.1. Comitative object The comitative participant is locationally close to the subject. It is a powerless undergoer of the same event the subject controls and

Table 2: Overview of applicatives.

Meaning	Marker	Applies to
comitative object	ø	intransitives
accompanying object	vowel change	intransitives, transitives
indirect object	-n-	intransitives, transitives
affected object	-la-	intransitives, transitives
objective object	-y-	intransitives

carries out. Roots that take this applicative are always intransitive and have an active interpretation. They are often movement verbs, but not necessarily so:

(7) a. *mala-ø*
go.SG-3
'He goes up'

b. *ka-mala-ø* *a-bëshë*
3SG-go.SG-3 3SG.P-clothing
'He brings his clothes'

c. *urup-ta-ø*
covered-MID-3
'He covers himself (e.g. with a blanket)'

d. *ka-urup-ta-ø*
3SG-covered-MID-3
'He covers himself together with someone else'

One could consider these verbs to be flexible with regard to their transitivity, that these are roots that can be considered to be either transitive or intransitive. However, I will consider these verbs to be extended intransitives because they behave differently in a number of respects.

First, and this is the main argument to consider these as subtypes of intransitives rather than transitives, consider the interaction with a system of participant-based verbal number that exists in Yurakaré. There is a small class of verbs that change stem according to the number of core participants. This system functions on an ergative basis: intransitive verbs change their stem according to the number of the subject participants, the transitive verbs change stems according to the number of object participants. The ergative pattern is characteristic of this type of verbal number in all languages investigated to date that have this system (cf. Corbett, 2000, p. 253), also the group of verbs that exhibits this suppletive opposition is similar across languages. This system of participant-based verbal number is observed in some geographically nearby languages such as Wari' (Everett & Kern, 1997, pp. 337–340), Itonama (Crevels, 2002, pp. 45–47) and Kwaza (Van der Voort, 2004, pp. 383–388).

The verb *mala* (with plural stem *bali*) is one of these verbs. But when it appears with the comitative participant it does not function as a transitive verb: it is still sensitive to the number of the subject participants:

(8) a. *mala-ø* 'He goes'
b. *bali-ø=w* 'They go'
c. *ka-bali-ø=w* 'They bring it'
d. *ma-mala-ø* 'He brings them'

This contrasts with 'real' transitives *chitta/kokko* 'throw away' and *mi(i)/pu(u)*[8] 'take', which have suppletive stems according to the number of object participants:

(9) a. *chitta-ø* 'He throws it away'
b. *ma-jokko-ø*[9] 'He throws them away'
c. *mii-ø* 'He takes it'
d. *ma-pu* 'He takes them'

The comitative *ka-* clusters with other applicatives, such as the affected object (cf. Section 2.2.4) in this respect:

(10) a. *ti-la-mala-ø*
1SG-AFO-go.SG-3
'He left me'

b. *ma-la-mala-ø*
3PL-AFO-go.SG-3
'He left them'

c. *ti-la-bali-ø=w*
1SG-AFO-go.PL-3=PL
'They left me'

A second difference between the transitive construction and the comitative object construction is that they have slightly different paradigms. Predicates with a comitative object consequently mark animate third person singular objects[10] with the prefix *ka-*, whereas transitive verbs cannot refer to such a referent with the cross-referencing prefix *ka-*, instead, they have to leave the animate third person singular object unmarked:

(11) *aysa* *müta-ø=ja* *ka-mala-ø=ya*
Aysa pull.out-3=SS 3SG-go.SG-3=REP
'When he had picked up Aysa, he took him with him.'

This is from a mythological story in which Aysa is one of the main characters. This animate being is the direct object of the verb *müta* 'pull out', here translated as 'picked up', and it is the comitative object of the verb *mala* 'to go (singular

---

[8] The vowel is doubled in some circumstances for prosodic reasons. For more discussion, cf. Van Gijn (to appear).
[9] Underlying /k/ changes to [h] (orthographically represented by *j*) when preceded by a vowel.
[10] Or any other third person singular object for that matter. I only mention animates here because with these nouns, the comitative object behaves morphologically different from the direct object.

stem)'. Aysa could not be referred to with *ka-* on the verb *müta*, while the same Aysa could not be referred to by the absence of object marking on the verb *mala*.

There are in fact a few verbs in Yurakaré that can be either intransitive or transitive without any additional marker:

(12) a. *daja-y*
hang-1SG.S
'I am hanging'

b. *daja-y* *ti-bëshë*
hang-1SG.S 1SG-clothing
'I hang up my clothing'

In (12) we can see that these verbs that are indeed flexible (even though I have found only two so far), indeed behave differently from the verbs with a comitative object.

The difference in paradigmatic structure is minimal, but real (Table 3).
The difference can be found in the third person singular, which is in boldface. Whereas third person singular direct objects are only referred to with *ka-* in a few cases (involving only a certain group of nouns, cf. footnote 7), third person singular comitative objects are always referred to with *ka-*. In this respect, the comitative also clusters with the other applicatives, which are all based on the paradigm in the right column of Table 3.

Finally, semantically there is a difference as well: the subject has to carry out the same event as the comitative object undergoes, whereas with transitive verbs this is not necessarily so. With transitives, the agent can affect the patient without undergoing the change in state or location him-/herself:

(13) *ma-jokko-y*
3PL-throw.away.PL-1SG.S
'I throw them away'

Table 3: Paradigms of direct objects and comitative objects.

	Direct object	**Comitative object**
1sg	ti-	ti-
2sg	mi-	mi-
3sg	**no mrk./ka-**	**ka-**
1pl	ta-	ta-
2pl	pa-	pa-
3pl	ma-	ma-

Because of these considerations I consider verbs of the type *kamala* 'bring/go with (sg)' to be extended intransitives rather than transitives.

2.2.2. Accompanying object The accompanying object is comparable to the comitative object, with the difference that the accompanying object is not a powerless undergoer. Instead, it is in many cases a controlling, voluntarily operating participant cooperating with, or accompanying the agent of the action. There is no separate morphological marker to mark the extension of the root valency, but this applicative is marked by means of a vowel change of the object prefixes, where *i* becomes *ë*, and *a* becomes *u*:

(14) object paradigm / accompanying object paradigm

	object paradigm		accompanying object paradigm
1SG	ti-	1SG	të-
2SG	mi-	2SG	më-
3SG	ka-	3SG	ku-
1PL	ta-	1PL	tu-
2PL	pa-	2PL	pu-
3pl	ma-	3pl	mu-

Examples of this marker are the following:

(15)

abëssë-ø	'he is playing'	*ku-bëssë-ø*	'he is playing with him'
abayla-ø	'he is dancing'	*ku-bayla-ø*	'he is dancing with her'
mala-ø	'he goes'	*ku-mala-ø*	'he follows him, goes with him'
dula-ø	'he makes it'	*ku-dula-ø*	'he helps him make it'
anënë-ø	'he is cooking'	*ku-nënë-ø*	'he helps him cooking'

Usually, the accompanying object is associated with the subject of the predicate. Occasionally, however, it is associated with the direct object rather than the subject:

(16) a. *ku-bu-che-ø*
3SG.AO-head.off-CAU-3
'Send him as well, make him go with another'

b. *ku-wili-che-ø*
3SG.AO-return-CAU-3
'He sends him back with him'

These are both causative versions of movement verbs; the accompanying object is associated with the agent of the underlying non-causative clause (the causee).

With a number of communication verbs, the accompanying object marker encodes the addressee, or conversation partner:

(17) a. *ku-ta-y*
3SG.AO-say-1SG.S
'I say to him'

b. *ku-ite-m*
3SG.AO-ask-2SG.S
'You ask him'

A verb like *ta* 'say' can also appear without a accompanying object, in which case the speaker's utterance is directed at no one in particular, as can be seen in the following fragment taken from a mythological text:

(18) *të-pshë* *li-tütü* *na* *ashaa=chi* *ta-ø=ya* *latiji*
what-entity DEL-sit-3 DEM above=DIR say-3=REP subsequently
na *samu*
DEM jaguar
'"What is it that up there?", the jaguar asked himself subsequently'

2.2.3. Indirect object Indirect objects are marked by the prefix *n-* (or its allomorph *m-*), which follows the object cross-reference markers and precedes the root of the predicate:

(19) cross ref. marker - **n** - root

This applicative encodes several different but related semantic roles. With movement verbs, or caused movement verbs such as 'giving', 'sending' and the like, it indicates the *goal* of that movement, or the *recipient*.

(20) a. *ti-n-wita-m*
1SG-IO-arrive.SG-2SG.S
'You came to me, arrived at my place'

b. *mi-m-bache-ni*
2SG-IO-send-IRR:1SG.S
'I am going to send it to you'

c. *a-mumuy* *pa-n-kaya-shti*
3SG.P- totality 2SG-IO-give-FUT:1SG.S
'I will give it all to you (pl)'

d. *ka-n-chaya-ø*
3SG-IO-feed-3
'She gave him food'

Related to this is the use of the indirect object applicative with communication verbs, where the indirect object encodes the *addressee*:

(21) a. *mi-n-dyuju-shti*
2SG-IO-inform-FUT:1SG.S
'I am going to inform you'

b. *ti-n-kama-ø*
1SG-IO-call-3
'He is calling me'

c. *ti-n-dyaju-ø*
1SG-IO-warn-3
'He warned me'

Apart from these two related meanings there is another function of this indirect object, and that is to indicate an interested party. This type of participant is often called ethical dative or dative of interest, especially in the romance tradition (cf. Van Hoecke (1996) for a number of other names given to this type of dative in Latin). Semantically, an ethical dative construction can be described as follows: 'the circumstances described by the predicate have significant consequences for the interested party, whose referent is not in control of the event' (Fried, 1999, p. 492). The type of interest that is indicated by this kind of construction can, in many languages be either beneficiary or adversative, without any formal distinction (cf. Shibatani,1994, p. 463). In Yurakaré, however, these two subtypes of the ethical dative have different formal encodings. The 'indirect object' encoding, with the prefix *n-*, is by and large associated with goal/approach and beneficiary, while the 'affected object' encoding, with the prefix *la-* which is discussed in Section 2.2.4, is associated with source/separation and adversity.

The *beneficiary* reading can be observed in the next examples:

(22) a. *ti-n-dula-ø* *ti-sibë*
1SG-IO-make-3 1SG-house
'He made me my house'

b. *ti-m-bëbë-ma* *ti-oshewo*
1SG-IO-search-IMP.SG 1SG-pot
'Find me my pot'

c. *ti-n-tëptë-ø* *ti-bëshë*
1SG-IO-wash-3 1SG-clothing
'She washed my clothes for me'

Verbs denoting states, processes or achievements also often carry this applicative. The indirect object denotes the person that benefits from the state of affairs that has arisen:

(23) a. *ti-n-duta-ø* *ayma*
1SG-IO-burn-3 fire
'The fire is burning (for me)'

b. *ti-n-kula-ø* *ti-chata*
1SG-IO-boil-3 1SG-food
'My meal is boiling'

c. *ti-n-sa-ø=naja* *ti-sawata*
1SG-IO-finished-3=CR 1SG-work
'My work is done'

d. *nij* *ti-n-suwita-ø* *ti-ballata*
NEG 1SG-IO-come.up-3 1SG-plant.seed
'My plants are not coming up'

2.2.4. Affected object The affected object marks a participant that is affected in a non-direct way by the event. Both transitive and intransitive roots can take this applicative marker. With transitive roots the affected object applicative can be either valency increasing, adding an argument to the existing valence, raising the valence from two to three, or it can be valency re-arranging. In the latter case, the patient of a transitive verb is expressed as an affected object, the applicative is a valence re-arranger instead of a valence increaser. The marker for affected object is the prefix *la-*, which is preceded by the cross referencing object prefixes:

(24) cross ref. marker - **la** - root

As mentioned, the indirect object and the affected object both mark participants that are affected by the event indirectly. The difference between them is the way in which the participant is affected. Indirect and affected objects are often opposite in this respect. For instance, where the indirect object-applicative marks the goal or recipient of a movement or caused movement verb, cf. the examples in (20) above, the affected object marks source or separation from an entity with this same class of verbs:

(25) a. *mi-la-bache-ni*
2SG-AFO-send-IRR:1SG.S
'I am going to send it away from you'

b. *ti-la-danda-ø*
1SG-AFO-go.up-3
'He went up on me (when he saw me he went up)'

c. *ti-la-shta-m*
1SG-AFO-throw.away.SG-2SG.S
'You took it from me and threw it away.'

d. *mi-la-bata-y*
2SG-AFO-head.off-1SG.S
'I am heading off, leaving you behind'

Another way in which the affected object is opposite to the indirect object is that where the indirect object participant benefits from some state of affairs, the affected object participant is detrimentally affected:

(26) a. *ti-la-che-m* *ti-chata*
1SG-AFO-eat-2SG.S 2SG-food
'You ate my food on me'

b. *ti-la-ense-ø* *ti-awaryente*
1SG-AFO-drink-3 1SG-alcohol
'He drank my alcohol on me'

c. *ti-la-dyinde-ø* *ti-nñu*
1SG-AFO-sad-3 1SG-baby
'My baby is sad (on me)'

Here we can see that there is some overlap with the separation-reading of the examples in (25), since examples (26a) and (b) also have a separation meaning, while the examples in (25) might also be explained as cases of detrimental affect. Many of these instances of malefactive/detrimental affect can be analyzed as external possessors. Core instances of external possession are described as "constructions in which a semantic possessor-possessum relation is expressed by coding the possessor as a core grammatical relation of the verb and in a constituent separate from that which contains the possessum" (Payne & Barshi, 1999, p. 3). Although external possession is certainly a sub-type of the affected object, this explains for instance the coreferentiality requirement in examples such as in (26), the affected object applicative has a wider use. Apart from the already mentioned use as a marker of entities undergoing separation, the affected object also marks locations which are used for some purpose, or entities being located in such a way that they are affected or bothered by the event, either consciously or unconsciously:

(27) a. *a-shuye=la* *ka-la-pëjta-jti-ø=ya* (...) *kummë*
3SG.P-crown=INS 3SG-AFO-place-HAB-3=REP tree
'He was placed with his head he placed against the tree'

b. *ti-la-dapa-ø* *papel*
1SG-AFO-stick-3 paper
'The paper sticks on me'

c. *ti-la-dele-ø* *ti-dala=y*
1SG-AFO-fall-3 1SG-head=LOC
'He fell on my head' (*tiladele* can also mean: 'it fell out of my hands')

d. *ti-la-bishmi-ø*
1SG-AFO-smoke-3
'He smoked on me (e.g. blew his smoke into my face)'

As can be seen in (27a), where the cross-referencing prefix licensed by the presence of the marker *la-* refers to *kummë* 'tree', the affected object does not need to be animate, something which makes the construction slightly different from the ethical dative, which is always animate (Fried, 1999, p. 492). Nevertheless, in Yurakaré too, affected objects are in almost all cases animate.

A final point to be made about the affected object is that, instead of adding an argument to the existing arguments, it can also rearrange the valency of a transitive root:

(28) a. *ti-duya-ø* *isuna*
1SG-sting-3 stingray
'The stingray stung me'

b. *ti-ma-la-duya-ø* *ti-ññu*
1SG-3PL-AFO-sting-3 1SG-baby:PL
'He stung my children on me'

c. *ti-la-duya-ø* *isuna*
1SG-AFO-sting-3 stingray
'The stingray is stinging me'

In (28a), the basic transitive use of the verb *duya* 'sting' is shown; in (28b), the applicative marker *la-* licenses another cross-reference marker, in addition to the direct object marker. In (28c), finally, the patient, which was expressed as a direct object in (28a) is expressed as an affected object. The semantic effect of expressing the patient of a transitive verb as an affected object is that the situation is in progression.[11]

2.2.5. Objective object Objective objects place the strictest restrictions on the root they can attach to. They only attach to intransitive predicates expressing movement. The objective object expresses the participant which is the ultimate

[11] It may be that these instances should be analyzed as bearing the progressive prefix *a-*: *ti-l-a-duya-ø*. The prefix *la-* reduces to *l-* before a vowel.

goal of the movement; it can often be translated with 'to get X'. The applicative marker is the prefix *y-*:

(29) cross ref. marker - **y** - root
Objective objects can be either animate or inanimate:

(30) a. *ka-y-danda-ma asha=chi toronja*
3SG-OO-go.up-IMP.SG above=DIR grapefruit
'Go up to get me a grapefruit'

b. *ka-y-mala-ma mi-bashti*
3SG-OO-go.SG-IMP.SG 2SG-wife
'Go and get your wife'

The objective object can also more generally refer to an entity which is associated with the goal of the movement without the intention of getting this entity:

(31) a. *të-y-yupa-ø*
1SG-OO-enter.SG-3
'He entered my house'

b. *pa-y-yajta-n-tu li-massë-p=chi*
2PL-OO-run;jump-IRR-1PL.S del-stand.pl-2pl s=dir
'Let's run to where you guys are!'

c. *ka-y-dele-ø bëjta-ø=ja*
3SG-OO-fall.SG-3 see-3=SS
'When he saw him_i, he fell where he_i was'

Like the comitative object versus accompanying object opposition, there is a marginally used distinction for objective objects as well in the same fashion:

(32) *pëpësu ka-la-pëjta-jti-ø kumme ku-y-dandu-ntu-ø=chi*
pëpësu 3SG-AFO-place-HAB-3 tree 3SG.AO-OO-go.up-DES-3=DUB
'Pëpësu placed himself against the tree, wanting to go up (to reach another person who went up already)'

2.2.6. Combinations of applicatives There can be maximally three arguments encoded on the verb. One of these is always the subject, the other two can vary. Combinations of the direct object with the applicative objects (except for comitative and objective object) are the following:

Direct object and accompanying object

(33) *ku-ma-weshwe-y*
3SG.AO-3PL-scrape-1SG.S
'I help him scraping them'

Direct object and indirect object

(34) *ti-ma-n-dula-ø*
1SG-3PL-IO-make-3
'He makes them for me'

Direct object and affected object

(35) *ti-ma-la-danche-ø*
1SG-3PL-AFO-go.up:CAU-3
'He put them up there on me'

In all three examples, the marker referring to the direct object comes in between the applicative marker and the cross referencing suffix associated with that applicative marker:

(36) AO - DO - root
IO
AFO

The question then arises how two applicatives and their cross-reference markers are situated on the verb. I have found the following combinations.

The comitative object behaves like a direct object as far as placement is concerned: before the applicative marker, after the cross-referencing prefix associated with that applicative marker:

Comitative object and indirect object

(37) *ti-ja-m-mala-ma* *ti-oshewo*
1SG-3SG-IO-go.SG-IMP.SG 1SG-pot
'Bring me my pot!'

Comitative object and affected object

(38) *ka-ja-la-mala-ø* *a-juyja-ta* *petche*
3SG-3SG-AFO-go.SG-3 3SG.P-fish-MID fish
'The fish went off taking his fishing line'

The objective object cross-referencing suffix follows the prefix referring to the indirect object. The applicative marker for the indirect object is lost, but the vowel in the prefixes with the vowel /a/ in them (third person singular and the plural participants) is changed to /u/,[12] in the example provisionally glossed 'VC' for 'vowel change':

[12] There is some speaker variation in these constructions in that some speakers also change the /a/ to /u/ in the prefixes referring to the objective object.

Indirect object and objective object

(39) a. *ti-ja-y-danda-ma* *toronja*
1SG-3SG-OO-go.up-IMP.SG grapefruit
'Go up and get me a grapefruit'

b. ***ku**-ja-y-mala-ma*
3SG(VC)-3SG-OO-go.SG-IMP.SG
'Go and bring it to him!'

c. ***tu**-ma-y-mala-ma* *dulsi=w*
1PL(VC)-3PL-OO-go.SG-IMP.SG orange=PL
'Go and get us some oranges'

Indirect object and affected object

In (40), where the indirect object and the affected object are combined, both applicative markers are present: *la-* for affected object, and *n-* for indirect object, in that order. The cross-referencing prefixes show the opposite order.

(40) *ti-ja-la-n-dyalala-ø*
1SG-3SG-AFO-IO-spit-3
'He spits at him for me'

The order of cross referencing prefixes can be schematized into the following template, whereby direct, comitative and objective objects are mutually exclusive, since they are only found in combination with intransitive roots. The accompanying object is placed in between the affected object and the indirect object, since we lack examples to be sure.

(41)

1		2	3	4	5
IO		AFO	DO	appl. (cluster)	root
	AO		COM		
			OO		

3. Dependent Marking: Postpositions

Apart from the system of affixing participants onto predicates, Yurakaré also has a system of marking the case relation on the dependent noun by means of postpositions. Yurakaré has the following postpositional markers:

(42) *=tina* comitative
=la instrument
=y locative
=chi directional
=jsha ablative

I consider these markers to be postpositions rather than case markers. Payne (1997, p. 100) describes case markers and adpositions as two extremes of a continuum, giving the following rule of thumb to distinguish them:

> Case marking is the morphosyntactic categorization of noun phrases that is imposed by the structure within which the noun phrase occurs. Adpositions are free of such configurational constraints.

None of the roles encoded by the markers in (42) are determined by the grammatical requirements of a verb. I furthermore consider the markers to be enclitics because they attach to different hosts, as long as it is the final element of the noun they mark:

(43) a. *inele* *elle=y*
inside earth=LOC
'in the ground'

b. *elle* *inele=y*
earth inside=LOC
'in the ground'

c. *danda-shti* *matata* *kummë=la*
go.up-FUT:1SG.S large tree=INS
'I am going to climb that large tree'

d. *danda-shti* *kummë* *matata=la*
go.up-FUT:1SG.S tree large=INS
'I am going to climb that large tree'

In (43a) and (43b), the postposition =*y* is the final element in both utterances. The only difference between (43a) and (43b) and between (43c) and (43d) is the order of the content words of the noun phrase, while the postposition remains where it is: at the end of the NP. In (43a), the postposition attaches to the head noun, in (43b) it attaches to the relator noun.[13] In (43c) and (43d), the same phenomenon is exemplified with another postposition: =*la* 'instrumental', which attaches to the head noun in (43c), while it attaches to the adjective in (43d).

In this section I will discuss each of the postpositions mentioned in (42) above: =*tina* 'comitative' (3.1), =*la* 'instrument' (3.2) and the local postpositions =*y* 'locative', =*chi* 'directional' and =*jsha* 'ablative' (3.3).

[13] There is a closed class of relator nouns which specify the meaning of the local postpositions. They are often, but not always, encoded as possessed items. A discussion of these relator nouns, although they play an important role in marking locations, falls outside the area of interest for this chapter. The reader is referred to Van Gijn (to appear).

3.1. =tina 'comitative'

The postposition *=tina* marks a comitative participant. Comitative can be defined as "expressing with whom an entity is located" (Blake, 2001, p. 197).

(44) a. *na* *yolojoto* *shunñe* *a-ye=tina* *li-tütü-ø*
DEM armadillo man 3SG.P-sister=COM DEL-be-3
a-sibë=y
3SG.P- house=LOC
'That armadillo-man was in his house with his sister'

b. *l-ati* *mi-ø=w=ja* *bobo-ø=w=ja* *sebbe=tina*
REF-DEM take.SG-3=PL=SS kill-3=PL=SS peanut=COM
dula-ø=w=ja *tiya-ø=w=ya* *ati*
make-3=PL=SS eat-3=PL=REP DEM
'They took it, killed it, prepared it with peanuts and ate it'

In (44a), the comitative participant is associated with the subject participant. They have exactly the same role and could in principle both be encoded as subjects. The fact that this is not the case has to do with topicality. The situation is presented from the viewpoint of the subject. The comitative participant is presented as being less topical. This becomes more clear in (44b), taken from a mythological text. The subject participants have captured a deer by making it trip over a rope. The topical direct object in this sequence of events is the deer, and that the peanuts are associated with it as a non-topical entity.

In the examples in (44), the comitative participant and the core participant they are associated with have the same semantic roles (agent and patient in (44a) and (44b), respectively). The roles of the core participant and the comitative object do not have to be exactly parallel, however:

(45) *yupa-ø=ya* *a-mumuy* *a-ballata=tina*
go.in.SG-3=REP 3SG.P- totality 3SG.P- seed=COM
'He went in with all of his seeds'

In the above example both the subject and the comitative object go in, but the subject actively, and the comitative object passively.

The interpretation of *=tina* can also have a reciprocal shade of meaning in the sense of cooperating:

(46) *lëtta* *më-jti=ja* *achu-ta* *dula-shta-m* *mi-ye=tina*
one you(SG)-LIM=S like.that-MID do;make-FUT-2SG.S 2SG-sister=COM
'Only you will do like that (i.e. making babies together), with your sister'

In some cases =*tina* encodes participants that are loosely associated with the event.

(47) *achu-ta dula-tijti shinama pa=tina att-ima*
like.that-MID do-HAB.1PL.S before you(PL)=COM other-COL
jente=w=tina
people=PL=COM
'That is what we used to do with you, and other kinds of people'

(48) *l-achu-ta ma-n-tütü-ø=w yarru=tina chajti=ya*
REF-like.that-MID 3PL-IO-be-3=PL chicha=COM always=REP
ma-ba=w
3PL-husband=PL
'Like that, they always awaited their husbands with chicha'

Grammatically, the comitative object postposition is in complementary distribution with head marking in the above examples. This is not always the case, however:

(49) a. *li-ma-tütü-ø=w=ya*[14] *a-ye=tina inele a-sibë=y*
DEL-3PL-be-3=PL=REP 3SG.P-sister=COM inside 3SG.P-house=LOC
'They were inside the house together, with his sister'

b. *më-jti lëtta-m ku-winani-shta-m mi-ye=tina*
you(SG)-LIM one-2SG.S 3SG.AO-walk-FUT-2SG.S 2SG-sister=COM
'You will be the only one that is going to live together with your sister'

The story from which this utterance was taken deals with two persons: a brother and a sister. The plural subject marker refers to both the brother and the sister, but the sister is additionally marked with the comitative marker. This is a construction that is often heard, also in local Spanish, as in e.g. *vamos conmigo* 'let's go with me', meaning 'let's go you and me'. The semantic effect of backgrounding an accompanying object or associated object is also visible here: the topical entity in (49) is the man. I will come back to this effect of the comitative marker below.

[14] The form *limatütüwya* seems to show double agreement, once in object position by means of the prefix *ma-*, and once with the suffixes *-ø=w* in the subject positions. The latter is indeed subject agreement in person and number, the former however seems to have a more derivational character, indicating that the action is performed together. This prefix only corresponds to the number of the subject, since there are instances such as the following, where the object prefix does not change to a 1PL-prefix:

a. *ma-daja-tu kummë=la*
3PL-hang-1PL.S tree=INS
'We are hanging in the tree together'

3.2. =la instrument

The postposition *=la* has a range of functions; the three most common are reason, path and instrument. I will gloss the suffix with INS for instrument, since this seems to be the basic notion from which the other uses are derived:

Instrument

(50) *wissa-jti-ø=w=ya ana a-dala katcha=la*
hit-HAB-3=PL=REP DEM 3SG.P-head ax=INS
'They hit his head with an ax'

Reason

(51) *ati a-bombo=la shama-ø=w=ya ta-ppë-shama=w*
DEM 3SG.P-smoke=INS die.PL-3=PL=REP 1PL-grandfather-DEC=PL
'Our ancestors died because of its (the fire's) smoke'

Path

(52) a. *danda-ø=w=ya kummë=la*
go.up-3=PL=REP tree=INS
'They went up the tree'

b. *ana sekure=la mala-ø ku-ta-jti-ø dojo=chi*
DEM secure=INS go.SG-3 3SG.AO-say-HAB-3 body=DIR
'They say he went upstream on the Secure-river'

The core function of the marker can be argued to be 'instrument' if you see instrument as something that you can use to get from situation (or location) A to B. In (51), *bombo* 'smoke' is the metaphorical instrument that causes a change of state in the ancestors: from living to dead; a path is an instrument that connects two places.[15] There are some other constructions where the postposition *=la* occurs:

(53) *mu-ta-jti-ø=w=ya tuwa ta-buybu=la*
3PL.AO-say-HAB-3-PL-REP we 1PL-language=INS
'They spoke with them in our language'

(54) *ma-ba=w ma-l-iwaja-jti-ø=w=ya*[16] *ma-yaru=la*
3PL-husband=PL 3PL-AFO-serve-HAB-3=PL=REP 3PL-chicha=INS
'They served their husbands chicha'

---

[15] Schlesinger (1995, pp. 15, 16, 78–84) links — among others — the three conceptual spaces of instrument, path and reason/cause.

[16] Verbs beginning with /i/ do not take direct objects. Patients are formally encoded as indirect object (with the prefix *n-*) and beneficiaries/goals are encoded as affected objects formally.

The use of =*la* in the examples (53) and (54) is related to the 'instrument' function: the language is used as an instrument to intermediate between two people; the serving is done with the help of chicha (manioc or corn beer).

The examples in (55) are related to the 'reason' function of =*la*. Water and food, or rather the lack of them, are the reasons for the starving and the almost dying of the subjects.

(55) a. *imi-ø=ya* *latiji* *samma=la*
starve-3-REP subsequently water=INS
'She was starving with thirst, craving for water'

b. *sey* *shamu=ya* *shama-shtu=y=chi*
almost die.PL:3PL.S=REP die.PL-FUT:3PL.S=REP=DUB
latiji *hata=la*
subsequently food=INS
'They almost died of hunger'

The path function, finally, also has some extensions:

(56) a. *ashaa=y* *li-ma-ssë-ø=w* *lëshie* *shunñe=w* *kummë=la*
above=LOC DEL-3PL-be-3=PL two man=PL tree=INS
'That man is hanging in the tree'

b. *a-tushu=la* *bushu-ø=ya*
3SG.P- chest=INS lie.down.SG-3=REP
'He lay down on his chest'

These are instances of what Talmy (2000, pp. 99–175) calls 'fictive' motion. I will come back to this phenomenon when discussing the local postpositions in the next section.

3.3. Local postpositions

There are three local postpositions in Yurakaré, leaving aside the path function of =*la*: =*y*, =*chi* and =*jsha*. The postposition =*y* marks the stative location of an object or person:[17]

(57) a. *basu* *tütü-ø* *mesa=y*
cup be-3 table=LOC
'The cup is on the table'

[17] Some of the examples in this section have been elicited with the help of two elicitation tools developed at the Max Planck Institute for Psycholinguistics in Nijmegen: Ameka, de Witte, and Wilkins (1999) and Bowerman (1993).

b. *mesa=y* *bushu-ø* *emme*
table=LOC lie.SG-3 meat
'The meat lies on the table'

c. *lampara* *daja-ø* *ashaa=y*
lamp hang-3 above=LOC
'The lamp hangs in the air'

The marker =*chi* marks the goal of a movement:

(58) a. *mala-y* *ti-sibë=chi*
go.SG-1SG.S 1SG-house=DIR
'I went to my house'

b. *ëshshë=ja* *ka-mala-ø* *inele* *sibbë=chi*
stone=S 3SG-go.SG-3 inside house=DIR
'The stone rolled inside the house.'

c. *juan=ja* *mala-ø* *adyum=chi*
juan=S go.SG-3 river=DIR
'Juan went to the river'

Finally, the marker =*jsha* marks the source of a movement:

(59) a. *juan=ja* *otto-ø=ja* *mala-ø* *a-sibë=jsha*
juan=S go.out-3=SS go.SG-3 3SG.P-house=ABL
'Juan went from his house'

b. *pelota=ja* *ka-mala-ø* *sibbë* *inele=jsha*
ball=S 3SG-go.SG-3 house inside=ABL
'The stone rolled out of the house'

c. *mitchi* *shopto-ø* *mesa* *a-dojo=jsha*
cat go.down-3 table 3SG.P-body=ABL
'The cat came down from the table.'

As with =*tina* and =*la* above, the use of these markers can be extended on the basis of these core functions. The stative location marker can sometimes also mark the goal of a movement:

(60) a. *patta* *yupa-ø* *Juan* *a-tewe=y*
thorn enter.SG-3 Juan 3SG.P-foot=LOC
'A thorn entered Juan's foot'

b. *sulayle=ja* *dele-ø* *mesa* *a-dojo=jsha* *elle=y*
leaf=S fall.SG-3 table 3SG.P-body=ABL ground=LOC
'The leaf fell from the table onto the ground'

c. *mitchi* *danda-ø* *mesa* *a-dojo=y*
cat go.up-3 table 3SG.P-body=LOC
'The cat climbed onto the table'

The marking of a goal with the stative locative marker occurs with predicates that have an inherent direction, like *yupa* 'to enter', *dele* 'to fall (down)', *danda* 'to go up'. This marking is optional: the complement can also be marked by the directional marker:

(61) *Juan* *yupa-ø* *a-sib=chi*
Juan enter.SG-3 3SG.P-house=DIR
'Juan entered his house'

With these movement verbs of inherent direction, there is no difference in the interpretation of a complement marked by =*y* and a complement marked by =*chi*.

The other way round is also possible; a directional marker marking a location:

(62) *Juan=ja* *li-tütü-ø* *a-sibë* *inel=chi*
Juan=S DEL-be-3 3SG.P-house inside=DIR
'John is inside his house'

The combination of a stative locative verb and the directional marker occurs when the speaker is not at the location talked about at the moment of speech.

The examples in (62) and, as mentioned, (56), are examples of fictive motion (Talmy, 2000, p. 104):

> (...) languages systematically and extensively refer to stationary circumstances with forms and constructions whose basic reference is to motion. We can term this constructional fictive motion.

There are more instances of fictive motion in Yurakaré; the ablative marker can mark a stative location, and even the goal of a movement in some circumstances:

(63) a. *Juan=ja* *li-tütü-ø* *ayma* *a-shëli=jsha*
juan=S DEL-be-3 fire 3SG.P-side=ABL
'Juan is lying next to the fire'

b. *ana=jsha* *i-tanti-ø=w*
DEM=ABL VBL-eye-3=PL
'They had eyes on this side (the back of their head)'

c. *ti-la-dyalala-ø* *ti-tanti=jsha*
1SG-AFO-spit-3 1SG-eye=ABL
'He spat into my eye'

If an entity is on the outside of some bounded location, it can be marked by the ablative marker. So in (63a), the entity *Juan* is on the outside of the location *fire*, and in (63b), on the outside of the back of the head, referred to by the demonstrative *ana*. Example (63c) has a predicate with inherent direction. As we have seen above, the affected object can sometimes refer to an involuntary recipient participant, so the complement may be marked as a stative location. The eye can be seen as having an outside where the spit is located; hence it can be marked with an ablative marker, even if it marks the goal of a movement.

Another extension of the locational markers *=y* and *=jsha* is that they can be used to derive temporal adverbs:

(64) a. *ati-kka=y*
DEM-MEA=LOC
'in, during that time'

b. *numa=jsha*
night=ABL
'tomorrow'

c. *l-ati=jsha*
REF-DEM=ABL
'then, after that'

The first example literally means 'in the size of that', but can be interpreted as a temporal adverb, meaning 'in that stretch of time'. *Numajsha* is the only term for 'tomorrow' and literally means 'from the night', interpreted as 'after the night'. *Latijsha*, finally, is a frequent discourse organizing adverb. The prefix *l-*, together with the demonstrative *ati,* refers to a previously mentioned event, the ablative marker indicates that the event that follows occurs after that.

4. Head Versus Dependent Marking

In the preceding sections we have seen that Yurakaré has a mixed system of head and dependent marking to encode grammatical relations. Governed relations (subject and object) are always head marked. There is no dependent-marking strategy for the indirect object either. For most of the uses of the other applicatives there are dependent strategies imaginable.

Now that we have seen the general outlines of the Yurakaré system of marking grammatical relations, we can direct our attention to the question what principles govern this split. Nichols (1986) in her paper on head and dependent marking of syntactic relations, mentions that languages that have a split (i.e. partly head-marking and partly dependent-marking) system for marking syntactic relations follow regular principles. She mentions a number of these principles, but the most important for the purposes of the present chapter is the following:

(65) MOST LIKELY LEAST LIKELY
Governed > Subcategorized > Inner adverbials > Outer adverbials

The hierarchy in (65) is taken from Nichols (1986, p. 78) and shows the relative propensity of nominals to trigger verbal marking. With 'governed' she refers to "dependents whose presence is required, and whose morphological form is determined, by their heads". Subcategorized is used for those dependents "whose presence is required, but whose form is not determined by the head". Inner and outer adverbials are adjuncts, which are ranked on purely semantic grounds: entities like locative, instrument, manner are more likely to be cross-referenced on the verb than adverbial notions like reason, condition, purpose and the like.

If we apply this hierarchy to Yurakaré, we could say that governed[18] arguments are always head-marked, and outer adverbials are always dependent marked. The two categories in the middle, subcategorized arguments and inner adverbials cannot be classified into either head-marking or dependent-marking. For instance, if you look at goals of movement verbs, sometimes they are marked on the head (as indirect object) sometimes on the dependent, by means of the postposition =*chi*. This distinction by and large coincides with animate (head-marked) and inanimate (dependent marked) goals. Animate goals can also be dependent marked, but inanimate goals cannot be head-marked as indirect objects. Comitatives can either be head-marked (*ku*-paradigm, *ka*-paradigm) or dependent marked (=*tina*). I will come back to the comitatives below. And even an outer adverbial-like purpose or reason can be marked on the head (objective object), even though the fact that this is restricted to movement verbs tells us that in Yurakaré we should not consider these kinds of purposive arguments as outer adverbials, since these are supposed to be relatively free to combine with any type of verb.

In short, the hierarchy in (65), cannot fully explain the head-dependent marking split in Yurakaré. The cut-off point for head marking and dependent marking seems to lie somewhere within the middle categories of subcategorized arguments and inner adverbials. There must be some additional principle involved.

With the comitative relation there is a choice as to whether it is head marked. Comitatives can be encoded as (plural) subjects, accompanying objects, comitative objects or as adjuncts, marked with the postposition =*tina*. Accompanying objects and comitative objects are distinguished from each other simply on

[18] With this I mean government at the level of the VP. I am not talking about governing relations within the PP between the postposition and the NP.

semantic grounds: they encode different semantic roles. Accompanying objects are most active and conscious participants, while comitative objects are affected and undergoing participants. Connected to this point the accompanying objects are animate, while comitative objects are either animate or inanimate. The comitative object and the subject are also distinguished from each other on semantic grounds: the subject is active and potentially affecting, the comitative object is passive and affected. It appears that in text usage, the accompanying object and the subject participant can also often be distinguished from each other on semantic grounds, in the sense that the two actions, one performed by the accompanying object, one by the subject, happen more or less independent from each other. They often have different event-times and the accompanying object is frequently a goal of a movement, albeit an active one, while the subject is the entity that directly controls the event:

(66) *otto-ø=ja* *mala-ø* *ku-mala-ø*
go.out-3=SS go.SG-3 3SG.AO-go.SG-3
'He came out and went, he followed him'

So, what we are left with, and what is of the greatest interest for the present chapter, what is the difference between head and dependent marking strategies for comitative? Note that we cannot, like with goals of movement verbs, explain the split in terms of animacy. Although accompanying objects are always animate, comitative objects can be either animate or inanimate, something which can also be said for dependent marked comitative objects.

Text examples suggest that the dependent marked comitatives are downplayed in terms of pragmatic salience. For instance, as an illustration, the following fragment comes from a mythological text about two protagonists: a man and his sister who escape from a devastating fire that threatens to extinguish all of mankind. The story is told from the perspective of the man, possibly because the story is told by a man. When the woman plays a role equivalent to that of the man, she is mentioned as an adjunct, marked with the postposition =*tina*:

(67) *li-ma-tütü-ø=w=ya* *a-ye=tina* *inele* *a-sibë=y*
DEL-3PL-be-3=PL=REP 3SG.P-sister=COM inside 3SG.P-house=LOC
shonko=y
hole=LOC
'They were inside his house with his sister, in the hole'

The woman is, in other words, downplayed in topicality relative to the man.

Another reason for using the marker =*tina* is to introduce a new participant. The following fragment is from a story about Noah's ark. The protagonist of the story is clearly Noah himself. The whole story is presented from his perspective.

If at a certain point in the story another entity joins Noah in the agent role, this participant is marked with the postposition=*tina*:

(68)	*l-achu*	*imbëtë-ø=ja*	*sawata-ø*	*an*	*noe*	*a-bontu=tina*
	REF-like.that	behave-3=SS	work-3	DEM	noah	3SG.P-son:PL=COM

'When he had done that, Noah worked with his sons'

This is the first mention of Noah's sons. Note that they join Noah in the protagonist's role; they do not replace him. They are introduced as adjuncts.

The opposition between head marking and dependent marking in the comitative semantic field then seems to be related to topicality. The more topical a participant, the more likely it is that this participant role is marked on the head.

Now consider data given in Table 4, which come from a count of five narrative texts.[19] There are a couple of differences that stand out. First, there is a difference between intransitive and transitive subjects in terms of the number of times they are overtly expressed. Intransitive subjects are overtly expressed in 36.5% of the cases, while transitive subjects, i.e. subjects of extended intransitives or transitives are overtly expressed in only about 18% of the cases.[20] This is a well-known phenomenon which can be attributed to information structure. New

Table 4: Percentages of overtly expressed and non-expressed head-marked arguments.

		%S overt	%S covert	%O overt	%O covert	%AP overt	%AP covert
%intr	43.17	36.46	63.54	*	*	*	*
%extintr	28.09	17.90	82.10	*	*	14.79	85.21
%tr	20.66	18.52	81.48	38.62	59.26	*	*
%other	8.09						

[19] The total number of head-marked arguments in these texts is 917. S = subject (both intransitive and transitive), O=direct object, AP=applied object.

[20] The category 'other' consists of nominal predicates and ditransitive verbs. The latter group, of course, is of interest for this chapter. However, the low frequency of ditransitive predicates in these texts (only 14 occurences) does not permit to draw any firm conclusions. Nevertheless, even these few instances show the same tendencies. Of the 14 occurences, 11 were extended transitives; of these 11 instances, the direct object was expressed overtly 9 times (81.8%), while applied objects were overtly expressed only twice (18.2%). There were three instances of doubly extended intransitives, one combination of an objective object and an indirect object, where both applied objects were not overtly expressed, and two combinations of a comitative object with an indirect object; a combination that is semantically close to an extended transitive. In both these occurrences, the comitative object was overtly expressed, while the indirect object remained unexpressed.

information is typically expressed in either intransitive subject or transitive object role, positions more salient than the transitive subject position. This is a universal tendency in spontaneous discourse (cf. Du Bois, 2003). This also fits with the fact that direct objects are overtly expressed in roughly the same number of cases as intransitive subjects.

More interesting for the present chapter is the difference between the percentage of overtly expressed applied objects (AP) in extended intransitives and the percentage of overtly expressed direct objects in transitive verbs: 15% against 39%. This suggests that applied objects of intransitive stems are more often topical than direct objects of transitive stems. If we take the non-expression of an argument to be a direct indication for its topical status, we must conclude that, as far as information structure is concerned, applied objects and transitive subjects group together as non-salient positions, which are used for old information, already introduced topics.

5. Conclusion

The facts summarized in Table 4 combined with the facts concerning the comitative participants suggest that the split between head marking and dependent marking in Yurakaré has to do with (i) argument structure of predicate roots and (ii) topicality. Arguments that are part of the core argument structure of the root (subject and object) are always head marked. The other head-marked arguments are applied arguments, which have the tendency to be highly topical.

According to Kuno (1976, cited in Payne, 1997, p. 151), there is an empathy principle of human communication: "Human beings tend to select as topics entities with whom they empathize." One could say that a location or a path for instance is also affected by an action, but locations and paths generally do not evoke so much empathy. Animate beings are more likely to be topical than non-animate beings. This tendency is also visible when we connect applicatives and animacy. A count of the five narratives used for the results in Table 4 implies that 85% of the entities that are encoded as applied objects are human, and 92% are animate, while among the dependent marked participants only 16% are animate.[21] The animate dependent marked participants are almost exclusively marked with =*tina* (91%). This can be explained by the fact that, as we have seen above, the postposition=*tina* is often used to downplay participants that are normally head marked.

[21] The total number of dependent marked-participants in these texts is 139.

All of the head-marked participants can be said to be affected in some way. This is most clear with the comitative object, objective object, indirect object and the affected object, but also in a sense the accompanying object. In the case of the accompanying object it is affected by the fact that it is joined by another agent (encoded as the subject), over which it has no control. This makes these participants analogous to direct objects in some sense. It is therefore no coincidence that they are treated as direct objects morphologically. If we combine this with the empathy principle we might say that affected participants are likely to be treated as direct objects if they evoke empathy.

These three criteria — argument structure of the root, topicality (and animacy) and affectedness — seem to determine which participants get to be marked on the head and which are dependent marked. This applies both synchronically — in the case of the comitative semantic field where there is competition and variation, or diachronically in the case of the indirect object, where there is no competing-dependent marking strategy.

Acknowledgements

I am thankful for the insightful comments of an anonymous reviewer, which helped me put the data of Yurakaré in a wider theoretical, typological and areal perspective. All remaining errors and shortcomings are mine. The data cited in this chapter come from the recordings and fieldnotes I collected during four trips to the field. I am thankful to my consultants who shared their knowledge with me generously.

References

Ameka, F., de Witte, C., & Wilkins, D. (1999). Picture series for positional verbs: Eliciting the verbal component. In: D. Wilkins (Ed.) *Manual for the 1999 field season, version 1.0* (pp. 48–56). Ms. Nijmegen: Language and Cognition group, Max Planck Institute for Psycholinguistics (and accompanying picture series).

Blake, B. J. (2001). *Case* (2nd ed.). Cambridge: Cambridge University Press.

Bowerman, M. (1993). Bowerman's topological relations picture series: Topological paths and static relations. In: *Space stimuli kit 1.2* (pp. 40–50). Ms. Nijmegen: Cognitive Anthropology Group, Max Planck Institute for Psycholinguistics (and accompanying picture series).

Corbett, G. G. (2000). *Number*. Cambridge: Cambridge University Press.

Crevels, M. (2002). Itonama o Sihnipadara: lengua no clasificada de la Amazonía Boliviana. *Estudios de lingüística, 16*, 5–56.

Dixon, R. M. W. (1999). Arawa. In: R. M. W. Dixon & A. Y. Aikhenvald (Eds), *The Amazonian languages* (pp. 239–306) Cambridge: Cambridge University Press.

Dixon, R. M. W., & Aikhenvald, A. Y. (2000). Introduction. In: R. M. W. Dixon & A. Y. Aikhenvald (Eds), *Changing valency: Case studies in transitivity* (pp. 1–29) Cambridge: Cambridge University Press.

Du Bois, J. W. (2003). Discourse and grammar. In: M. Tomasello (Ed.), *The new psychology of language: Cognitive and functional approaches to language structure* (Vol. 2, pp. 47–87) Nahwah, NJ: Erlbaum.

Everett, D. L., & Kern, B. (1997). *Wari': The Pacaas Novos language of Western Brazil.* London: Routledge.

Frajzyngier, Z., & Shay, E. (2003). *Explaining language structure through sytems interaction.* Amsterdam/Philadelphia: John Benjamins.

Fried, M. (1999). From interest to ownership: A constructional view of external possessors. In: D.L. Payne & I. Barshi (Eds), *External possession.* (pp. 473–504). Amsterdam/Philadelphia: John Benjamins.

Kuno, S. (1976). Subject, theme and the speakers empathy — a reexamination of relativization phenomena. In: C. N. Li (Ed.), *Subject and topic* (pp.137–153). New York: Academic Press.

Metzger, R. G. (1998). *The morpheme KA- of Carapana (Tucanoan).* SIL Electronic working papers 1998–003.

Nichols, J. (1986) Head-marking and dependent-marking grammar. *Language*, *62*(-1), 56–119.

Palmer, F. R. (1994). *Grammatical roles and relations.* Cambridge: Cambridge University Press.

Payne, D. L., & Barshi I. (1999). External possession: what, where, how and why. In: D. L. Payne & I. Barshi (Eds), *External possession* (pp. 3–29). Amsterdam/Philadelphia: John Benjamins.

Payne, T. E. (1997). *Describing morphosyntax.* Cambridge: Cambridge University Press.

Schlesinger, I. M. (1995). Cognitive space and linguistic case: Semantic and syntactic categories in English. Cambridge: Cambridge University Press.

Shibatani, M. (1994). An integrational approach to possessor raising, ethical datives, and adversative passives. In: S. Gahl, A. Dolbey & C. Johnson (Eds), *Proceedings of the twentieth annual meeting of the Berkeley Linguistics Society. General session, dedicated to the contributions of Charles J. Fillmore.* Berkeley: Berkeley Linguistics Society.

Talmy, L. (2000). *Toward a cognitive semantics (Vol. I): Concept structuring systems.* Cambridge, MA/London: MIT Press.

Van der Voort, H. (2004). *A grammar of Kwaza.* Berlin/New York: Mouton de Gruyter.

Van Gijn, R. (2004). Number in the Yurakaré noun phrase. In: L. Cornips & J. Doetjes (Eds), *Linguistics in the Netherlands 2004.* (pp. 69–79). Amsterdam/Philadelphia: John Benjamins.

Van Gijn, R. (to appear). Grammatical description of Yurakaré. Ph.D. thesis, Radboud University, Nijmegen.

Van Hoecke, W. (1996). The Latin dative. In: W. van Belle & W. van Langendonck (Eds), *The dative, volume I: descriptive studies* (pp. 3–38). Amsterdam/Philadelphia: John Benjamins.

Appendix: Abbreviations

abl, ablative; afo, affected object; ao, accompanying object; cau, causative; col, collective; com, comitative; cr, current relevance; dec, deceased; del, delimiter; dem, demonstrative; des, desiderative; dir, direction; dub, dubitative; ds, different subject; fut, future; hab, habitual; ideo, ideophone; imp, imperative; ins, instrument; io, indirect object; irr, irrealis; lim, limitative; loc, locative; mea, measure; mid, middle voice; neg, negative; oo, objective object; p, possessive; pl, plural; ref, referential; rep, reportative; s, subject; sg, singular; ss, same subject; vbl, verbalizer; vc, vowel change

Chapter 4

Case Pattern Splits, Verb Types and Construction Competition

Andrej L. Malchukov

1. Introduction: Transitivity Splits and Verb-Type Hierarchies

At the beginning of the 1980s two important studies appeared, both advocating a prototype approach to the notion of transitivity: Hopper and Thompson (1980) and Tsunoda (1981). In a highly influential paper Hopper and Thompson argued that transitivity is a gradable and multi-factorial notion. Among the features contributing to high transitivity they mention both parameters relating to participants of the event, such as the subject's volitionality and the object's affectedness and definiteness, as well as properties of the event itself, such as perfectivity, affirmativity and reality. In their article Hopper and Thompson convincingly demonstrate that if some of the semantic features of transitivity are lacking (e.g. O is only partially affected, the verb is imperfective or negated), it can result in a formally less transitive construction (construction with an oblique object, passive or antipassive transformation, etc.).

A similar approach has been independently proposed by Tsunoda (1981), who suggests that a two-argument clause will receive a transitive encoding, if it satisfies the Effectiveness Condition (EF-CON). EF-CON is again seen as a multi-factorial notion including such parameters as impingement of action on O (O is affected and/or attained), O-individuation (specificity, definiteness), completedness of the verbal action, actuality and telicity of the event. As compared to Hopper and Thompson's article, Tsunoda's paper is somewhat restricted in scope as it primarily

Competition and Variation in Natural Languages: The Case for Case
Edited by M. Amberber and H. de Hoop

ISBN: 0-08-044651-5

focusses on ergative languages (alternations of the ergative and intransitive patterns). On the other hand, it is broader in scope in that it pertains not only to transitivity alternations, also discussed by Hopper and Thompson, where the same verb takes alternative case-frames depending on the properties of the clause (tense/aspect/mood (TAM) properties), but also to verb splits, where different lexical classes of verbs subcategorise for different case-frames. Tsunoda makes an important point that "TAM splits (i.e. transitivity alternations, A.M.) and verb splits are fundamentally not different from each other, their semantics and case-marking mechanisms being governed by the common principles" (Tsunoda, 1981, p. 391). The principles alluded to in the quote above pertain to EF-CON. For example, it has been observed that incompleteness/imperfectivity can condition a transitivity alternation in some languages; not surprisingly, the verbs, which like 'look for' include the property of non-completeness into their lexical meaning, tend to select for an intransitive pattern. Similarly, since dynamicity/telicity has been identified as a factor contributing to high transitivity in TAM-conditioned transitivity alternations, it is not surprising that intrinsically stative verbs such as 'like', which lack these characteristics, often fail to take a transitive case-frame.

In this chapter I shall focus on verb-splits in case-marking patterns and propose a set of constraints that accounts for the cross-linguistic variation in case patterns for different verb classes (constraints on encoding of transitivity alternations have been proposed in Malchukov forthcoming). As a starting point I shall take the verb-type hierarchy proposed in Tsunoda's original publication (Tsunoda, 1981) and elaborated on in Tsunoda (1985). In these articles Tsunoda suggested the following hierarchy of verb types that predicts distribution of intransitive and transitive patterns in individual languages:

Effective action >> Perception >> Pursuit >> Knowledge >> Feeling >> Relation

The hierarchy above represents a scale stretching from the more transitive verb types on the left to the less transitive verb types on the right. It is called a hierarchy since it predicts that if a verb type lower in the hierarchy allows for a transitive case-frame (NOM–ACC in accusative languages or ERG–ABS in ergative languages), so do the verb types higher in the hierarchy. The hierarchy is semantically grounded in that the verb classes higher in the hierarchy conform to the transitivity prototype (in Tsunoda's terms, satisfy the EF-CON), while those further to the right fail EF-CON on one or several dimensions. For example, verbs of feeling (cf. 'like', 'fear') are lower on the transitivity hierarchy since an object of liking (unlike an object of killing or breaking) is less affected and the event itself is atelic.

Some verb types in the hierarchy are subdivided into further subclasses. Thus, verbs of effective action (i.e. verbs referring to an action that actually affects the patient), which are highest on the transitivity hierarchy, come in two types: the resultative subtype that includes canonical transitives like 'kill' and 'break' as well as the non-resultative subtype including verbs of contact like 'hit' and 'touch'. Similarly, perception verbs also include two subtypes depending on whether the act of perception is described as necessarily resultative, the visual image of O being attained (the 'see'-class), or not (the 'look'-class) (on further distinctions between these classes of perception verbs see Section 3.4). The Relation subtype, which is lowest in the hierarchy, is heterogeneous, including apart from possession (as originally proposed in Tsunoda, 1981) also other types of static relations (cf. *consist*, *correspond*, etc.). Table 1, taken in an abridged[1] form from Tsunoda (1985), summarizes the cross-linguistic data on valency patterns ('case-frames') of individual verb types adduced in support of the hierarchy above.

As is clear from the table above, in general, the presented data lend support for the hierarchy. The verbs of effective action select invariably for a transitive case-frame (NOM–ACC or ERG–ABS), but as we go down the hierarchy we find this case-frame alternating with or replaced by one of the intransitive case-frames. The table also demonstrates that languages allow for different extensions of the transitive pattern along the hierarchy. For example, while Avar allows the extension of the transitive pattern only as far as the pursuit predicates, in Basque all verb types instantiate a transitive construction (albeit mostly in alternation with an intransitive pattern).

On the other hand, the table above demonstrates a great deal of variation in valency patterns within the same verb type. First, some verb types may be split in that different items select for different case-frames. For example, in English some predicates from the feeling and knowledge classes are transitive (cf. *like,-know*), while some other (cf. *fond of*, *aware of*) are intransitive (note however that the latter are non-verbal; see Section 4.1 for further discussion). Second, the same verb may take different patterns alternatively (as in the case of *hit/hit at* alternation discussed below).[2] Still all these noise factors cannot obscure the fact that verb types show a gradual reduction in transitivity as one moves down the verb type hierarchy.

Tsunoda's work has been assessed in the typological literature as a major contribution to the cross-linguistic research on the valency patterns for particular verb classes (cf. Drossard, 1991; Lehmann, 1991, p. 186; Lazard, 1998, p. 61). Nevertheless, some problems with the proposed hierarchy have been noted as

---

[1] I have left out the lowest verb type "Ability" (for which data are lacking for many languages in the table), as well as the data on Samoan that show a similar pattern to Tongan and Warrungu that shows a similar pattern to Djaru.

[2] Furthermore, variation in case-frames may depend on the class of the nominal, like in many Australian languages, where nouns pattern ergatively while pronouns pattern accusatively. Such cases will not be discussed here.

Table 1: Transitivity splits (Tsunoda 1985).

	effective action		perception		pursuit	knowledge	feeling	relation
	+result (kill, break)	−result (hit, shoot)	+attained (see, hear)	−attained (look, listen)	search, wait	know, understand	like, fear, fond of,	possess, similar, consist
English	Nom/Acc	Nom/Acc Nom/PP	Nom/Acc	Nom/PP	(Nom/Acc) Nom/PP	Nom/Acc (Nom/PP)	Nom/Acc Nom/PP	Nom/Acc Nom/PP
Japanese	Nom/Acc	Nom/Acc Nom/Dat	Nom/Acc Dat/Nom	Nom/Acc	Nom/Acc	Nom/Acc Dat/Nom	Nom/Acc Nom/Dat Dat/Nom Nom/Nom	Nom/Acc Nom/Dat Nom/Obl Dat/Nom
Basque	Erg/Abs	Erg/Abs Erg/Dat	Erg/Abs	Erg/Abs Erg/Dat	Erg/Abs Erg/Dat Abs/Dat	Erg/Abs Erg/Dat Dat/Abs	Erg/Abs Dat/Abs	Erg/Abs
Tibetan	Erg/Abs	Erg/Obl	Erg/Abs, Erg/Obl		Erg/Abs Erg/Obl	Erg/Abs	Erg/Abs Abs/Obl Obl/Abs	Obl/Abs
Avar	Erg/Abs		Loc/Abs		Erg/Abs Abs/Dir	Loc/Abs	Dat/Abs	Gen/Abs
Tongan	Erg/Abs		Erg/Abs Abs/Dat Abs/Loc		Erg/Abs Abs/Dat	Erg/Abs Abs/Dat Dat/Abs	Erg/Abs Abs/Dat Abs/Loc	Abs/Abs
Djaru	Erg/Abs		Erg/Abs		Erg/Abs Abs/Dat	Erg/Abs Abs/Dat	Erg/Abs Abs/Dat Abs/Loc	Erg/Abs Abs/Dat Abs/Abs
Eskimo	Erg/Abs		Erg/Abs		Erg/Abs	Erg/Abs	Erg/Abs	Erg/Abs

well. Lehmann (1991, p. 234) notes that the hierarchy above conflates several semantic dimensions. Lazard expresses some doubts as to whether Tsunoda's ranking indeed constitutes a hierarchy, since "the hierarchy is only evident at the two extremes" (Lazard, 1998, p. 60). Indeed, as can be seen from the table above, some of the intermediate types are not strictly ordered. For example, while perception predicates outrank pursuit predicates in Djaru (the former are transitive, the latter allow for variation of transitive and intransitive patterns), the reverse is true for Avar (where the former invariably show the intransitive LOC–ABS pattern, while the latter can be either transitive or intransitive). Similarly, pursuit verbs outrank knowledge verbs in Japanese (the latter allow for an inverse pattern, as well), but the reverse is true of Tibetan (the former allow for a variation of a transitive and an ERG–OBL pattern, the latter are only transitive).

Furthermore, the hierarchy in the present form only predicts that verb types down in the hierarchy may deviate from the transitive pattern in some way, but it does not predict what case-frames would be selected. Yet, there seem to be cross-linguistic preferences for particular verb types to select for particular case-frames. It is true that some languages use the same pattern for different verb types deviating from the canonically transitive construction. For example, in Finnish, verbs with less affected objects including verbs of emotion and cognition as well as pursuit will take the object in the partitive case. However, other languages will use different case-frames for different verb types. For example, in German both pursuit verbs (cf. *warten auf* 'wait for') and mental verbs (e.g. *gefallen* 'like') can be intransitive but only the latter can take the inverse DAT–NOM pattern. In Basque, as reported in Tsunoda's table, ERG–DAT is used as an alternative case-frame to the transitive one for the contact and pursuit type, while DAT–ABS is an alternative for the knowledge and feeling types. In Avar perception and cognition verbs take the inverse OBL–ABS case-frame, while the pursuit verbs an ABS–OBL frame. Similarly, in Ingush, another (North-East) Caucasian language, pursuit verbs take an ABS–OBL pattern, while mental verbs take the DAT–ABS frame (Nichols, 1994, p. 118–119):[3]

[3] The following abbreviations are used in the glosses: A=transitive subject; ABS=absolutive case; ACC=accusative case; ADVL=adverbializer; AGRabs=absolutive (S/O) agreement paradigm; AGRdat=agreement with indirect (dative) object; AGRs=agreement with S; AGRo=agreement with O; ALL=allative case; AOR=aorist; CL=class/gender marker; COND=conditional; DAT=dative case; DECL=declarative (mood); DU=dual; ERG=ergative case; F=feminine (gender); FUT=future tense; IF=illocutionary force marker; IMPFV=imperfective aspect; INCH=inchoative aspect; LOC=locative case; M=masculine gender; N=neutrum (gender); NOM=nominative case; O=(direct) object; OBJ=object marker; OBL=oblique object; PL=plural; POST-EL=post-elative case; PRES=present tense; PROG=progressive aspect; REP=repetitive aspect; REFL=reflexive marker; S=intransitive subject; SG=singular; TAM=tense/aspect/mood marker; TOP=topic marker.

Ingush (Nichols, 1994, p. 118–119)

(1) *İz* *suoga* *he*
he.ABS me.ALL wait
'He is waiting for me'

(2) *Suona* *Iz* *viez*
I.DAT he.ABS like
'I like him'

The presented data strongly suggest that the choice of the valency pattern is not random and particular verb types show certain predilections for certain case-frames. Clearly, Tsunoda's hierarchy as it stands does not allow for any principled predictions concerning the distribution of non-transitive case-frames across the verb types.

Actually, some typologists have expressed scepticism concerning a possibility of establishing a proper hierarchy of verb classes since it will presuppose an in-depth research into the whole verbal and adjectival vocabulary in a wide range of languages (Lehmann, 1991, p. 187). This can be a reason why this line of research has not been systematically pursued in the subsequent typological literature, and typologists addressing this topic such as Drossard (1991) and Lazard (1998) did not attempt to elaborate on the verb-type hierarchy or present a new hierarchy with more predictive power. Moreover, the results of cross-linguistic research on verbal valency patterns are not always compatible with each other. For example, in a recent volume dealing with non-canonical case-marking of arguments (Aikhenvald, Dixon, & Onishi, 2001), the authors of two cross-linguistically oriented contributions addressing the typology of verb splits have arrived at somewhat different conclusions (they do not discuss how their findings relate to Tsunoda's hierarchy). Thus, Onishi (2001), who draws in particular on the contributions to the above volume, concludes that sensation verbs are more prone for non-canonical subject marking as compared to emotion and perception predicates. Haspelmath (2001), on the other hand, referring to the results of Bossong's (1998) research of experiencer constructions in European languages, concludes that experiencers of cognition/perception predicates show a strong predilection for being encoded as a (canonical) subject, while experiencers of emotion predicates show an equally strong predilection for being encoded as objects (in particular, dative objects), and sensation predicates are intermediate in that respect. Note that the ranking of sensation and emotion predicates is reversed here as compared with Onishi's hierarchy.

Nevertheless, in this chapter I will show that constructing a universally valid hierarchy is feasible. Moreover, on the basis of such a hierarchy specific predictions can be made concerning the preferred patterns of case-frames for particular semantic classes of verbs. The remainder of the chapter is structured as follows. In Section 2, I propose to decompose Tsunoda's hierarchy, recasting it in the form of a two-dimensional semantic map. Furthermore, I present some data in support of the particular 'routes' (sub-hierarchies) on the semantic map. In Section 3, I propose a set of constraints that can predict which case-frames would be preferred by different verb classes. In Section 4, some other factors contributing to case marking are considered. Finally, in Section 5, I integrate the proposed two-dimensional hierarchy into a more comprehensive semantic map showing further connections between the transitive and intransitive domains.

2. Verb Classes on a Semantic Map

2.1. Decomposing Tsunoda's Hierarchy

There is a general consensus in the functional-typological literature as to what constitutes a semantically transitive construction (cf., e.g. Hopper & Thompson, 1980; Comrie, 1989; Givón, 1985; Dixon, 1994; Palmer, 1994; Lazard, 1998). Thus, Givón (1985, p. 90) identifies the following properties as contributing to semantic transitivity:

(a) *Agent-related*: The prototypical transitive clause has a visible, salient, volitional, controlling *agent-cause* which initiates the event;
(b) *Patient-related*: The prototypical transitive clause has a visible, salient, non-volitional, non-controlling *patient-effect* which registers the bulk of change associated with the event;
(c) *Verb-related*: The prototypical transitive clause has a compact, perfective, realis verb or verbal tense-aspect-modality.

Thus, the transitive prototype is defined in terms of the role properties of its core arguments, as well as the properties of the verb itself (the latter properties corresponding to the TAM properties described by Hopper and Thompson and Tsunoda will not concern us here). While there is general consensus that the transitivity prototype should appeal to semantic roles of its arguments, there is much less agreement how the semantic roles themselves should be defined. In some approaches going back to the classic Fillmorian Case Grammar tradition semantic roles are used as labels or supplied with informal definitions (cf. Givón's definition above), in some other they are characterized in terms of binary features (e.g. Rozwadowska, 1988), or derived from the position of semantic arguments in the event structure or — in somewhat different terms — in the lexical-semantic representation of the verb's meaning (cf. van Valin & Lapolla, 1997; Rappoport &

Levin, 1998; Croft, 1998; Wunderlich, 1997). The latter approach seems to be the most promising, but definitions have not been offered so far for all argument roles that will concern us here. Therefore, I shall use traditional labels for Semantic (Thematic) Roles as familiar from the literature, and further view them as multifactorial and, consequently, gradient concepts (cf. Palmer, 1994).

While a canonical transitive construction should conform to a certain semantic transitive prototype, no such prototype is available for an intransitive construction. In fact, the intransitive construction is rather defined in negative terms, as a clause not conforming in formal and semantic terms to the transitive prototype. However, deviations from the prototype may be numerous. To begin with a verb may have only one argument (which may additionally be more similar either to A or O). Additionally if a verb has two arguments, O may not represent a typical (affected) Patient and A may not be a typical (controlling) Agent.

Now, it is clear from looking at Tsunoda's verb hierarchy, that we are dealing with different deviations from the transitivity prototype. If we compare the canonical transitive (effective action) verbs with the pursuit type, we witness a difference in affectedness: while O is affected (undergoes some change) in the former case, it is not affected in the latter case, as an action is merely intended but not realized (cf. 'wait for', 'search'). In this respect, verbs of contact like 'hit' are intermediate between the effective action and pursuit types, as they refer to an action that has taken place but not necessarily yielded a result (change of state of O). Thus, one can set up a hierarchy of decreased Patienthood (affectedness of the O participant) where 'break' ranks higher than 'hit' and 'hit' ranks higher than 'search'. In other words, 'break'-verbs and 'search'-verbs differ in argument structure: the argument structure of the former is Agent–Patient, while the argument structure of the latter is Agent–Goal. Still clearer instantiations of the Goal role are objects of (two argument) motion verbs, which constitute a maximal deviation from transitivity along this dimension.

On the other hand, the difference in argument structure between canonical transitives and mental verbs such as 'see' and 'like' is more profound and complex, as the differences relate not only to properties of O but more importantly to properties of A as well. Indeed, also here as in the case of pursuit verbs, we are envisaging decreased patienthood on the part of O. Still more importantly, these verbs instantiate a deviation from the agentive prototype on the part of A: the A of emotional predicates is not a controlling but rather an affected participant. In other words, the argument structure of mental verbs is Experiencer–Stimulus, rather than Agent–Patient. The shift in argument structure along this dimension is also gradual, inasmuch as perception predicates of the 'see' type are arguably intermediate between 'break' and 'like' types. On the one hand, one can follow Kemmer (1993, p. 137) in her suggestion that experiencers of perception verbs are less typical than those of emotion verbs since the former are less affected. On the other hand, even though the object of perception is physically unaffected, as long as a visual image is obtained, the action can qualify as resultative and the O as more Patient-like (see Tsunoda, 1981). Finally, sensation predicates (such as 'freeze', 'be sick') deviate arguably even further from transitivity prototype than emotion predicates, since Experiencer is their only argument, while many emotion predicates ('like', 'fear') take two arguments (admittedly, the situation is less clear with predicates like 'be worried', 'be sad', which fall in between two- and one-argument predicates).

Let us sum up the discussion. Above, I have argued that the verb-type hierarchy proposed by Tsunoda conflates two different dimensions: a (sub-) hierarchy of decreased patienthood on the part of O argument (that leads from 'break' to 'go'), and another (sub-) hierarchy that additionally involves decreased agentivity on the part of the A participant (that leads from 'break' to 'sick'). The new — two dimensional — hierarchy is represented below (verb types absent on Tsunoda's hierarchy are in parentheses).

Figure 1 represents a 'semantic map' inasmuch as the adjacent verb types show semantic affinities (see Haspelmath, 2003, for discussion of the semantic map approach). On the assumption of form–function iconicity, which underlies the semantic map methodology, it is expected that the map should be contiguous. That is, if two categories (here, verb types) on the map share a certain case-frame this will hold for intermediate categories as well. For example, if emotion verbs share the transitive case-frame with the effective action verbs, the intermediate types — perception and cognition — should allow for the transitive pattern as well. On the other hand, inasmuch as the proposed semantic map preserves the initial insight of Tsunoda's hierarchy in being oriented from more transitive to less transitive verb types, it can also be called a two-dimensional hierarchy. As noted by Haspelmath (2003), hierarchies have more predictive power than semantic maps, since apart from predictions based on contiguity requirements, they generate predictions based on the directionality of the map: if categories lower on a hierarchy display a certain pattern (the transitive pattern, in our case), categories higher in the hierarchy should display this pattern as well.

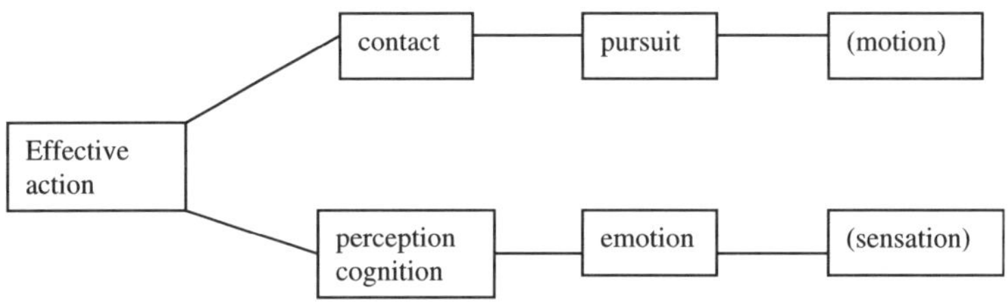

Figure 1: Two-dimensional verb type hierarchy.

An important qualification is in need here, though. An implicational hierarchy dealing with semantic classes should be formulated in existential terms (for some member of the class X) rather than universal terms (for every member of the class X); see Cristofaro (2003) for a recent discussion of 'quantified' implicational universals. That is, if some member(s) of the semantic class X displays a particular morpho-syntactic characteristic, the hierarchy would predict that some member(s) from the semantic classes higher in the hierarchy will display this characteristic as well (given, naturally, that this characteristic is associated with the high ranking in the hierarchy). This is a common assumption in typological research dealing with semantic classes. Thus, Dixon (1977) in his well-known typological study of adjectives, formulates a generalization to the effect that if (some of the) non-basic property words (e.g., human propensity items) pattern as adjectives in a particular language, (some of the) basic property words (referring to value, size, age and colour) should do the same (while the opposite does not hold of course; that is, languages having a closed class of adjectives may well confine this class to the basic property words). Thus, although the verb-type hierarchy cannot predict for every lexical item which case-frame it selects,[4] it can predict that if there are some verbs from a type lower in the hierarchy that take a transitive pattern there should also be found some verbs from types higher in the hierarchy that do the same.

Generally, I assume with functional linguists such as Givón, Croft, Dixon, Tsunoda, Lazard that while in all languages clauses, which conform to the transitivity prototype, will receive a transitive encoding, languages will differ in the extent to which a transitive construction will be extended to other clause types departing from the transitive prototype. Such extensions can be conceived as metaphoric extensions from one semantic domain to the other, as suggested by Givón (1984) and Rice (1987). (For example, the notion of object affectedness clearly needs reinterpretation when one shifts from a domain of physical actions to the domain of mental events.) Alternatively, it can be regarded as assimilation of minor sentence types to the major construction, as suggested by Lazard, who views the transitive clause as a major construction type for two-actant action verbs, serving as a model for other two-actant patterns (Lazard, 1998, p. 40).[5] In any case

[4] In fact, Lucassen (1985, p. 258) expresses doubt in feasibility of cross-linguistic generalizations about valency patterns of verb types referring to the fact that in Abkhaz 'see' is transitive while 'hear' is not.

[5] There is ample diachronic evidence in support of this view. For example, some experiencer verbs such as *thyncan* 'seem' that took the inverse pattern in Old and Middle English have been assimilated to the transitive class in Modern English (see Blake, 1994, p. 148). In Russian, the verbs of the pursuit class such as *iskatj* 'look for' which originally took a genitive object are transitive in the modern language. The class of affective verb in North-east Caucasian languages is on decline and in some languages like Udi has been nearly ousted by the ergative pattern (Klimov & Alekseev, 1980; Schulze, 1994). In Kayardild some verbs of emotion, which historically derive from middle verbs, have been assimilated to the transitive pattern (Evans, 1995, p. 329).

a language may confine a transitive construction to some domain in the hierarchy presented above, or extend it to some further point down the hierarchy. In the next section, I shall present evidence for the different cut-off points for such an extension on the two-dimensional hierarchy, drawing on the data presented in Tsunoda's publications as well as in subsequent typological and descriptive studies. The data pertaining to particular sub-hierarchies will be presented separately; in Section 2.2, I present evidence for the ranking of the verb types on the sub-hierarchy leading from 'break' to 'go', while in Section 2.3, I shall present evidence for the ranking of the verb types on the sub-hierarchy leading from 'break' to 'sick'.

2.2. From 'Break' to 'Go'

Evidence for this sub-hierarchy is easy to obtain, in fact ample evidence was provided already in Tsunoda's initial publication. Below I shall present evidence for the ranking on this (sub-)hierarchy focussing on cases where a verb type higher in the hierarchy takes the transitive case-frame, while the verb types lower in the hierarchy take the intransitive case-frame.

2.2.1. break > hit Most languages seem to assimilate contact verbs like 'hit' and 'touch' to transitives. Yet, already in European languages we can observe that 'irresultative' verbs like 'hit' and 'touch' may diverge from canonical transitives (the 'resultative' subtype of effective action verbs) in behaviour. As noted by Tsunoda (1981), many of these verbs in English show an alternation between a transitive and prepositional O construction, while 'resultative' transitives do not (cf. *hit at* and **break at*; see Levin (1993, pp. 41–43) for further discussion of the 'conative alternation'[6]).

Some other languages offer more straightforward evidence for the reduced transitivity of the contact verbs. Thus, it is well known that in many Caucasian languages "verbs of surface contact" including such items as 'hit', 'bite', 'kiss', 'pinch' take an oblique O (Klimov & Alekseev, 1980, p. 180). Consider the following example from Chechen (Daghestanian):

Chechen (Nichols, 1984, p. 188)

(3) *Da:s* *woʕa:* *b-iett*
father-ERG son.DAT CL-beat(PRES)
'The father beats his son'

[6] Admittedly, not all groups of contact verbs participate in the "conative alternation": thus, while Levin's (1993) *hit*-verbs allow such alternation, *touch*-verbs do not. However, there are other syntactic characteristics that indicate lower transitivity of contact verbs from either class, for example, neither permits the "middle alternation" (cf. *it breaks easily*, **hits easily*, ** touches easily*); see (Levin, 1993, pp. 149, 155).

As illustrated in (3), in Chechen contact verbs take the patient in the DAT rather than absolutive case thus deviating from the ergative pattern. Admittedly, the origin of this case-marking pattern in Daghestanian languages is due to the fact that ditransitives take the notional instrument as a direct O (cf. Klimov & Alekseev, 1980). This is particularly evident for Chechen, where an example like (3) can still be interpreted as a ditransitive construction involving the understood (omitted) absolutive O *bi:* 'fist' (see Nichols, 1984, p. 188). However, this interpretation cannot be generalized — at least in synchronic terms — to other Caucasian languages, Kartvelian and West-Caucasian where contact verbs pattern intransitively as well. Consider an example from Abkhaz (West-Caucasian):

Abkhaz (Lucassen, 1985, p. 260)

(4) *D-sə-sə-yL*
3sg/AGRabs-1sg/AGRdat-beat-TAM
'S/he beats me'

In Abkhaz, encoding of grammatical relations manifests itself in agreement rather than morphological case. The verb in (4) cross-references both arguments, but differs structurally from the transitive in that it involves 'absolutive' and 'dative' AGR prefixes rather than 'ergative' and 'absolutive'; also the order of AGR markers is different from the transitive pattern.

In many other languages, such as Tibetan mentioned in Tsunoda's table, contact verbs pattern intransitively as well. In Amele, which makes a three-way distinction between transitive (with a full paradigm of object AGR), half-transitive (with a restricted AGRo), and the intransitive (lacking AGRo) verbs, includes *q-oc* 'hit' in the half-transitive class (Roberts, 1987, p. 285). In Trumai (language isolate, South American) the verb for 'bite' takes the ABS-DAT rather than ergative pattern, just as other 'aiming verbs' (Lazard, 1998, p. 147). A more complex case is represented by Marathi, a split ergative Indo-Arian language. Marathi, makes no formal distinction between 'break' and 'hit' verbs in imperfective tenses, since ACC is identical to DAT. However, the difference between the two becomes obvious once one regards case-marking in perfective tenses. Since Marathi is a split ergative language with an aspect-based split, in perfective tenses the direct O of 'break' verbs appears in the NOM case, while the O of 'hit' retains its case just as other indirect objects.

2.2.2. hit > look for As is also clear from Tsunoda's table, some languages treat pursuit verbs as intransitives, while some other assimilate them to the transitive class. The former case can be illustrated by European languages, where many pursuit verbs take oblique (prepositional) objects; cf. English *look for*,

German *warten auf* 'wait for'; see Christol (1998, p. 474 ff.) for more examples. The latter case is attested in Japanese where all pursuit predicates are transitive:

Japanese (Jacobsen, 1992, p. 46)

(5)	*tomodati*	*o*	*matu*
	friend	ACC	wait
	'wait for a friend'		

(6)	*apaato*	*o*	*sagasu*
	apartment	ACC	look.for
	'look for an apartment'		

Similar variation in the treatment of pursuit predicates is attested for ergative languages. While Eskimo treats pursuit verbs as transitives, for Australian languages pursuit verbs constitute a cut-off point on the transitivity hierarchy: rather than taking the transitive ERG-ABS pattern, they opt for the ERG–OBL or ABS–OBL patterns (Tsunoda, 1981; cf. Blake, 1977). Consider the following example from Djaru:

Djaru (Tsunoda, 1981, p. 407)

(7)	*Mawu-nu*	*ngal-ø-la*	*jaji-wu*	*jarra+nyang-an*
	man ERG	C-3sgAGRs-3sgAGRdat	kangaroo-DAT	wait-PRES
	'A man waits for (looks for) a kangaroo'			

Note that contact verbs are treated as transitives in Djaru, indeed most Australian languages do not distinguish lexically between 'break' and 'hit' (Dixon, 1980, p. 103).[7]

On the other hand, for Caucasian languages, where contact verbs pattern intransitively, the hierarchy predicts that (some) pursuit verbs will pattern intransitively as well. This prediction is borne out; see the Ingush example with the 'wait' predicate in (1). Sometimes, pursuit verbs share the same pattern with verbs of contact. Thus, in Lezgian *QeQün* 'look for' and '*galuq'un* 'hit (against)' govern the same 'postessive' case (Haspelmath, 1993, p. 274). And Basque, as is clear from Tsunoda's table, uses the same ERG-DAT pattern (in alternation to the transitive) for both the pursuit and the contact type.

Lazard (1998, p. 144) includes both verbs of pursuit and contact into his class of 'aiming' verbs denoting "actions directed towards an object but without necessarily attaining or affecting it". Note, however, that Lazard's notion of 'aiming verbs' extends even further to include verbs of directed perception (look at), emotional attitudes (worry about) and of social interaction (help, speak to). As evidence for taking 'aiming verbs' as a natural class, Lazard refers to the fact that in

[7] Yet some of them distinguish these meanings grammatically through an antipassive alternation.

many Oceanic languages they constitute a group of "middle" verbs selecting an ABS–DIR frame (cf. Chung, 1978, p. 47).[8]

2.2.3. search > go Most languages including the ones discussed above treat motion verbs as intransitives, although they may admit some few into the transitive class (cf. English *enter*, *leave*; see also Dixon, 1991, p. 281 for a discussion of 'preposition omission' with motion verbs). Admittedly, languages differ in the extent to which they extend the transitive pattern to the motion type. As compared to English, Japanese (not surprisingly, given that pursuit verbs take invariably a transitive pattern) allows more motion verbs to appear in a transitive construction (Jacobsen, 1992, p. 50).

Japanese (Jacobsen, 1992, pp. 32, 46)

(8) *rooka* *o* *hasiru*
hall ACC run
'run down the hall'

(But see Spencer & Otoguro (this volume) for a challenge to the traditional assumption that Japanese has a case system.) This construction, however, is mostly reserved to cases when the location is traversed completely and unidirectionally, elsewhere it alternates with the intransitive NOM–LOC frame:

(9) *rooka* *de* *hasiru*
hall LOC run
'run in the hall'

The preference of motion verbs for intransitive constructions pertains to languages of different alignment type, at least to those that have case morphology. In languages, which lack the case category, the distinction can be blurred, however. Thus, Mathews and Yip (1994, p. 136) note that in (Cantonese) Chinese many verbs of motion and posture ('go to', 'sit on') that are intransitive in other languages take an O like transitive verbs.

2.3. From 'Break' to 'Freeze'

In this section, I shall present evidence from typologically diverse languages for the ranking of particular verb types on the second sub-hierarchy leading from 'break' to 'freeze'.

---

[8] Cf. also Blume (1997) for a comparative study of 'interaction verbs' in European and Polynesian languages.

2.3.1. break > see As repeatedly noted in the typological literature, many languages — both accusative and ergative — distinguish 'affective verbs' from transitives. In some languages, verbs of perception are treated as belonging to this class on a par with the verbs of cognition and emotion. Thus, in Japanese perception predicates can pattern either transitively or intransitively: the former pattern is found with verbs of attentive perception the latter with inactive perception, which take the DAT–NOM pattern:

Japanese (Jacobsen, 1992, pp. 30, 31)

(10) (*Watashi* *wa)* *kokuban* *o* *mita*
(I TOP) blackboard ACC look-at-PAST
'I looked at the blackboard'

(11) (*Watashi* *ni* *kokuban* *ga* *mieta*
(I DAT) blackboard NOM see/visible-PAST
'I saw the blackboard'

Not surprisingly, emotion predicates that are lower in the hierarchy also take the intransitive pattern:

Japanese (Shibatani, 2001, p. 312)

(12) *Mami* *ni* *(wa)* *Hata-sensei* *ga* *osorosii* *(sooda)*
Mami DAT (TOP) Hata-professor NOM fear (ful)
'Mami is afraid of Professor Hata'

Similarly, in some ergative languages verbs of perception pattern intransitively. Again Caucasian languages, which display an inverse pattern with perception predicates, can serve as an example (cf. Tsunoda, 1981; Drossard, 1991; Lazard, 1998 for discussion and further exemplification). Consider the following examples from Avar (contrasted with the transitive construction), where cognition and perception predicates take an experiencer in the locative case, while emotion predicates take an experiencer in the dative case:

Avar (Blake, 1994, p. 121 from Ebeling, 1966):

(13) *Ínssucc-a* *j-as* *j-écc-ula*
(M)father-ERG F-child F-praise-PRES
'The father praises the girl'

(14) *Ínssu-du* *j-as* *j-ix-ula*
(M)father-LOC F-child F-see-PRES
'The father sees the girl'

(15) *Ínssu-je* *j-as* *j-óλ'- ula*
(M)father-DAT F-child F-love-PRES
'The father loves the girl'

Thus, in all these languages verbs of perception share intransitive pattern with other mental verbs, although they do not necessarily display the same case-marking of arguments. A similar situation occurs in many Indic and Dravidian languages, where perception predicates also take a non-canonical subject (Onishi, 2001).

2.3.2. see/know > like/fear Languages cited above do not provide evidence for the relative ranking of perception and emotion predicates, since both take an intransitive pattern. For perception verbs, however, it is rather an exception than a rule, since in most languages (inactive) perception verbs as 'see' and 'hear' pattern transitively (cf. Tsunoda, 1981; Blake, 1994, p. 57; Palmer, 1994, p. 26). This has been generally confirmed by Bossong's (1998) study of the experiencer construction in European languages which showed that 'see'-verbs show a strong predilection for a transitive construction with a subject experiencer, while 'like'-verbs show an equally strong predilection for the 'inverse' object experiencer construction (Haspelmath, 2001, p. 61). The European languages from Bossong's sample are mainly nominative, but the same reference for perception verbs to pattern transitively can be observed in ergative languages as well (as can be readily seen from Tsunoda's table). Thus, Tsunoda notes that 'see' is used as an exemplary transitive predicate in many descriptions of Australian languages, while emotion predicates often select for some other case-frame, typically ABS-OBL like in Djaru:

Djaru (Tsunoda, 1981, p. 407)

(16) *Ngali* *nga-li-nyanta* *minyirri* *ngumpirr-a*
we.ABS C-1duS-3sgLOC shy.ABS woman-LOC
'We are shy of the woman'

Also 'split intransitive' (or split-S) languages, which consistently differentiate between agentive and patientive subjects, are instructive in that respect. It seems that few of these languages treat 'see'-verbs as having a patientive subject, as is the case in Oneida which groups 'see' together with 'like' and 'sick'. Most split-S languages, however, assimilate 'see'-verbs to semantic transitives with agentive subjects and patientive objects. Thus, Ika treats 'see' as agentive, while 'know', 'afraid' and 'tired' are non-agentive. Similarly, in Guarani 'see' is transitive, while emotion verbs take a patientive subject. The same holds for Acehnese, where 'see' is cross-referenced as an agentive subject verb and needs to take a special detransitivizing marker (the 'accidental action' prefix *teu-*) in order to shift to the patientive subject class (Durie, 1985, p. 60).

The presented evidence indicates that perception predicates of the 'see' class (for a discussion of the 'look' predicates see Section 3.4) rank higher in the hierarchy than emotion predicates. Cognition predicates seem to fall in between the two types. Note that Tsunoda's hierarchy suggests that cognition predicates are more similar to emotion predicates (they are adjacent in the hierarchy). However, in Daghestanian languages perception ('see') and cognition ('know') predicates are invariably treated in the same manner, while emotion predicates can receive a distinct treatment (for example, the former take the A in the affective/locative case, the latter in the dative, as in Avar (14)–(15); see Comrie and van der Berg (2003), Ganenkov (forthcoming) for an overview of case-marking of experiencers in Daghestanian languages). In general, however, these data are consistent with Tsunoda's hierarchy inasmuch as cognition predicates take an intermediate position between perception and emotion. As a final illustration consider the situation in Acehnese (Durie, 1985, p. 63ff.). While perception verbs are treated as agentive, emotion verbs are for the most part either patientive or 'variable' (i.e. can cross-reference their subject with either agentive or patientive clitics). Cognition verbs fall in-between these two classes, as some of them belong to the agentive class, while some others belong to the 'variable' class.

2.3.3. like/fear > freeze/be cold Tsunoda does not include sensation predicates into his classification, apparently because his hierarchy is concerned with two place verbs, while sensation predicates are predominantly one place. Still there is a class of emotion predicates like 'be sad' which seem to be similar in terms of the number of valencies to sensation predicates like 'be cold' (see Kemmer (1993, p. 128ff.) on the distinction between two- and one-participant mental events). Thus, Onishi (2001) assigns both groups into the same class of "one- or two-place verbs with affected S/A". Interestingly, even within this class sensations sometimes pattern differently from verbs of emotion. Consider the case from Quechua, where sensation predicates take the Subject in the ACC case:

Quechua (Hermon, 2001, p. 151)

(17) *ñuka-ta* *chiri-wa-rki-mi*
me-ACC cold-AGRO.1sg-PAST3-IF
'I was cold'

(18) *ñuka-ta-ka* *uma-ta* *nana-wa-n-mi*
me-ACC-TOP head-ACC hurt-AGRO.1sg - PRES3-IF
'My head hurts me'

As noted by Hermon (2001, pp. 151–152) other mental verbs including those of emotion display the canonical pattern with the nominative experiencer (with the exception of *muna-* 'want' that takes the accusative subject).[9] In Tariana (Amazonian) there

[9] Note that predicates with the desiderative suffix also take the subject in the accusative.

is a small class of verbs (S_{io} verbs in Aikhenvald's (2001) classification) that takes the S in the objective case, rather than in the unmarked case (Aikhenvald, 2001). As noted by Aikhenvald (2001, p. 180), this class mostly includes predicates of physical state. In Amele (Papuan) sensation predicates pattern as impersonal constructions with object-experiencers. As noted by Roberts (1987, p. 316), the majority of impersonal constructions are used to refer to a physiological state ('tired', 'itchy', 'cold', 'hurt', 'breathless'), although some few refer to emotional states ('sorry', 'willing') and one verb — to the mental state ('aware/understand').[10]

These data prompted Onishi's (2001) conclusion that sensation predicates are most prone among the verb types for non-canonical subject marking. Below we shall see how this observation can be reconciled with the results of Bossong's (1998) study showing that in European languages emotion predicates seem most prone for non-canonical subject marking, a result which seems to contest the lowest position of the sensation predicates on the verb-type hierarchy.

2.4. Conclusions on the Verb Hierarchy

Above I have shown that the verb-type hierarchy proposed by Tsunoda indeed conflates several semantic dimensions.[11] Once these dimensions are disentangled into separate sub-hierarchies we find how exceptions to the hierarchy mentioned in Section 1 fall into place. Note that all these exceptions concern the relative ranking of the pursuit predicates vis-à-vis mental verbs. As argued above, pursuit type cannot be reasonably ranked with respect to mental verbs since the two types display a deviation from a transitivity prototype in a different way; in particular, only the latter hierarchy involves a decrease of agentivity of the A participant. With these amendments, the two-dimensional hierarchy can be set up as a universal hierarchy having predictive power. If (some members of) the verb types lower in the hierarchy display the transitive pattern, (some members from) the verb types higher in the hierarchy will do so as well.

[10] Another instructive example comes from Malayalam. As noted by Jayaseelan (2004, p. 230), while mental psych-verbs allow for both NOM and DAT subjects, physical sensations take only the latter. Interestingly, if a verb referring to a physical sensation (e.g., *vedaNicc-* 'feel pain') exceptionally takes a NOM subject it is interpreted as (metaphorically) referring to mental suffering.

[11] The same conflation is found in the typology of verb types proposed by Onishi (2001), where mental predicates and pursuit/interaction verbs are treated as subclasses of the same group (Class II in his classification: 'two place verbs with less agentive A and less affected O'). More generally, it is doubtful if a one-dimensional universal hierarchy of semantic (thematic) relations can be constructed. Indeed, Experiencer and Instrument are opposed to Agent along different dimensions (the former characterized as affected, the latter lacking the agent's sentience) and thus cannot be ranked in relation to each other on a principled basis.

As a further illustration of independence of the sub-hierarchies on the two-dimensional map, consider extensions of transitivity along the hierarchy in English and Japanese. If one operates in terms of a one-dimensional hierarchy as proposed by Tsunoda it hardly makes sense to ask which of the languages is 'more transitive', that is more liberal in extension of the transitive pattern. However, such a comparison makes perfect sense in terms of a two-dimensional hierarchy. Indeed, Japanese is more permissive in extension of the transitive pattern along the first sub-hierarchy: as noted above it treats pursuit predicates (and even many motion verbs) as transitive. On the other hand, English is more liberal than Japanese in extending of the transitive pattern along the second sub-hierarchy as it assimilates mental verbs to the transitive pattern. This is consistent with Jacobsen's (1992) conclusion that English (unlike Japanese) downplays the distinction between the verb types in agentivity/volitionality, while Japanese (unlike English) does not make a consistent distinction between intended (potentially resultative) and accomplished (actually resultative) actions.[12]

3. Constraining Case Frames for Verb Types

3.1. Introductory Remarks

In the previous sections, I have introduced a two-dimensional hierarchy of verb-splits and demonstrated how verb types can be located on the semantic map ranging from prototypical transitive to intransitive constructions. In this section, I shall suggest constraints that will account for the cross-linguistic variation in case-marking of arguments for individual verb types.

Let us first briefly consider the motivation behind case-marking in natural languages (see De Hoop & Narasimhan, this volume for further discussion). In functional-typological literature (Comrie, 1989; Mallinson & Blake, 1981; Kibrik, 1985), two main functions of case-marking have been identified: the indexing function, that is cases are used to encode semantic roles, and the discriminating function, that is the need to distinguish between the core arguments (subjects and objects). First recognition of these functions have given rise to two analyses of case-marking that were viewed as alternative and even competing, nowadays it is generally acknowledged that both are indispensable to account for

[12] Further, as noted by Jacobsen (1992, p. 47), Japanese is also more restrictive than English in extending a transitive pattern to verb types involving a symmetrical relation between the participants ('marry', 'resemble'). This is still another dimension of the transitivity hierarchy, which we shall briefly address in 5.

a cross-linguistic variation in case patterns (cf. Mallinson & Blake, 1981, pp. 91–96; Song, 2001, pp. 156–167). For example, the indexing approach provides a better account of case-marking of oblique arguments, as well as of marking of core arguments in 'role-dominated' languages (in terms of van Valin & LaPolla, 1997), where case reflects semantic roles of arguments. On the other hand, the indexing approach on its own cannot account for the well-known tendency attested both in accusative and ergative languages to leave a single argument of an intransitive verb (S) and one of the core arguments of the transitive verb (A or O) unmarked. To account for this one usually invokes other functional factors, such as the need to differentiate between the two arguments (that is, the discriminating function mentioned above), as well as economy that disfavours overt marking of (core) arguments. Note that due to the latter factors both intransitive and transitive clauses contain one unmarked argument (in the nominative case in accusative languages and in absolutive case in ergative languages). Based on these observations Tsunoda (1981), following some earlier suggestions of Shibatani (1977) and Dixon (1979), has proposed the UNMARKED CASE CONSTRAINT, which reads as follows:

> UNMARKED CASE CONSTRAINT (UCC): In a non-elliptical sentence at least one NP must be in the unmarked case (nominative or absolutive).

Tsunoda used this constraint to account for the fact why case patterns lacking an unmarked argument (such as the ERG–DAT pattern lacking the unmarked absolutive argument and impersonal constructions lacking nominative subject in accusative languages) are generally dispreferred. In (Malchukov, forthcoming) I relied on the UCC to account for certain asymmetries between accusative and ergative languages in the encoding of the transitivity parameters suggested by Hopper and Thompson (1980). In particular, I proposed two constraints on the encoding of transitivity parameters: (i) the Relevance Principle that predicts that a parameter should be encoded on an argument to which it pertains (e.g. agentivity on the A argument and affectedness on the O argument), and (ii) the PRIMARY ACTANT IMMUNITY PRINCIPLE (PAIP) derived from Tsunoda's UCC, which prohibits encoding of a transitivity parameter on the primary (unmarked) argument. PAIP helps to explain why differential object marking is mostly confined to accusative languages (where O is not a primary argument and therefore permits manipulation on its case-marking), while differential subject marking is mostly found in ergative languages (where ERG argument is not a primary argument and therefore permits manipulation on its case-marking). However, in this chapter I shall return to the original form of the UCC constraint suggested by

Tsunoda, since it provides a more appropriate formulation than PAIP for an account of transitivity splits, where no case alternation with the same verb is involved.

More recently the typological variation in case-marking systems has been addressed in Optimality-Theoretic literature, which attempts to provide a principled account of cross-linguistic variation in case-marking systems (see, e.g. contributions by de Hoop, Koeneman, Mulders, & Wearman, 2001; Legendre, Grimshaw, & Vikner, 2001). Within Optimality-Theory (OT) (Prince & Smolensky, 2004), the grammar of a natural language is viewed as the result of interaction of conflicting and violable constraints. Since the functional factors discussed above can be viewed as competing motivations in the domain of case-marking it is clear that they are amenable to an OT-style analysis. Here I shall briefly regard three optimality-theoretic approaches to case-marking, proposed by Primus (1999), Wunderlich and Lakämper (2001) and De Hoop and Narasimhan (this volume), which are most similar to my approach (see Nakamura, 1999; Woolford, 2001 for alternative approaches).

Although these approaches are based on somewhat different theoretical assumptions, they are similar in a number of respects, in particular with regard to a set of constraints used to account for variation in case-marking. Firstly, all these approaches assume some version of the UCC, termed 'Obligatory NOM Requirement' by Primus, 'Default Linking' by Wunderlich and Lakämper and PAIP (Malchukov, forthcoming) by De Hoop and Narasimhan. Secondly, all approaches propose a constraint (called Distinctness by Primus, Distinguishability by De Hoop and Narasimhan, and Uniqueness by Wunderlich and Lakämper) that is used to rule out constructions with doubling of a particular case (NOM–NOM, ACC–ACC, etc.). The functional motivation for this constraint lies in the discriminating function of cases, as explicitly acknowledged by Primus and De Hoop and Narasimhan. Thirdly, both Primus and Wunderlich and Lakämper assume some default rules assigning structural cases to core arguments. In the analysis of Wunderlich and Lakämper, structural cases are defined in terms of features [+hr] 'there is a higher role' and [+lr] 'there is a lower role' on the argument structure. In this system, the feature [+hr] licenses ACC-marking, the feature [+lr] licenses ERG-marking, the feature combination [+hr/+lr] licenses DAT-marking, while NOM lacks a feature specification. Primus introduces structural case defaults in the form of NOM–ACC preference for two-argument verbs in an accusative language. I shall follow Primus in her assumption that assignment of structural cases in a transitive pattern is best viewed as the result of the assimilation of two-argument verbs to the transitive construction (cf. Lazard, 1998, for a similar view).

Most differences between these analyses concern the question of how the indexing function of cases is captured. In the approach of Primus this is captured most directly since the choice of case is motivated by the 'Proto-Role' properties, as suggested by Dowty (1991). NOM in the accusative system (and ERG in the ergative system) is assigned to an argument that displays (the greater number of) Proto-Agent properties, while ACC (and ABS in the ergative system) case are assigned to an argument that displays (the greater number of) Proto-Patient properties. DAT is assigned by default (DAT-Default constraint) to an argument displaying a small number of Proto-Agent properties. Thus, in this system the choice of cases is ultimately semantically motivated. Wunderlich and Lakämper, on the other hand, argue for an approach that treats assignment of structural and semantic case in a different fashion and further assume that semantic and syntactic case assignment can compete with each other. Structural cases are assigned through feature specification [+hr] and [+lr], as mentioned above. Semantic cases, by contrast are sensitive to semantic roles (this is ensured by their Mark Semantics constraint). Generally, Wunderlich and Lakämper argue that structural case assignment is oblivious to semantic information, even though they admit that there is a connection between the semantic features [+hr] and Proto-Patient properties (affectedness) and the feature [+lr] and Proto-Agent properties (control).[13] Finally, within the 'uniform model of case-assignment' advocated by de Hoop and Narasimhan, the indexing function of cases is directly captured through the Identify constraint.

I shall follow Wunderlich and Lakämper in their assumption that structural and semantic case assignment can be in competition (cf. also Woolford, 2001), but shall assume with Primus and de Hoop & Narasimhan that (in a semantically transitive construction) there is a semantic motivation for assignment of structural cases as well, as long as the A argument represents an Agent and the O argument a Patient. The requirement that semantic roles are properly encoded can be viewed as instantiations of Faithfulness to semantic input, where the semantic roles are specified (or can be deduced from lexical decomposition of verbal semantics). Furthermore, I shall assume that Faithfulness ensures existence of the default (iconic) mapping from semantic roles to cases such that the agent will be iconically encoded through a NOM case in an accusative system (and the ERG case in the ergative system), the patient will be encoded through ACC (or ABS in the ergative system), the experiencer by DAT[14] case and local roles by oblique cases (see de Hoop & Narasimhan for a similar approach).

[13] The semantic import of structural cases is used by Wunderlich and Lakämper to account for the licensing of DAT case beyond ditransitive constructions.

[14] Dative is systematically notationally distinguished from obliques (DAT vs. OBL), when it encodes the Experiencer, elsewhere (e.g., when marking Goal arguments) it is subsumed under OBL.

Thus, building on the previous work in the functional-typological and optimality-theoretic traditions, I shall assume the following constraints on case-marking:

(a) cases encode semantic roles (FaithRole);
(b) core arguments are assigned default structural cases (Trans(itive)Def(ault));[15]
(c) each clause contains an unmarked Case[16] (UCC);
(d) doubling of (structural) cases is prohibited (Uniq(ueness)).

In the following sections, we shall see how these constraints can be applied to predict the cross-linguistically available case-marking patterns for the different verb types on the verb-type hierarchy. In accordance with previous OT approaches, variation in case-marking patterns is viewed as resulting from a different ranking of potentially conflicting and violable constraints. Of course, these constraints do not need to be in conflict in every particular case. Thus, in the case of canonically transitive constructions all these constraints are satisfied. Indeed, the FaithRole constraint ensures that Agent and Patient roles are encoded (in an accusative language) by NOM and ACC respectively.[17] Similarly, the transitive

[15] As noted above, I regard TransDef as a markedness constraint, as I take the transitive construction to be the unmarked pattern for polyvalent verbs (cf. Primus, 1999). Alternatively, it can be viewed as a conjunction of structural case assignment to both subject and object (which is motivated by the [+hr] and [+lr] features in LDG). On Woolford's (2001) approach the choice of the transitive pattern is determined by the case markedness hierarchy, where NOM and ACC are the two least marked cases. On this view one has to specify further that A receives the NOM case (in the accusative system) and O receives the ACC case, rather than the other way around. As far as I can judge, taking one of the latter two alternatives would not change anything in the computation of the optimal case-frames below.

[16] I leave here open the question whether UCC itself should not be better viewed as a result of interaction of some (faithfulness and markedness) constraints, as suggested by an anonymous reviewer. In the present chapter, UCC is taken as a stipulation informed by the typological data, much like the Default Linker constraint in Wunderlich's LDG or — in a somewhat different perspective — Obligatory Case Requirement in Bobaljik (1993). Generally, however, the relation between UCC and markedness is indisputable; one way to derive UCC from markedness constraints has been suggested by Woolford (2001), who derives UCC from the case markedness hierarchy (*DAT >> *ACC >> *NOM).

[17] The question whether the NOM case marking of subjects satisfies the FaithRole constraints or is rather due to (un)markedness, as suggested by an anonymous reviewer, depends on the assumptions about the candidate set. Here I assume that the candidate set is restricted to case forms available in a particular language. Therefore, say, ERG case markers do not enter into the set of competitors in an accusative language (hence also the notational distinction between NOM case in an accusative language and ABS case in an ergative language). Furthermore, I follow Primus (1999) in her view that there is a correlation between Proto-agent properties and nominative marking of subjects in an accusative language; otherwise it is difficult to account for non-canonical subject marking in cases when the subject lacks (some) of the Proto-agent properties. If a set of competing case-frames is assumed to be universal and the computation itself determines the inventory of cases, the absence of ergative case in an accusative language must be due to some higher-ranking Economy constraint prohibiting

pattern will be preferred by virtue of the Transitive Default. The transitive pattern further satisfies UCC and Uniqueness constraints. As can be seen from the following Generalized Tableau, all the alternative patterns violate one or more of these constraints (this is indicated by asterisks in the respective boxes).

Tableau1 (Table 2) above is called 'generalized' since it is designed to predict a cross-linguistically preferred pattern given a certain set of constraints, rather than to account for (un)grammaticality of certain patterns in individual languages in terms of language particular constraint ranking. No universal ranking among the constraints is assumed here, rather constraint reranking gives rise to a cross-linguistic variation in case-patterns. Thus, Tableau 1 accounts for the fact that a semantically transitive construction will be consistently encoded through a transitive pattern since the constraints reinforce each other rather than conflict. For the other verb types from Tsunoda's hierarchy that deviate from the transitive prototype the patterns will be more diversified since different patterns would be favoured by different constraint rankings. In what follows we shall see how the proposed set of constraints can predict possible case-frames for the other verb types from Tsunoda's hierarchy drawing on the data introduced in the previous sections.

3.2. Case Patterns of Pursuit Verbs

Since FaithRole constraints play a crucial role in my analysis let us first consider the argument structure of the pursuit verbs. Above I have argued that pursuit verbs differ from semantic transitives in that the argument roles of the former are

Table 2: Tableau 1: Case-marking in the canonical transitive construction in an accusative language.

	FaithRole	**UCC**	**TransDef**	**Uniq**
NOM–ACC				
NOM–OBL	*		*	
DAT–ACC	*	*	*	
NOM–NOM	*		*	*
ACC–ACC	*	*	*	*

case-marking of As (maybe it can be identified with the UCC, as in the approach of Nakamura, 1999). On this view, indeed, marking of A by the NOM case (instead of ERG) will incur a FaithRole violation. This move however will not have any consequences for the competition among the case-frames in Tableau 1, since all the other competitors will incur similar violations. Hence the NOM–ACC pattern will be optimal on either account.

Agent–Goal rather than Agent–Patient. In an accusative language mapping of this argument structure onto cases is unproblematic. The agent is iconically encoded by NOM, while the goal is iconically encoded by an oblique case. The only alternative seems to be to assimilate this pattern to the transitive pattern, which would account for the fact why some languages treat (certain) pursuit verbs as transitive. Indeed both patterns are attested; as discussed in Section 2.2, pursuit verbs in English generally prefer the NOM–OBL pattern while Japanese usually opts for the NOM–ACC pattern (see (5)).

For ergative languages mapping of this argument structure is less convenient, though. The mapping rules based on semantic faithfulness would predict the ERG–OBL case-frame which lacks the unmarked argument, and hereby incurs a violation of UCC. In such a case several options are available, as represented in Tableau 2 (Table 3). Henceforth I disregard case-frames violating the Uniqueness Constraint since those are considered sub-optimal.[18] Generally, only these candidates that are optimal under different constraint rankings are represented in the Tableau 2.

Thus, the following case-frames would be optimal under different constraint rankings:

- ERG–OBL, if FaithRole >> {UCC, TransDef};
- ERG–ABS, if TransDef >> {UCC, FaithRole};
- ABS–OBL, if UCC >> FaithRole >> TransDef.

The last option is in need of further justification, as it seems to follow from Tableau 2 that if UCC is the highest ranking constraint, the ABS–OBL will always lose to the ERG — ABS pattern as only the latter satisfies the TransDef constraint. This is, however, a consequence of considering FaithRole as a unitary feature "checked" for both arguments simultaneously, rather than assuming two

Table 3: Tableau 2: Case-frames for pursuit verbs in ergative languages.

	FaithRole	**UCC**	**TransDef**
ERG–OBL		*	*
ERG–ABS	*		
ABS–OBL	*		*

[18] Some languages like Japanese seem to permit the 'double nominative' pattern, in an apparent violation of Uniqueness constraint. Note, however, that the 'double nominative' construction seems to presuppose a separate topic position that is distinct from the subject position and thus can be seen as "parasitic" on the latter.

separate Faithfulness constraints pertaining to each of the arguments individually: FaithRole/A and FaithRole/O. On the latter approach, FaithRole/A is violated if A is not assigned the ERG case in accordance with its Agent role and FaithRole/O is violated if O is not assigned the OBL case in accordance with its Goal role. Now, a different ranking of FaithRole/A and FaithRole/O with regard to UCC will yield different results. If FaithRole/A dominates UCC and UCC dominates FaithRole/O, ERG–ABS emerges as a winning candidate. In this case, the Faithfulness constraint for A is satisfied, while the Faithfulness constraint for O is violated. This violation however is 'minimal', given that it is more important for the optimal candidate to satisfy the higher-ranking UCC, which requires that one argument in a transitive frame will be unmarked. An opposite ranking of Faithfulness constraints with respect to UCC (FaithRole/O >> UCC >> FaithRole/A) will yield ABS–OBL as an optimal candidate, provided that TransDef is a low-ranking constraint. The ABS–OBL pattern will also be optimal under the constraint ranking UCC >> FaithRole/O >> FaithRole/A. Thus, the crucial factor for the choice of the ABS–OBL pattern as the optimal pattern is that both UCC and FaithRole/O dominate FaithRole/A.

All case-marking options listed in Table 3 are indeed attested. The ERG–OBL pattern exists as a minor pattern in many Australian languages, as illustrated above for Djaru in (7). The transitive pattern ERG–ABS is attested, as reported in Tsunoda's Table, in Eskimo and Basque among others. Finally, the ABS–OBL pattern is found in many Caucasian languages and has been illustrated above for Ingush in (1).

3.3. Case Patterns of Emotion Verbs

Mental verbs (also called psych verbs) have attracted much attention in the linguistic literature since they allow for alternative mapping from semantic roles to grammatical relations and thus challenge those theories which (like Perlmutter and Postal's Universal Alignment Hypothesis) assume that there is a universal alignment between the two (see, e.g. Grimshaw, 1990, p. 19ff; Croft, 1991, p. 213 ff. for discussion). As noted by Croft (1991), such versatility is particularly characteristic for the class of mental state verbs, which can show alternative alignment both cross-linguistically and within one language: for example, in English *like* takes a subject-experiencer, but *please* takes an object-experiencer.[19] Croft (1991) attributes this variation to the fact that mental states cannot be easily interpreted in

[19] Some other classes of mental verbs are universally aligned, according to Croft (1991): thus, causative mental verbs like *frighten* invariably take an experiencer as their object, while activity verbs (cf. *think*) and inchoative verbs (cf. *get mad*) invariably take an experiencer as their subject.

terms of the causal chain of events including an agent-causer and an affected Patient. Here the causal chain is rather bi-directional, as the experiencer must first direct his attention to the stimulus, and then the stimulus causes a change in the mental state of the experiencer.

Let us consider how my analysis can account for the cross-linguistic variation found in the case-marking of arguments of mental state verbs, in particular, of emotion verbs. As stated above, the semantic role of A is traditionally defined as Experiencer, and the role of O as Stimulus. Since the second argument is not an affected Patient, while the first is not a controlling Agent, the iconic mapping from argument roles to cases will yield the DAT–OBL frame. As is clear from the discussion above, this case-frame is problematic for both accusative and ergative languages, since it lacks an unmarked argument and thus incurs a violation of UCC. Tableau 3 (Table 4) shows the competition of case-frames of Experiencer-verbs in accusative languages.

Tableau 3 predicts which patterns would be optimal under different constraint rankings:

(a) DAT–OBL survives, if FaithRole >> {UCC, TransDef};
(b) NOM–ACC is a winning candidate, if TransDef >> {FaithRole, UCC};
(c) both NOM–OBL and DAT–NOM may be optimal candidates if UCC >> FaithRole >> TransDef or if UCC is interpolated between the Faithfulness constraints.

As in the previous case, the choice between the two patterns in (c) is determined by ranking of the FaithRole/A and FaithRole/O with regard to UCC and to each other. Thus, if both UCC and FaithRole/A dominate FaithRole/O, DAT–NOM is the optimal candidate, if UCC and FaithRole/O dominate FaithRole/A, NOM–OBL wins.

All patterns are indeed attested. The (a) pattern is attested as a minor pattern in some European languages (cf. Icelandic *Mér likar vel við henni* 'I like her', or German *Mir graut von dir* 'You horrify me' cited by Primus, 1999, p. 70). The tran-

Table 4: Tableau 3: Experiencer-verbs in accusative languages.

	FaithRole	**TransDef**	**UCC**
DAT–OBL		*	*
NOM–OBL	*	*	
NOM–ACC	**		
DAT–NOM	*	*	

sitive (b) pattern illustrated by the English *like* is typical for 'Standard Average European' languages (Haspelmath, 2001). The two patterns in (c) are also common in European languages (cf. Bossong, 1998, for an overview): NOM–OBL is attested for example in Finnish, where the O of emotion verbs (such as *rakastaa* 'love' and *pelätä* 'fear') is in the Partitive case, while the inverse DAT–NOM is found in many other European languages such as Russian, German and Italian (cf. Russian *Mne nravitsja eta kniga* (me.DAT like.REFL this book) 'I like this book').

Now consider case-marking options for experiencer verbs in ergative languages as shown in Tableau 4 (Table 5).

Tableau 4 shows case-frames that would be optimal under different constraint rankings:

(a) the iconic DAT–OBL survives, if FaithRole >> {UCC, TransDef};
(b) transitive ERG–ABS is the optimal candidate, if TransDef >> {FaithRole, UCC};
(c) both ABS–OBL and DAT–ABS may be optimal candidates, if UCC >> FaithRole >> TransDef or if UCC is interpolated between the Faithfulness constraints.

The transitive (b) pattern is attested in many languages such as Basque and Eskimo, referred to in Tsunoda's Table. Both case patterns in (c) are common and show a particular areal distribution: ABS–OBL is attested in Polynesian and Australian languages (cf. e.g. (16) from Djaru), while DAT–ABS is found in many Caucasian languages (see (15) from Avar). The iconic pattern in (a) seems to be less frequent, but crops up in some (East-)Caucasian languages, such as Tabasaran.

Tabasaran (Ganenkov, forthcoming)

(19) *Uzu-z Xu-jir-i-q-an gučura*
I-DAT dog-PL-OBL-POST-EL be.afraid
'I am afraid of dogs'

A similar pattern has been reported for Budukh (Alekseev, 1994, p. 285), Lezgian (Haspelmath, 1993, p. 284) and Udi (Schulze, 1994, p. 495). Interestingly,

Table 5: Tableau 4: Experiencer-verbs in ergative languages.

	FaithRole	**TransDef**	**UCC**
DAT–OBL		*	*
ABS–OBL	*	*	
ERG–ABS	**		
DAT–ABS	*	*	

in Udi both Experiencer and Stimulus are marked by the 'dative' cases, but it does not incur a Uniqueness violation since the two dative cases are different ('dative-1' and 'dative-2' in Schulze's notation).

3.4. Case Patterns of Perception Verbs

Among perception predicates two classes are traditionally distinguished: inactive perception verbs like *see* and *hear* in English and active (alias 'attentive') perception verbs like *look* and *listen*. As already observed by Tsunoda (1981), both classes deviate from a transitivity prototype albeit in a different way. Inactive perception predicates characterize their O as attained, but are not necessarily agentive (as one can see something accidentally). On the other hand, attentive perception predicates are agentive but not (necessarily) resultative (as one can look at something without seeing). Now, Tsunoda (1985) observed that cross-linguistically 'see'-verbs are always more transitive than 'look'-verbs (and used this observation as evidence that it is affectedness rather than volitionality that is crucial for the transitive encoding). Næss (2004), however, noted that in some ergative languages (East Futunian, Udi) 'look'-verbs are more transitive than 'see'-verbs since the former take an ergative pattern while the latter do not. Her explanation is that ergative languages are more consistent in encoding control, while accusative languages are more consistent in encoding affectedness. Note however, that Tsunoda's evidence for 'see' being more transitive than 'look' is in no way restricted to accusative languages: for example, in Tsunoda (1985) he refers to Chepang (Tibetan) where 'see' takes the transitive ERG–ABS pattern while 'look' does not. On the other hand, we have also witnessed accusative languages, like Japanese, where 'see' patterns intransitively and 'look' patterns transitively (see (10)–(11)). Perhaps most interestingly, in some ergative languages both 'look' and 'see' are intransitive, but display different patterns. Such a situation is common in (East-)Caucasian languages. This is illustrated below for Lezgian, where 'see' takes the DAT–ABS pattern, while 'look' takes the ABS–DAT pattern:

Lezgian (Haspelmath, 1993, pp. 89, 281)

(20) *Zun mad wa-z akwa-n qhiji-da-č*
I.ABS still you-DAT see-PER REPET-FUT-NEG
'You will not see me again'

(21) *Xtul č'exi bubadi-z kilig-na*
grandchild.ABS great father-DAT look-AOR
'The grandson looked at his grandfather'

Similar patterns have been reported for perception predicates in Ingush (Nichols, 1994, pp. 118–119). In Khinalug (Kibrik, 1994, p. 399) *zaği* 'see' takes a DAT–ABS pattern and *l'äkši* 'look' takes a ABS–LOC pattern. In most Andic languages, except for Bagvalal where 'listen' takes a transitive ERG–ABS pattern, 'hear' takes an inverse DAT/AFF–ABS pattern, while 'listen' an ABS–OBL pattern (Comrie & van der Berg, 2003). In Svan (Kartvelian) 'listen' takes the ERG–DAT pattern, and 'hear' the DAT–ABS pattern (Testelets, 1998). How does the proposed analysis fare when confronted with such variation in case-marking patterns?

Let us consider first the predictions of my constraint-based analysis for the two classes of perception predicates. Note first that 'look' predicates are similar in argument structure to pursuit verbs: both have a controlling A (Agent) and non-affected O (Goal). In fact, Dixon (1991) attributes *look for* and *look at* to the same semantic class of 'Attention'-verbs, while Lazard includes both into his class of 'aiming verbs'. Then, evaluation of case-frames for the 'look' verbs would be predicted to proceed similarly to pursuit verbs (see Tableaux 3 and 4). In the accusative languages the two preferred patterns would be NOM–OBL or the transitive pattern. These predictions are borne out: the former pattern is attested, for example, in English, while the latter is attested, for example, in Japanese (see (10) above). For ergative languages, the optimal case-frames would be ERG–OBL, ABS–OBL, ERG–ABS. Again all these patterns are attested, in fact, all occur in Caucasian languages cited above: ERG–OBL occurs in Svan, ERG–ABS occurs in Bagvalal, while ABS–OBL is attested in most other Andic languages.

For 'see' predicates predictions are similar to those for 'like' predicates, as they are similar in argument structure. However, the O of 'see' predicates is characterized as attained which makes it more similar to Patient (in this respect 'see' verbs are like 'find' verbs). Therefore, I shall assume that the iconic pattern for the 'see'-predicates is DAT–ACC rather than DAT–OBL. On these assumptions, the evaluation of competing case-frames is represented in Tableau 5 (Table 6).

Thus the optimal candidates are:

(a) DAT–ACC, if FaithRole >> {UCC, TransDef};
(b) NOM–ACC, if TransDef >> {FaithRole, UCC};
(c) DAT–NOM is optimal under constraint ranking {UCC, FaithRole/A} >> {FaithRole/O, TransDef}. Under the opposite ranking of Faithfulness

Table 6: Tableau 5: 'See'-verbs in accusative languages.

	FaithRole	**TransDef**	**UCC**
DAT–ACC		*	*
NOM–ACC	*		
DAT–NOM	*	*	

constraints {UCC, FaithRole/O} >> {FaithRole/A, TransDef}, the transitive pattern emerges.

Tableau 5 accounts for the fact that assimilation to the transitive pattern would be more frequent for (inactive) perception verbs as compared to emotion verbs; indeed, as noted in Section 2.2 the transitive pattern is cross-linguistically most common. DAT–NOM pattern in (c) is found, for example, in Japanese (see (11)). Finally, the iconic DAT–ACC pattern seems to be rare, but is attested, for example, in Sinhala:

Sinhala (Gair & Paolillo, 1997, p. 33)

(22) *MeṭE dæŋ aliyawE peenEwa*
I.DAT now elephant.ACC see.PRES
'I now see the elephant'

For ergative languages the default mapping from roles to cases based on semantic faithfulness will yield the DAT–ABS frame. The only alternative to this case-frame is the transitive pattern. The choice of the optimal candidate among these two patterns will depend on the relative ranking of FaithRole (more specifically, FaithRole/A) and TransDef, as both patterns satisfy UCC.

Again, the predictions of Tableau 6 (Table 7) seem to be corroborated by the data available. As can be readily seen in Tsunoda's table, ERG–ABS pattern is used for 'see' predicates in many Australian languages, while DAT–ABS pattern is found in Caucasian languages (see (20) from Lezgian).

Thus, my account correctly predicts that both 'see' and 'look' verbs can pattern either transitively or intransitively, but in the latter case the alternative case-frames would be different: 'see' verbs will have the DAT–NOM (or DAT–ABS) pattern as an alternative to the transitive pattern, while the 'look' verbs will have the NOM–OBL (or ABS–OBL) pattern as an alternative. As is clear from the discussion above, these predictions are generally borne out (however, as we shall see in Section 4.2, counterexamples can be found in languages where classes of inactive and active perception verbs are not lexically differentiated).

Table 7: Tableau 6: 'See'-verbs in ergative languages.

	FaithRole	**TransDef**	**UCC**
DAT–ABS		*	
ERG–ABS	*		

3.5. Case Patterns of Sensation Predicates

Finally, let us consider sensation predicates like 'freeze, be cold'. As noted above, these predicates are distinguished from other classes by the fact that they are normally one-argument. The sole argument inasmuch as it is conceptualised as Experiencer, should be iconically encoded through the DAT case (but see below). Tableau 7 (Table 8) shows the optimal case-marking patterns given the familiar set of constraints (note since these verbs are one-argument TransDef is out of force/satisfied vacuously):

The former pattern resulting from FaithRole dominating UCC is familiar from European languages, which regularly use this pattern, in particular, with non-verbal predicates: cf. German *Mir ist kalt*, Russian *Mne xolodno* (me.DAT cold.ADVL) 'I am cold'; see Bossong (1998) for more examples from European languages. Among ergative languages this pattern occurs, for example, in Lezgian (Haspelmath, 2001, p. 61): *Zaz meqi-da* (me.DAT cold-COP) 'I am cold'.

Interestingly, with verbal predicates the dative-experiencer pattern competes with the accusative-experiencer pattern; consider the following examples from Icelandic and Russian:

Icelandic (Andrews, 1982, p. 461)

(23) *Mig* *kelur*
I.ACC is.freezing
'I am freezing/getting frost-bitten'

(24) *Mer* *kolnar*
I.DAT is.getting.cold
'I am getting cold'

Russian

(25) *Mne* *nezdorovitsja*
me.DAT sick.REFL.PRES.3sg
'I feel sick'

Table 8: Tableau 7: Case-marking of sensation-predicates.

	FaithRole	UCC
DAT _		*
NOM/ABS _	*	

(26) *Menja* *toshnit*
I.ACC nauseat.PRES.3sg
'I am nauseated'

The occurrence of the ACC-marked experiencers is unexpected under the present account. Note, however, that in both Icelandic and Russian cases the ACC-marked experiencers are conceived as more affected and more Patient-like as compared to Dative experiencers. This has been noted for Icelandic by Andrews (1982, p. 463; cf. Onishi, 2001, p. 27), who discussing examples (23) and (24) writes that the former, with the ACC-marked experiencer, suggests a physical damage (frostbite), while the latter, with the DAT-marked experiencer, speaks merely of being cold. The same point can be made with regard to the Russian examples in (25) and (26). Thus ACC marking of experiencers of sensation verbs can be attributed to FaithRole, inasmuch as S is conceptualized as a Patient. Interpretation of the sole argument of sensation verbs as Patients is also uncontroversial in the case of Papuan languages mentioned in Section 2.3, where constructions with sensation predicates pattern like an impersonal clause with an object-experiencer.

The first pattern motivated by the FaithRole constraint is also found in languages with 'split intransitivity', where subjects of sensation predicates pattern with non-actors rather than actors (see Mithun, 1991; Dixon, 1994, for discussion of these languages). The distinction between actors and non-actors in these languages manifests itself in agreement rather than morphological case. Sensation verbs regularly take non-actor agreement to cross-reference the experiencer. Some languages, which like Koasati (Muskogean) make a further distinction between 'object'-agreement and 'dative'-agreement paradigms, may use both paradigms with sensation verbs. (Note that the nominative case on the pronominal subject remains unaltered in the examples below, since case-marking of nouns, which operates on the nominative-accusative basis, is totally independent of the choice of agreement.)

Koasati (Kimball, 1991, pp. 253–254)

(27) *(Anó-k)* *am-hó·p*
(I-NOM) I.ACC-sick
'I am sick'

(28) *(Anó-k)* *ca-hó·p*
(I-NOM) I.DAT-sick
'I feel sick/I am hurt'

The choice between the two paradigms is largely lexically determined (hence Koasati qualifies as a predominantly 'split-S' language, in terms of Dixon, 1994). When the same verb may alternatively chose AGR_{acc} or AGR_{dat} paradigms, as in the examples above, the choice between the two paradigms is determined not by the degree of affectedness but rather depends on whether the sensation is viewed as permanent or temporary (Kimball, 1991, p. 253; cf. Payne (1982, p. 356) for the discussion of Chickasaw, another Muskogean language). In most Split-S languages, however, these two AGR series are not differentiated, which is consistent with the view that these languages do not consistently differentiate between experiencers and patients (Drossard's (1991, p. 169) observation that languages of the 'active' type normally lack a distinct class of experiencer verbs).

However, in most other languages — both nominative and ergative — sensation verbs are assimilated to the intransitive pattern with the sole unmarked argument. On my account this is due to the fact that UCC outranks FaithRole. This pattern with subject experiencers of sensation verbs is predominant in European languages (as long as predicates are verbal), as clearly shown by Bossong's (1998) study. My account also provides a straightforward explanation for the fact that even though sensation predicates generally have a stronger predilection for the object-experience construction, in European languages they are less prone for inversion as compared to emotion predicates. Note that in an inverse construction with a (two-place) emotion verb Stimulus is encoded by the unmarked nominative case, hence UCC is satisfied. But for one-place sensation predicates encoding Experiencer as the subject is the only way to satisfy UCC: they simply lack another argument to "invert" with. This account relating the possibility for inversion to the number of arguments is corroborated by the fact that one-argument emotion predicates on Bossong's list such as 'be glad' are less prone for inversion and pattern similarly to sensation predicates.[20]

3.6. Conclusion

In this section, I have argued that cross-linguistic variation in case-frames of individual verb types is best viewed as resulting from interaction of a few functionally based constraints. Notably, the same variation in case-marking arising from competition of several case-frames can often be attested for a single language as well, as each of the competing case-frames may be optimal under different conditions. First, there may be a lexical split within the verb type, some verbs opting for one case pattern while other verbs choosing another one (see, for example the situation in Amele discussed in Section 2.2.). Second, the same lexical item may allow for alternative case patterns (cf. the 'conative alternation' for contact verbs in English).

[20] Haspelmath (2001) provides the following percentages for Bossong's sample showing predilection of particular predicates for object-vs. subject-like experiencer constructions: 'like' shows a clear predilection for the object-experiencer construction (21% vs. 79%), while distribution of these patterns for 'glad' and 'cold' predicates is about the same (52% vs. 48% for 'glad' and 54% vs. 46% for 'cold').

Yet another manifestation of competition between several case-marking options is a 'mixed' pattern when different case-marking strategies (say, morphological case and agreement) opt for different case-patterns. We have seen an example of such mixed patterns in Koasati in (27)–(28), where AGR is chosen on the semantic basis (that is, FaithRole >> {UCC, TransDef}[21]) while case-marking is determined syntactically (that is, {UCC, TransDef} >> FaithRole).

4. Other Factors

Above I have argued for an approach where different functional factors that determine case-marking are viewed as conflicting constraints and I showed how case-frames for individual verb types can be predicted on the basis of these constraints. In this section, I shall briefly discuss for the sake of completeness, two other factors that can influence the case-marking pattern, the lexical class of the predicate, and polysemy and pattern inheritance.

4.1. Non-Canonical Marking of Arguments and Non-Verbal Predication

As noted by Drossard (1991, p. 172) who further attributes this insight to Hansjakob Seiler, the origin of transitivity splits is often rooted in the non-verbal character of certain predicate types. The evidence for this claim is not difficult to come by. For example, in Russian mental predicates regularly take an inverted pattern if the predicate is adverbial: *Ja znaju* 'I know' vs. *Mne izvestno* lit. 'me.DAT (it) is known'. In German, only nonverbal intransitive predicates can take a DAT subject (cf. Primus, 1999); *Mir ist (es) kalt*, etc. In English, verbal types low on Tsunoda's hierarchy (knowledge, feeling, relation) usually pattern transitively if verbal (cf. *know*, *like*, *have*), but intransitively if non-verbal (cf. *aware of*, *fond of*, *similar to*). In Japanese, mental predicates take a transitive pattern if they are verbs but an inverted or a 'double-nominative' pattern if they are adjectives (see Sugamoto, 1982; Shibatani, 2001, for discussion and exemplification). In Tamil, the only productive class of predicates that allow dative subjects are adverbialized nouns:

Tamil (Asher, 1985, p. 106)

(29) *Enakku* *daakamaa* *irukkutu*
I.DAT thirst.ADVL be.PRES.3sn
'I am thirsty'

[21] Of course, for one-argument verbs such as those in (27)–(28) TransDef is out of force.

In Tariana the largest class of sensation predicates that take a non-canonical subject is nominal:

Tariana (Aikhenvald, 2001, p. 180)

(30) *Amiri-ka-mha* *du-na*
drunk(enness)-DECL-PRES 3sgnf-OBJV
'She is drunk'

In many Australian languages, the ABS–DAT pattern is regularly found with nominal predicates denoting emotion or will (see, e.g. Crowley, 1983, p. 342, on Uradhi; cf. also a similar ABS–LOC pattern with the non-verbal predicate in (16) from Djaru). Notably, even when the nominal predicate is verbalized it inherits this case-pattern. Compare the following example from Nyawaygi where the verbal predicate is derived from the nominal through the use of an 'inchoative verbalizer':

Nyawaygi (Dixon, 1983, p. 455)

(31) *ŋayba* *walŋgambiNa* *yagugu*
1sg.S thirsty.INCH.TAM water.DAT
'I am thirsty for water'

Of course, the fact that non-verbal predicates often select a non-canonical subject or object marking has not gone unnoticed in the previous literature. For example, Tsunoda (1981) also notes that the likelihood of a non-verbal expression increases as one moves down the hierarchy. Still, the relation of non-canonical marking to the verb class has not been systematically investigated. Sometimes this relation is conceived as indirect and attributed to the fact that non-verbal predicates are normally stative, and stativity is one of the parameters correlated with low transitivity (see, e.g. Drossard, 1991; Onishi, 2001). Although the correlation between non-verbal predication and stativity is undeniable, not all such cases can be accounted for in semantic terms: after all, mental verbs such as 'like' and 'know' are stative as well, but still select for a transitive pattern in many languages. In general, it seems that a predilection for non-canonical marking related to word class of the predicate is much more robust since it is ultimately rooted in syntactic valency of the respective word classes. For example, a non-nominative encoding of Experiencer in the Russian and Tamil examples cited above can be straightforwardly accounted for by the fact that predicative adverbs do not subcategorise for a subject. Similarly, non-canonical marking of the subject and/or object in two-argument clauses with adjectival (and nominal) predicates can be attributed to the fact that adjectives normally are more restricted in their transitive potential than verbs. Although adjectives in some languages (e.g. in Quechua) can govern a direct object, in most languages they cannot. In this respect, adjectives in Japanese and Korean which take an object in the NOM rather than ACC case provide a good example. As these examples demonstrate, encoding of arguments is severely constrained by the syntactic valency of individual word classes.

How can this factor be integrated into my model? I shall assume that non-verbal and verbal predicates share the same argument structure as long as they belong to the same predicate type, but differ in syntactic constraints. TransDef would be out of force with respect to non-verbal predicates since it pertains only to two-argument verbs. For adverbial predicates that do not subcategorise for a subject either, UCC is further irrelevant. The fact that adjectival predicates in many languages disallow a direct object can be captured through an additional constraint prohibiting assignment of accusative case to the second argument (cf. the common view in the generative linguistics that (transitive) verbs are structural case assigners, while adjectives are not). Thus, case-marking patterns for non-verbal predicates can be straightforwardly accounted for in terms of (additional) syntactic constraints pertaining to individual word classes.

4.2. Polysemy and Pattern Assimilation

Another factor that can influence case-marking is polysemy. Notably, polysemous items often share the same case pattern even if their different meanings represent different verb types. Perception predicates provide a good example in this respect. In some languages a verb meaning both 'see' and 'look' can select for different case-frames hereby differentiating these meanings. For example in Even (Tungusic), the same verb *it-* means 'see' when it takes an accusative object (*d'uu-v itten* 'saw a house(-ACC)), but 'look' when it takes an allative object (*d'uu-tki itten* 'looked at a house(-DIR)). However, in other cases of such polysemy one pattern can be assimilated to another. A telling example is found in Tongan (Polynesian), mentioned by Tsunoda (1981), where 'see' takes an ABS OBL pattern (ABS-DAT/LOC). As I have argued in Section 3.4, this pattern is more appropriate for the 'look' class. Importantly, in this case 'see' can also mean 'look at', as in the following example from Chung (1978):

Tongan (Chung, 1978, p. 217)

(32) *'Oku sio ki tahi 'a e sianá na*
PROG see to ocean ABS the man that
'That man was looking at the ocean'

An opposite situation is found in Caucasian languages, where inactive perception predicates display an inverse (affective) pattern (see (20) for Lezgian). Interestingly, in Lezgian, the verb 'see' can also mean 'look', when used in the imperative sentence, without changing its case-frame (DAT-ABS).

Lezgian (Haspelmath, 1993, p. 283)

(33) *Gila* *kwe-z* *[za* *wuč-da-t'a]* *aku!*
now you-DAT [I.ERG do.what-FUT-COND] see.IMPFV
'Now look (lit. see) what I am going to do!'

Thus, both in Tongan and Lezgian perception verbs are ambiguous between inactive and attentive perception, but in Tongan verbs inactive perception follow the pattern of verbs of attentive perception, while in Lezgian the situation is reverse.

Similar examples of pattern assimilation due to polysemy can be found elsewhere. For example, Dixon (1980, p. 103) notes that many Australian languages do not differentiate lexically between intended and accomplished action and therefore use the same item (with the same transitive frame) in both meanings, citing Dyirbal verb *balga-l* 'hit (with a club etc.); kill by hitting' as an example. Similarly, the fact that psych-verbs from the Levin's (1993) 'amuse' class take Object-Experiencer may be due to the fact that many of them (including *amuse* itself) also allow an agentive interpretation (Grimshaw, 1990, p. 23), while some others (e.g. *depress*, *strike*) can also refer to a physical (agentive) action (Levin, 1993, p. 191).

In the above cases, polysemy can be observed on a synchronic level. However, in many cases it can be observed only diachronically, when in the course of semantic evolution a verb can shift its verb type. Now, if the original meaning providing motivation for a particular case-frame is lost we are not dealing with a case of polysemy any more; for such cases it would be more appropriate to speak of pattern inheritance. For example, Haspelmath (2001) suggests a historical explanation for the preference of subject-experiencer constructions in 'Standard Average European' languages. He notes that many emotion predicates historically arise through metaphorical extension from the verbs denoting a physical action; thus, *worry* derives its meaning from 'strangle; seize by the throat', *stun* from 'deprive of consciousness with a blow', etc. (Haspelmath, 2001, p. 79). Mithun discussing idiosyncrasies of subject marking in the split intransitive Haida, where a subject of a 'know' predicate is treated as a non-agent while the subject of 'believe' as an agent, remarks (1999, p. 217): "It might be hypothesized that belief requires a stronger will or instigation than knowledge, but a better explanation is rooted in idiomaticity and lexicalization. Levine notes that to 'believe' in Haida is literally 'make true for oneself' (1977, p. 135)". Similarly, Nichols (1984, p. 199) states that it is impossible to provide a synchronic explanation for idiosyncratic marking of indirect objects of two-place verbs in Nakh-Daghestanian languages, but for most cases a diachronic explanation is readily available.

It is more difficult to integrate constraints associated with lexical assimilation into my model. As long as we are dealing with cases of synchronic polysemy, these can be dealt by introducing an additional constraint requiring unification of valency patterns of a polysemous verb. However, once the source meaning which could account for the deviant case-frame is lost, it appears to be impossible to account for the deviant valency patterns in terms of universal constraints; here the valency pattern should be specified in the lexicon.

5. Conclusion: Towards a Comprehensive Map for Transitivity Splits

Above I have set up a two-dimensional semantic map for transitivity splits and showed how a small set of functionally based constraints can account for variation in case-frames for particular verb types on the map. However, in the present form this map is incomplete. As mentioned in the introduction, while there is only one way to comply with the transitive prototype (see Givón's definition above), there are numerous ways in which a construction can deviate from it. Above we have considered two strategies for violating the transitivity prototype, as manifested with the respective sub-hierarchies on the proposed two-dimensional map: the first resulting in a decrease in affectedness on the part of O, the other primarily concerned with reduction in agentivity on the part of A (the sub-hierarchy from 'break' to 'freeze'). These two routes lead into the domain of agent-like and patient-like intransitives, respectively.

However, there are more conceivable routes between the semantic domains of transitivity and intransitivity. A route that we have not considered so far pertains the loss of referential distinctness and asymmetrical relation between A and O arguments. This dimension of transitivity have been addressed by Kemmer (1993) in her study of middle voice, although her study concerns the issue of voice alternations as much as the issue of verb classification. As far as the latter is concerned we can integrate the semantic map suggested by Kemmer (1993, p. 202) for the middle situation types with the semantic map of transitivity splits presented above. On the one hand, semantic middles (I use this term to refer to inherent reflexives like 'wash' and inherent reciprocals like 'meet') are connected with syntactic reflexives, which are closer to the transitive pole of the map. On the other hand, middle verbs are linked to the intransitive domain via the verbs of spontaneous action ('burst', 'be born', etc.). A further connection between semantic middles to experiencer verbs is corroborated by the fact that the latter often display middle morphology (Kemmer, 1993).

Further the map is incomplete, as there are other verb types not represented on the map. First, we have seen evidence that cognition verbs are intermediate between the perception and emotion verbs, as originally proposed by Tsunoda. Furthermore, transitive verbs involving an affected subject (like 'eat', 'put on' or 'take') are arguably intermediate between canonical transitives and perception verbs.[22] The distinction between verbs with the affected subject and experiencer verbs can be conveniently captured in the analysis proposed by Testelets (1998) who distinguishes between different verb types in terms of control and affectedness but additionally 'weights' A and O with regard to these features. On this view, both mental verbs and verbs from the 'eat' class involve an affected A, but for the latter verb class O is arguably more affected than A. There is also an obvious connection between the class of affected subject verbs to inherent reflexives Kemmer's (1993, p. 54) class of "grooming" verbs). Another verb type missing on the map is the verbs of social interaction, studied by Blume (1997). This is again a rather heterogeneous group including such items as 'follow', 'help', 'obey' and 'speak'. On the one hand, interaction verbs show similarities with pursuit verbs (note that both are treated as subclasses of 'aiming' verbs by Lazard, 1998).[23] On the other hand, interaction verbs show an affinity with symmetric predicates (inherent reciprocals); cf. 'agree' or 'marry' which can be interpreted as members of both types. In terms of Testelets (1998), the difference between the two types lies in the degree of control ascribed to the O participant: with symmetric predicates A and O are equal in control, with interaction verbs O is lower in control than A (with canonical transitives O has no control whatsoever). The connection between aiming verbs and symmetric predicates is corroborated by the data from Kartvelian languages where both classes are encoded as middle verbs (Testelets, 1998). Finally, as we have witnessed above (in Sections 2.3 and 3.5), sensation predicates additionally show links that lead beyond one-argument intransitives to the domain of impersonal constructions.

The connections between individual verb types discussed above can be tentatively summarized in the form of semantic map shown in Figure 2 for verb classes, which subsumes the two-dimensional map introduced in Section 2:

Needless to say, much cross-linguistic research is needed to substantiate the claims embodied in this map. Again, the map is still incomplete in that further semantic types can be introduced. Thus, my map does not mention Tsunoda's 'Relation' type, which in itself is a rather heterogeneous class. Furthermore, some other classes may need further decomposition; for example, it is instructive, that both Tsunoda's 'pursuit' verbs and Blume's 'interaction verbs' are distributed across several subclasses in Levin's (1993) comprehensive study of English verb classes. Generally, much more descriptive and theoretical work (along the lines

[22] Ample cross-linguistic evidence for the reduced transitivity of verbs with affected subjects (in particular, 'ingestive verbs') has been presented by Saksena (1980), Amberber (2002) and Næss (2004).
[23] In Chukchee, for example, interaction verbs (e.g. 'call', 'help') and pursuit verbs ('wait for', 'stalk') belong to the same class in that both are labile and allow for the alternation of the ERG–ABS and ABS–OBL patterns (Inenlikej & Nedjalkov, 1967, p. 254).

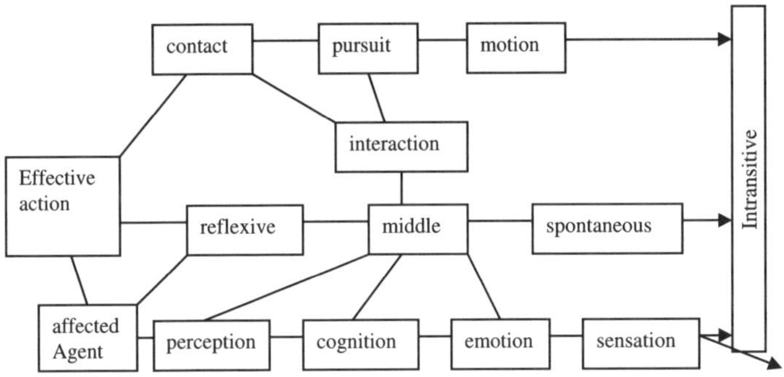

Figure 2: A Comprehensive semantic map for transitivity splits.

of Apresjan (1969) for Russian and Dixon (1991) and Levin (1993) for English) is needed to identify verb classes in individual languages, as well as to provide a proper definition and diagnostics for particular argument roles. Still, as I tried to show in this chapter, constructing a cross-linguistically viable hierarchy of verb types is feasible, and this hierarchy can be used to constrain cross-linguistic variation in case-marking patterns for particular verb types.

Acknowledgements

I am grateful to H. de Hoop, P. de Swart, T. Tsunoda and two anonymous reviewers for the helpful comments on the earlier version of this chapter. I acknowledge the Netherlands Organization for Scientific Research (NWO) for financial support, grant 220-70-003 for the PIONIER project "Case cross-linguistically".

References

Aikhenvald, A. (2001). Verb types, non-canonically marked arguments and grammatical relations: a Tariana perspective. In: A. Aikhenvald, R. W. M. Dixon & M. Onishi (Eds) *Non-canonical marking of subjects and objects* (pp. 177–201) (Typological studies in languages, 46) Amsterdam: Benjamins.

Aikhenvald, A., Dixon, R. W. M., & Onishi, M. (Eds), *Non-canonical marking of subjects and objects* (pp. 177–201) (Typological studies in languages, 46). Amsterdam: Benjamins.

Alekseev, M. E. (1994). Budukh. In: R. Smeets (Ed.), *North East Caucasian Languages* (Vol. 2, pp. 259–297). New York: Caravan Books.

Amberber, M. (2002). Quirky alternations of transitivity: the case of ingestive predicates. In: M. Amberber & P. Collins (Eds) *Language universals and variation.* (pp. 1–20). Westport, Connecticut: Prager.
Andrews, A. (1982). The representation of case in modern Icelandic. In: J. Bresnan (Ed.), *The mental representation of the grammatical relations* (pp. 427–503). Cambridge: MIT Press.
Apresjan, Ju. D. (1969). *Eksperementaljnoe issledovanie russkogo glagola.* Moskva: Nauka.
Asher, R. E. (1985). *Tamil.* London: Croom Helm.
Blake, B. J. (1977). *Case marking in Australian languages.* (*Linguistic Series*, 23). Canberra: Australian Institute of Aboriginal Studies.
Blake, B. J. (1994). *Case.* Cambridge: Cambridge University Press.
Blume, K. (1997). A contrastive analysis of interaction verbs with dative complements. *Linguistics, 36*(2), 253–280.
Bossong, G. (1998). Le marquage différentiel de l'objet dans les langues d'Europe. In: J. Feuillet (Ed.), *Actance et valence dans le langues de l' Europe* (pp. 259–294). Berlin: Mouton de Gruyter.
Christol, A. (1998). Marquage oblique des actants. In: J. Feuillet (Ed.), *Actance et valence dans le langues de l' Europe* (pp. 457–525). Berlin: Mouton de Gruyter.
Chung, S. (1978). *Case marking and grammatical relations in Polynesian.* Austin: University of Texas Press.
Comrie, B. (1989). *Language universals and linguistic typology.* Chicago: University of Chicago Press.
Comrie, B., & van der Berg, H. (2003). Experiencer constructions in Daghestanian languages. Handout of talk presented at the PIONIER workshop on *Case, valency and transitivity.*
Cristofaro, S. (2003). *Subordination.* Oxford: Oxford University Press.
Croft, W. (1991). *Syntactic categories and grammatical relations: The cognitive organization of information.* Chicago: University of Chicago Press.
Croft, W. (1998). Event structure in argument linking. In: M. Butt & W. Geuder (Eds), *The projection of arguments* (pp. 21–65). Stanford: CLSI Publications.
Crowley, T. (1983). Uradhi. In: R. W. M. Dixon & B. Blake (Eds), *The handbook of Australian languages* (Vol. 3 pp. 307–430). Amsterdam: Benjamins.
Dixon, R. M. W. (1977). Where have all the adjectives gone? *Studies in Language, 1*, 19–80.
Dixon, R. M. W. (1979). Ergativity. *Language, 55*, 59–138.
Dixon, R. M. W. (1980). *The languages of Australia.* Cambridge: Cambridge University Press.
Dixon, R. M. W. (1983). Nyawaygi. In: R. W. M. Dixon & B. Blake (Eds). *The handbook of Australian languages* (Vol. 3, pp. 431–527). Amsterdam: Benjamins.
Dixon, R. M. W. (1991). *A new approach to English grammar, on semantic principles.* Oxford: Clarendon Press.
Dixon, R. M. W. (1994). *Ergativity.* Cambridge: Cambridge University Press.
Dowty, D. R. (1991). Thematic proto-roles and argument selection. *Language, 67*, 547–619.
Drossard, W. (1991). Verbklassen. In: H. Seiler & W. Premper (Eds), *Partizipation: das sprachliche Erfassen von Sachverhalten* (pp. 150–182). Tubingen: Narr.
Durie, M. (1985). *A grammar of Acehnese.* Dordrecht: Foris.

Ebeling, C. L. (1966). Review of Chikobava and Cercvadze's "The grammar of literary Avar". *Studia Caucasica, 2*, 58–100.

Evans, N. (1995). *A grammar of Kayardild*. Berlin: Mouton.

Gair, J. W., & Paolillo, J. C. (1997). *Sinhala*. München: LINCOM.

Ganenkov, D. (forthcoming). Experiencer coding in Daghestanian languages. In: L. Kulikov, A. Malchukov & P. de Swart (Eds), *Case, valency and transitivity: A cross-linguistic perspective*. Amsterdam: Benjamins.

Givón, T. (1984). *Syntax: A functional-typological introduction*. Vol. 1. Amsterdam: Benjamins.

Givón, T. (1985). Ergative morphology and transitivity gradients in Newari. In: F. Plank (Ed.) *Relational typology* (pp. 89–107). Berlin: Mouton.

Grimshaw, J. (1990). *Argument structure*. Cambridge, MA: MIT Press.

Haspelmath, M. (1993). *A grammar of Lezgian*. Berlin: Mouton.

Haspelmath, M. (2001). Non-canonical marking of core arguments in European languages. In: A. Aikhenvald, R. W. M. Dixon & M. Onishi (Eds), *Non-canonical marking of subjects and objects* (pp. 53–85). (Typological studies in languages, 46). Amsterdam: Benjamins.

Haspelmath, M. (2003). The geometry of grammatical meaning: Semantic maps and cross-linguistic comparison. In: M. Tomasello (Ed.), *The new psychology of language* (Vol. 2, pp. 211–243). New York: Lawrence Erlbaum Associates Publishers.

Hermon, G. (2001). Non-canonically marked A/S in Imbabura Quechua. In: A. Aikhenvald, R. W. M. Dixon & M. Onishi (Eds), *Non-canonical marking of subjects and objects* (pp. 149–177). (Typological studies in languages, 46). Amsterdam: Benjamins.

Hopper, P. J., & Thompson, S. A. (1980). Transitivity in grammar and discourse. *Language, 56*, 251–299.

de Hoop, H., Koeneman, O., Mulders, J., Weerman, F. (Eds). (2001). Effects of morphological case. *Lingua*, 111.

de Hoop, H., & Narasimhan, B. this volume. Differential case-marking in Hindi.

Inenlikej, P. K., & Nedjalkov, V. P. (1967). Iz nabljudenij nad ergativnoj konstrukciej v chukotskom jazyke (Some observations on the ergative construction in Chukchee). In: V. Z. Zhirmundskij (Ed.), *Ergativnaja konstrukcija predlozhenija v jazykax raznyx tipov* (pp. 247–261). Leningrad: Nauka.

Jacobsen, W. M. (1992). *The transitive structure of events in Japanese*. Tokyo: Kurosio Publishers.

Jayaseelan, K. A. (2004). The possessor-experiencer dative in Malayalam. In: P. Bhaskararao & K. V. Subbarao (Eds), *Non-nominative subjects* (Vol. 1, pp. 227–245). Amsterdam: Benjamins.

Kimball, G. D. (1991). *Koasati grammar*. University of Nebraska Press.

Kemmer, S. (1993). *The middle voice*. (Typological studies in language, 23). Amsterdam: Benjamins.

Kibrik, A. E. (1985). Toward a typology of ergativity. In: J. Nochols & A. Woodbury (Eds), Grammar inside and outside the clause (pp. 268–324). Cambridge: Cambridge University Press.

Kibrik, A. E. (1994). Khinalug. In: R. Smeets (Ed.), *North East Caucasian languages* (Vol. 2, pp. 367–407). New York: Caravan Books.

Klimov, G. A., & Alekseev, M. E. (1980). *Tipologija kavkazskix jazykov*. Moskva: Nauka.

Lazard, G. (1998). *Actancy*. Berlin: Mouton.

Legendre, G., Grimshaw, J., Vikner, S. (Eds). (2001). *Optimality-theoretic syntax*. Cambridge, MA: The MIT Press.

Lehmann, C. (1991). Predicate classes and PARTICIPATION. In: H. Seiler & W. Premper (Eds), *Partizipation: das sprachliche Erfassen von Sachverhalten* (pp. 183–239). Tubingen: Naar.

Levin, B. (1993). *English verb classes and alternations*. Chicago: University of Chicago Press.

Levine, R. (1977). *The Skigate dialect of Haida*. Ph.D., University of Columbia.

Lucassen, W. (1985). On the verbal structure of Abkhaz. In: F. Plank (Ed.) *Relational typology* (pp. 257–269). Berlin: Mouton.

Malchukov, A. (forthcoming). Transitivity parameters and transitivity alternations: constraining co-variation. In: L. Kulikov, A. Malchukov & P. de Swart (Eds), *Case, valency and transitivity: A cross-linguistic perspective*. Amsterdam: Benjamins.

Mallinson, G., & Blake, B. (1981). *Language typology: Cross-cultural studies in syntax*. Amsterdam: North-Holland.

Mathews, S., & Yip, V. (1994). *Cantonese: A comprehensive grammar*. London: Routledge.

Mithun, M. (1991). Active/agentive case marking and its motivations. *Language*, *67*, 510–546.

Mithun, M. (1999). The languages of native North America. Cambridge: Cambridge University Press.

Næss, Å. (2004). *Transitivity: From semantics to structure*. Doctoral dissertation, University of Nijmegen.

Nakamura, W. (1999). Functional optimality theory: Evidence from split case systems. In: M. Darnell et al. (Eds). *Functionalism and formalism in linguistics* (Vol. 2, pp. 253–277). Amsterdam: Benjamins.

Nichols, J. (1984). Direct and oblique objects in Chechen-Ingusgh and in Russian. In. F. Plank (ed.) *Objects: towards a theory of grammatical relations* (pp. 183–211). London: Academic Press.

Nichols, J. (1994). Ingush. In: R. Smeets (Ed.), *North East Caucasian languages* (Vol. 2, pp. 79–147). New York: Caravan Books.

Onishi, M. (2001). Non-canonically marked subjects and objects: Parameters and properties. In: A. Aikhenvald, R. W. M. Dixon & M. Onishi (Eds), *Non-canonical marking of subjects and objects* (pp. 1–53). (Typological studies in languages, 46). Amsterdam: Benjamins.

Palmer, F. R. (1994). *Grammatical roles and relations*. Cambridge: Cambridge University Press.

Payne, D. (1982). Chickasaw agreement morphology: A functional explanation. In: P. Hopper & S. Thompson (Eds), *Studies in transitivity*. (*Syntax and Semantics*, Vol. 15, pp. 351–379). New York: Academic Press.

Primus, B. (1999). *Cases and thematic roles*. Tübingen: Niemeyer.

Prince, A., & Smolensky, P. (1993). *Optimality theory. Constraint interaction in generative grammar*. Malden: Blackwell.

Rappoport Hovav, M., & Levin, B. (1998). Building verb meaning. In: M. Butt & W. Geuder (Eds). *The projection of arguments* (pp. 97–135). Stanford: CLSI Publications.

Rice, S. (1987). *Towards a cognitive model of transitivity*. Ph.D, thesis University of California.

Roberts, J. (1987). *Amele*. London: Croom Helm.

Rozwadowska, B. (1988). Thematic restrictions on derived nominals. In: D. Wilkins (Ed.), *Thematic relations*. (*Syntax and Semantics*, (Vol. 21, pp. 147–165). San Diego: Academic Press.

Saksena, A. (1980). The affected agent. *Language, 56*, 812–825.

Schulze, W. (1994). *Udi*. In: R. Smeets (Ed.), *North East Caucasian languages* (Vol. 2, pp. 447–503). New York: Caravan Books.

Shibatani, M. (1977). Grammatical relations and surface cases. *Language*, *53*, 789–809.

Shibatani, M. (2001). Non-canonical constructions in Japanese. In: A. Aikhenvald, R. W. M. Dixon & M. Onishi (Eds), *Non-canonical marking of subjects and objects* (pp. 307–355). (Typological studies in languages, 46). Amsterdam: Benjamins.

Song, J. J. (2001). *Linguistic typology: Morphology and syntax*. London: Longman.

Spencer, A., & Otoguro, R. (this volume) Limits to case — a critical survey of the notion.

Sugamoto, N. (1982). Transitivity and objecthood in Japanese. In: P. J. Hopper & S.A. Thompson (Eds), *Studies in transitivity*. (*Syntax and Semantics*, Vol. 15, pp. 423–449). New York: Academic Press.

Testelets, J. (1998). On the two parameters of transitivity. In: L. Kulikov & H. Vater (Eds), *Typology of verbal categories*. Papers presented to V. Nedjalkov on the occasion of his 70th birthday (pp. 29–45). Tübingen: Niemeyer.

Tsunoda, T. (1981). Split case-marking in verb types and tense/aspect/mood. *Linguistics*, *19*, 389–438.

Tsunoda, T. (1985). Remarks on transitivity. *Journal of Linguistics*, *21*, 385–396.

van Valin, R., & Lapolla, R. (1997). *Syntax: Structure, meaning and function*. Cambridge: Cambridge University Press.

Woolford, E. (2001). Case patterns. In: Legendre, G. et al. (Eds). *Optimality-theoretic syntax* (pp. 509–545). Cambridge, etc.: MIT Press.

Wunderlich, D. (1997). Cause and the structure of verbs. *Linguistic Inquiry*, *28*, 27–68.

Wunderlich, D., & Lakämper, R. (2001). On the interaction of structural and semantic case. In: H. de Hoop, O. Koeneman, J. Mulders, F. Weerman (Eds), *Effects of morphological case*. *Lingua*, 377–417.

Chapter 5

Limits to Case — A Critical Survey of the Notion

Andrew Spencer and Ryo Otoguro

1. Introduction — The Problem of Case

In his wide-ranging and insightful survey of case, Blake (1994, p. 1) defines 'case' as "... a system of marking dependent nouns for the type of relationship they bear to their heads." Blake terms this a 'central definition' (p. 7) but also allows the possibility of deviations from this central type, including case systems which are expressed by adpositions rather than marking on nouns, and systems which do not realize head-dependent relationships.

It is common to see the same case labels used in grammatical descriptions. It is a very noteworthy fact that in language after language, the same clustering of syntactic and semantic properties are expressed by a single type of form, a case. Some (see, for instance, Goddard, 1982; Mohanan, 1994) take this to imply that linguistic theory has to appeal to a feature of case with a variety of values such as nominative, accusative, ergative, absolutive, genitive, dative and so on, in which each case has a canonical function, in addition to peripheral ones possibly. Undoubtedly, there are languages in which we would lose generalizations if we failed to appeal to such a property. But there is also the assumption that whenever a morphosyntactic construction is used to realize the more common grammatical or semantic functions then that construction deserves the label of 'case'.[1]

[1] This impulse is taken to its logical conclusion, of course, in the doctrine of 'abstract case' (Chomsky, 1981), under which all languages have a case system even if it receives absolutely no morphological realization whatever.

Competition and Variation in Natural Languages: The Case for Case
Edited by M. Amberber and H. de Hoop

ISBN: 0-08-044651-5

Moreover, there is frequently the further assumption that this label is enshrined in the grammar itself as some kind of feature.

However, it is not entirely clear whether this is the right view to take. The purpose of this chapter is to take a critical look at the notion of case and ask to what extent such a feature is needed in the grammars of individual languages. A leitmotiv running through our discussion will be the necessity to consider both formal, morphological aspects of case and the functional, syntactico-semantic aspects. We shall elaborate on this in Section 3 but for the present, the consequence is that we must consider both the way that case markers are realized (as affixes, clitics, adpositions or whatever) and the grammatical properties that they express. One of the puzzling aspects of case is that a language can have a morphological (inflectional) case system that is not completely congruent to its system of syntactic case relationships. We take the view that this fact (which is not always fully appreciated in the literature) is of importance to our understanding of the nature of case.

Little of what we have to say will be new, rather, our aim is to tie together a number of observations which are not usually considered together, with a view to determining what we can safely call a 'case system'.

The broad conclusions that we shall draw are the following: A language can have grammatical markers (inflections, function words) to which we can conveniently give a pre-theoretical label of 'case', but this does not necessarily mean that the grammar of that language needs to appeal to a feature 'case'. In many instances, no generalizations are lost if we simply refer directly to form-based categories such as 'the *-de* suffixed form of the word' or 'phrase bearing final particle *ga*' or whatever. In languages with declensional classes or other types of morphological complication, it is often necessary to set up a morphological feature of case in order to generalize across different types of word forms. In such languages, the syntactic principles that make appeal to case can often be couched directly in terms of the morphological category. For instance, we might find that the syntax selects a phrase whose lexical noun head has to be in a certain morphological form. This is true of languages in which case does not enter into an agreement system, for instance. For such languages, there seems to be no motivation for setting up a special syntactic feature of case. However, there are languages that show case agreement or other syntactic processes appealing to a syntactic property of case. This syntactic case is then a property of the noun phrase as a whole, and not a property of its head (though it may be realized by selecting a specific morphological form of the lexical head, the corresponding morphological case). A clear sign of the need to distinguish phrase-level syntactic case of this sort from word level morphological case is the phenomenon of 'non-autonomous case' or 'virtual case', where we can define a syntactic case that has no unique morphological correspondent.

2. Beard's Criterion

Case systems represent an advanced stage in the grammaticalization of lexical concepts, in which lexical meaning is all but lost, and the cases function as exponents or signals of other grammatical categories such as grammatical functions. This makes case a somewhat odd category, in that the more indirect the relationship between form and function, the more canonical the case category will be. This point is taken to its logical conclusion by Beard (1995) who argues that we can not really speak of (morphological) case at all unless we have inflectional classes. Only in that type of grammar will it be necessary to generalize over distinct forms and hence only in truly inflectional systems will we need morphological case labels. We shall refer to this as 'Beard's Criterion'. Where a language such as Turkish uses exactly the same case affixes with every noun (modulo predictable phonological allomorphy such as vowel harmony) then we need only speak of the *-In* form, the *-dA* form or whatever, without providing labels which make reference to morphosyntactic properties such as 'genitive' or 'ablative'. Blake (1994, p. 25) points out that Australianists often effectively adopt Beard's thesis, what he calls the 'formal approach'. Australian languages resemble Turkish in that they lack inflectional classes of the Indo-European kind. Although we often find distinct allomorphs of case markers, the distribution of these allomorphs is usually determined by phonology (for instance, the prosodic properties of the stem, or whether the stem ends in a vowel or a consonant). In true inflectional class systems there is no phonological, morphological or semantic justification for class membership and hence the distribution of case allomorphs. For this reason, linguists working on those languages will often refer to cases by their form, not by some abstract label for their (canonical) function. A similar descriptive tradition is found among the native Dravidian grammarians. Beard's Criterion in the simple form that we are applying here presupposes a clean cut between inflectional and 'agglutinating' languages. This is unrealistic of course. An important issue of analysis arises when a language has just a few nouns or even just a single noun that has a case form differing from the norm. Dixon (1980, p. 295) provides just such a case from Yidin^y, in which one word, *bama* 'person', takes the ergative suffix *-lu* rather than the expected *-ngu*. However, it should be pointed out that such analytic conundrums are found throughout morphosyntax. For instance, do we have to say that all English verbs have a 1SG form and a singular/plural distinction in the past tense, simply because the verb 'be' retains such forms?

There is an important point lying behind the formal approach to case, namely that morphological features have the principal function of regulating the construction of inflected word forms. Thus, in the first instance, the reason why we need a label such as 'ablative case' for Turkish or Latin is so that the morphology can affix *-dA* to a Turkish noun or *-a:*, *-o:*, *-i:*, *-i:s*, *-ibus*, etc. to a noun or adjective in Latin.

For Latin, we will miss generalizations if we do not use an abstracted label such as 'ablative (case) form)'. We need to be able to say that word forms such as *nauta:* 'sailor', *domino:* 'lord', *turri:* 'tower', *pueri:s* 'boys', *hominibus* 'men' are all 'in the ablative case', and for this it makes sense to say that rules affixing *-a:*, etc. appeal to that property. However, for Turkish nothing is lost, apparently, by setting up a noun category of '*-dA* form' and letting principles of morphology and of morphosyntax to appeal to the label '*dA* form'. This point is all the more valid for languages such as Japanese (see below) or Indo-Aryan in which the 'case markers' are actually postpositions, particles or phrase-final clitics. In such languages, there is absolutely no need to refer to any kind of case label in the morphology, since all statements can be couched in terms of the postposition, particle or clitic.

3. Syntactic Reflexes of Case

As is well known, in languages with reasonably rich morphological case systems (at least as rich as that of Turkish, say) the cases may express purely semantic concepts such as 'instrument', 'location', and so on or they may be used to realize grammatical relationships, specifically grammatical functions (GFs) such as subject, object, possessor. In addition, the features in a canonical case system can be expected to participate in grammatical processes of various kinds that are defined in purely formal, grammatical terms, without semantic conditioning. This principally means semantically arbitrary morphosyntactic processes of agreement and government. The most well-known type of case agreement is that of attributive modifiers, determiners, and so on within the NP (although it is not always entirely clear whether to treat this as agreement in case or multiple marking of case within the noun phrase). However, we also find occasional instances of case agreement in secondary predications and similar constructions. Latvian has a debitive mood in which the verb is prefixed by *jā-* and the subject appears in the dative case. With the debitive form of the copular verb, a predicative adjective agrees in all features, including dative case, with the subject (Mathiassen, 1997, p. 130):[2]

(1) *Studentiem* *jābūt* *uzmanīgiem*
student.DAT.PL be.DEBITIVE attentive.DAT.PL
'The students have to be attentive'

---

[2] Abbreviations used are as follows:
ACC 'accusative', ALLAT 'allative', CL '(noun) class' (Swahili); 'classifier' (Japanese quantified nouns), COP 'copula', DAT 'dative', EMPH 'emphatic (particle)' (Japanese), ERG 'ergative', FEM 'feminine', GEN 'genitive', HON 'honorific', INSTR 'instrumental', LOC 'locative', MASC 'masculine', NOM 'nominative', PL 'plural', SG 'singular', SUBJ 'subject', TOP 'topic', VOC 'vocative'.

In a number of Slavic languages, and especially Polish, the instrumental case marks predicatively used nominals, especially where temporary or stage-level predication is involved. There is absolutely no semantic connection between the core meaning of the instrumental and stage-level predication, of course.

Constructions such as these more or less demand that we set up a syntactic attribute of case, over which agreement/government relations can be defined in the syntax. Further indications of the need for a syntactic attribute of case, over and above morphological case, is found when the list of cases manipulated by the syntax is apparently greater than that provided for by the morphology, what Spencer (2005) has called 'virtual case'. (Mel'čuk (1986) discusses a variant of this under the heading 'non-autonomous case'; see also Brown, Baerman & Corbett, (forthcoming) for further exemplification.) A well-known example of this is provided by the Finnish syntactic accusative, which is expressed by either the morphological nominative or genitive (except in pronouns). In Estonian, even pronouns lack a dedicated accusative. Ackerman and Moore (2001, Chap. 4) claim that the grammar of Estonian would be greatly complicated if there were no way of referring to the notion 'noun phrase marked CASE ACCUSATIVE'.

A less well-known but also very clear instance of virtual case is provided by Chukchee. Nouns fall into two inflectional classes, with distinct sets of case suffixes. The language has a very clear ergative morphosyntax, with all nouns and pronouns showing ergative–absolutive case marking and no splits with respect to tense/aspect or any other property. However, no single noun or pronominal has a case form which can be uniquely identified as purely ergative. In one inflectional class, the ergative is always identical to the locative, and in the other class ergative is always identical to the instrumental. In this respect, Chukchee nouns exhibit both of the common case syncretisms found in Australian languages (Dixon 1980, pp. 303, 304), but the Chukchee case is particularly striking because two distinct inflectional classes are involved, thus satisfying 'Beard's Criterion' for a canonical case system (see Spencer, 2005 for full justification of this claim). We must therefore set up an ergative case for the syntax to appeal to while recognizing that the morphological realization of ergative is always syncretic.

A slightly more complex example is provided by the animate accusative in Russian (see Fraser & Corbett, 1995 for detailed discussion). Only nouns of the predominantly feminine gender *–a* class (2nd declension) have a dedicated accusative singular form (e.g. *Marinu* 'Marina.ACC'). Adjectives which agree with such a noun take the ending *-uju*: *moloduju Marinu* 'young.ACC Marina.ACC'. No noun has a dedicated accusative in the plural. Inanimates which are not in the *–a* declension have an accusative which is identical to the nominative. Animate nouns in the plural and 1st declension animates in the singular have their accusatives identical to the genitive form. An interesting situation arises with regard to a largish class of 2nd declension (*-a* class) nouns which are masculine, such as *mužčina* 'man'. Being 2nd declension nouns, these have an accusative singular in *–u*: *mužčinu*. However, an adjective modifying such a noun can not appear in the *–uju* form since this is reserved for feminine gender nouns. Instead, the adjective has to adopt the form it would have if the masculine gender noun were in the expected inflectional class, the exclusively masculine gender 1st declension. This form is identical to the genitive singular, ending in *-ogo*: *molodogo mužčinu* 'young-ACC/GEN man.ACC'.

Syntactically speaking, the animate genitive/accusative is really an accusative case. For instance, when a transitive verb taking an accusative direct object is passivized the object is promoted to subject and marked with nominative case as in (2):

(2) a. *My* *priglasili* *èt-u* *molod-uju*
we invited this-FEM.ACC.SG young-FEM.ACC.SG
ženščin-u
woman.FEM-ACC.SG
'We invited this young woman'

b. *Èt-a* *molod-aja* *ženščin-a*
this-FEM.NOM.SG young-FEM.NOM.SG woman.FEM-NOM.SG
byla *priglašena*
was invited
'This young woman was invited'

This is also true of animate nouns that appear morphologically to be in the genitive case, as seen in (3):

(3) a. *My* *priglasili* *èt-ogo* *molod-ogo*
we invited this-MASC.GEN.SG young-MASC.GEN.SG
čelovek-a
person.MASC-GEN.SG
'We invited this young person'

b. *Èt-ot* *molod-oj* *čelovek*
this-MASC.NOM.SG young-MASC.NOM.SG person.MASC.NOM.SG
byl *priglašen*
was invited
'This young person was invited'

However, the case labelling in (3a) is wrong. There is a small number of verbs which selects complements in non-accusative cases, amongst them *trebovat´* 'demand' which takes a genitive complement. However, verbs of this sort lack a passive construction because only accusative-marked objects can be passivized.

Thus, the complement phrase *ètogo molodogo čeloveka* in (3a) must be in the accusative case, despite appearances. This is a little clearer when we substitute *mužčina* for *čelovek* in (3), as in (4):

(4) a. *My priglasili èt-ogo molod-ogo*
we invited this-MASC.ACC.SG young-MASC.ACC.SG
mužčin-u
man.MASC-GEN.SG
'We invited this young man'

b. *Èt-ot molod-oj mužčina*
this-MASC.NOM.SG young-MASC.NOM.SG man.MASC.NOM.SG
byl priglašen
was invited
'This young man was invited'

The only way to describe such a situation adequately is to say that the forms which are apparently genitive are actually accusative. If there were no dedicated 2nd declension accusative singular form, we would be dealing with a virtual accusative case here.

Australian languages furnish an even more complex instance of this type of mismatch. We mentioned earlier potential problems with Beard's Criterion, where a language with otherwise agglutinating case forms has occasional lexemes with unexpected morphology. More complex instances of this problem often arise, where pronouns and other nominals exhibit split case marking behaviour. Blake (1994, p. 27) discusses this in the context of Australian languages. He cites example (5) from Pitjantjatjara:

(5) *Tjilka-ngku ngaya-nya tjina waka-rnu*
prickle-ERG 1SG-ACC foot.ACC pierce-PAST
'A prickle pierced my foot'

Pronouns, but not nouns, are distinctively marked for accusative case. The neatest way to describe case marking is to assume that the entire NP is in one and the same (syntactic) case. If we permit ourselves an abstract label 'accusative', generalizing over the overtly marked pronoun form *ngaya-nya* and the unmarked noun form *tjina* then we avoid having to describe the morphosyntax with appeal to disjunctive statements along the lines 'the *–nya* form of the pronoun and the unmarked form of the noun'.

A consequence of this reasoning we is that must assume that the unmarked form of a noun is ambiguous with respect to (syntactic) case marking. On the one hand, it realizes the absolutive case, for objects and intransitive subjects, in contrast with the ergative case (on nouns). On the other hand, it realizes the accusative case, which stands in contrast to the unmarked nominative case realizing (transitive and intransitive) subjects. As Blake points out (1994, p. 27), this means that the unmarked form of the pronoun, when in construction with an ergative marked subject noun, has to be analysed as realizing ergative case. Hence, ultimately all transitive subjects are simultaneously in the ergative and the nominative, all intransitive subjects are simultaneously nominative and absolutive, while all objects are simultaneously absolutive and accusative. It is not entirely obvious how such a paradoxical situation should best be resolved. However, we can agree with Blake that simply labelling the morphological forms will not be of much help.

These examples of virtual or non-autonomous case all involve primary grammatical functions of subject/object, but we find the same phenomenon with the semantic locative case in Latin. For first and second declension nouns the locative is identical to the genitive in the singular and to the dative/ablative in the plural. In third declension nouns it is identical either to the dative or the ablative in the singular (*Cartha:gini:* or *Cartha:gine* 'at Carthage') and to the dative/ablative in the plural (Allen & Greenhough, 1983, pp. 34, 269; see also Blake, 1994, pp. 23f). Notice that the only justification for the case value of 'locative' is the meaning, 'at such and such a place', not the syntactic behaviour.[3]

Phenomena such as those sketched in this subsection are typical of the morphosyntactic constructions which motivate a syntactic notion of case. Without such a feature for the syntax to refer to, it is very difficult to see how the facts of case government and agreement could ever be stated. This fact makes it easier to see the close dependency between morphological form and syntactic function. If cases are not realized morphologically, by head-marking on the noun, then we have fewer opportunities for case government, and there is virtually no way in which we could see case agreement. For this reason, then, we should be very suspicious of descriptions which appeal to case features where no straightforward head-marking morphology is involved.

4. Cases at the Periphery

One of the features of case systems is that they accrue gradually over time through grammaticalization and they decay over time, so that certain cases may

[3] The search for virtual case can sometimes mislead. Traditionally, Latvian grammars have distinguished an instrumental case, which is identical to the accusative in the singular and the dative in the plural. This case is governed by the preposition *ar* 'with'. Unfortunately, the instrumental does not exist. In the singular, different prepositions in Latvian take different cases, but in the plural all prepositions, without exception, govern the dative (Fennell, 1975; Mathiassen, 1997, p. 41).

be restricted to very limited grammatical contexts or specific lexical items. The vagaries of grammatical change can make it very difficult to characterize 'the' case system of individual languages and raises interesting questions about the relationship between morphology and syntax and the formalization of grammars generally. In this section, we just present a handful of examples to illustrate some of the puzzles that tend to be swept under the carpet.

4.1. Minor/moribund Contrasts

It may well be a universal of morphosyntax that wherever we have a stable inflectional case system, it will include outliers that are hard to integrate into the system as a whole. There are three ways in which cases may be peripheral:

(i) They may be recently grammaticalized, and hence not fully integrated into the case system as a whole, either morphologically or syntactically or both;
(ii) They may be moribund;
(iii) They may be cases which have unusual functional or formal properties.

In addition, there may be types of constructions that resemble case marking formally and may interact with case marking, but may appear to be something different. Particles marking discourse functions fall into this category. As an instance of the third type, we shall discuss the vocative in the next subsection. Here we will just briefly sketch instances of the first two types.

Instances of type (i) due to the partial grammaticalization of cases are not hard to come by. For instance, at the morphophonemic level we find that some of the cases of Hungarian fail to take part in vowel harmony (Kenesei, Vágo & Fenyvesi, 1998). At the level of syntax we find that in Estonian the 'final four cases' fail to trigger agreement (Blake, 1994, p. 103; Viitso, 1998). At the purely morphological level we often find that some cases seem to be composites, whose first component is often comparable to the genitive case marker. This is found in Uralic generally, the Dravidian languages (for instance, Kannada, Schiffman, 1983, p. 27f), the language isolate Ket (Werner, 1997, p. 102). In Hungarian, we find a slightly different morphological reflex of grammaticalization, in that some of the case endings appear to be cognate with stems that can themselves take nominal inflections. Thus, the formative *–nek/nak* functions as a dative case suffix (*ember-nek* 'to the man', *László-nak* 'to László), but it can also take possessor agreement suffixes: *nek-em* 'to me', where *–em* is the 1sg possessor marker (cf. *kép-em* 'my picture'). Dunn (1999, p. 97f) discusses examples of recently grammaticalized case markers in Chukchee that are cognate with postpositions and whose behaviour is still very similar to that of postpositions (and which have not been recognized as case forms in previous grammatical descriptions).

Case inflections typically develop from postpositions, and so it is not surprising that they may retain some of the properties of postpositions in incomplete grammaticalization. A simple example of this is found in languages such as Turkish (Lewis, 1967, p. 35) in which a case affix can take scope over a coordinated noun construction without having to be repeated, much like the preposition in the English gloss to (6b):

(6) a. *sıhhat-te* *ve* *afiyet-te*
health-LOC and well-being-LOC

b. *sıhhat* *ve* *afiyet-te*
health and well-being-LOC
'in health and well-being'

Exemplifying type (ii) it is particularly common to see moribund residual case systems in languages that have lost an earlier inflectional case system but in which traces of that system survive with high frequency function words such as pronouns. This is found in English, Scandinavian, Romance, Bulgarian and Macedonian, for instance, and gives rise to the frequently explored, but not solved, conundrum 'can a language like English be said to have a nominative/accusative distinction in the nominal system on the basis of such a distinction in five pronouns?' (see, for instance, Hudson, 1995). A rather different version of this conundrum arises in the well-known situation in which a nominative-accusative language switches to an ergative-absolutive organization but in which the pronouns retain the old nominative-accusative patterning, as is common in Australian languages. We shall not discuss these familiar cases in detail, but rather will outline a number of other, less familiar examples, to illustrate the scope of the problem.

In reference grammars of Uralic languages, we typically find a long list of case forms with a caveat about the last few, to the effect that they are lexically restricted. Finnish (Karlsson, 1987, p. 113) is typical. The abessive, instructive and comitative are expressed by suffixes *-tta/ttä*, *-n* and *-ine* and they "appear mainly in fixed expressions like idioms". The comitative includes the plural suffix *-i* and therefore neutralizes singular and plural, and it is also invariably followed by a possessive suffix. Restriction to specific morphosyntactic contexts of this sort is not uncommon with moribund cases.

Slavic languages[4] which have retained the Indo-European case system generally retain the distinction between dative and locative, though the latter is usually only found in construction with prepositions (and for this reason is often called the

[4] See de Bray (1980) and Comrie and Corbett (1991) for comparative grammars of the Slavic languages.

prepositional case). However, the two cases are formally similar and are neutralized in some declensions. In Serbian and Croatian, this neutralization is almost complete (even though pedagogic grammars tend to reproduce both cases as though they were in general distinct). In the plural there are no nouns, adjectives or pronouns which distinguish dative from locative. In the singular, all adjectives, pronouns and most nouns fail to distinguish dative from locative. The two cases are distinguished in just a small number of nouns such as *grad* 'town' or *stvar* 'thing'. Segmentally the two cases are identical: *gradu*, *stvari* but in the dative the stem has a falling accent while the prepositional it has a rising accent.

A slightly different situation is found with German. The noun phrase system exhibits a clear four-case contrast of nominative, accusative, genitive and dative. However, this is only visible with phrase-initial determiners and adjectives. Nouns have a much depleted system. So-called weak nouns only have two forms, a nominative singular (*Student* 'student) and a form in *-en* (*Studenten*). So-called strong nouns with masculine or neuter gender (or feminine proper names and close kin terms) have a genitive singular in *-(e)s* and most nouns have a dative plural in *-(e)n* (Spencer, 2005). Because of considerable affixal homophony the actual case of a phrase can sometimes only be deduced by performing a set of inferences over the forms of the determiner, attributive adjective and noun (see Durrell, 1979 for extensive discussion of this and its theoretical implications).

Although Bulgarian has largely lost its case system, there are interesting vestiges. Bulgarian has a definite article, which suffixes to the first adjective or noun in a noun phrase. In the literary language (though less so in the spoken language), with masculine singular nouns, the definite article is found in a "long" form *-ət* and a "short" form *-ə*. The long form is used with subjects and the short form with non-subjects, so that the article effectively signals a nominative/oblique case distinction. No such distinction is observed with neuter or feminine nouns or with plural nouns. A further vestigial case in Bulgarian is the vocative, which we leave for special discussion in the next subsection.

4.2. The Vocative

Languages frequently have markers on nouns or noun phrases which realize discourse-related functions such as information structure (topic-focus, old-new information and so on). Nahuatl lacks noun morphology except for plural formation but it has a prenominal particle, *in*, whose function, while poorly understood, seems to be related to topicality, definiteness, specificity and so on (Andrews, 1975). This marker would never be mistaken for a case formative. However, in other languages, things are not so clear. We discuss the situation with Japanese in some detail in Section 5.1.

One very common discourse function which lies outside the normal range of grammatical functions, adverbials or information structuring is the vocative. This is a very understudied phenomenon, even though many languages are said to have a distinct vocative case and even though it is clear that in some languages the vocative is fully integrated into the nominal declension system. We devote this section to the vocative case, noting that we are just scratching the surface of a rather complex and controversial phenomenon. We begin with the Slavic languages.

With the exception of the pronominal accusative and dative, and the special oblique masculine singular definite form mentioned above, Bulgarian has lost all the cases of Old Slavonic except for a residual vocative. Masculine and feminine nouns in the singular have special suffixed forms (Scatton, 1984, pp. 140–142, 1993, pp. 199–201): *sin* 'son' *sine*, *grad* 'town' *grade*, *knjaz* 'prince' *knjaže*, *səprug* 'husband' *səpruže*, *učitel* 'teacher' *učitelju*, *məž* 'man' *məžo*, *graždanin* 'citizen' *graždanino*, *žena* 'woman' *ženo*, *carica* 'tsarina' *carice*, *sədija* 'judge' *sədijo*.

Unlike in other Slavic languages, the vocative case in Bulgarian has to be treated as a genuine syntactic case rather than just an adverbial or other peripheral form that has been integrated into the noun morphology. This is because adjectives agree with masculine (though not feminine) nouns in vocative case (Scatton, 1984, p. 27, 1993, p. 207): *nov učitel* 'a new teacher', *novi učitelju!* 'new teacher!'. This special form is an innovation in Bulgarian arising from the breakdown of case and also the loss of the short/long adjective forms of Old Slavic.[5] Macedonian, which is very closely related to Bulgarian, has a similar system of noun morphology except that the 'count form' and the vocative are now optional (and tend to be avoided, Friedman, 1993, p. 263f). Optionality, of course, raises additional problems of interpretation and analysis.

The vocative in Slavic is a rather interesting case. All the Slavic languages except Bulgarian and Macedonian have preserved a very full case system for nouns and adjectives. Old Church Slavonic (often taken as a surrogate for Common Slavonic) had a vocative case, but like other Slavonic languages did not distinguish a plural vocative form. A survey of the grammatical descriptions of the modern languages in Comrie and Corbett (1993) reveals that in Czech, Polish, Serbian/Croatian, Upper Sorbian, Kashubian (Cassubian), and Ukrainian

[5] As an anonymous reviewer reminds us, agreement in vocative case is found in Ancient Greek, too, for instance, in the address to Odysseus recurring throughout the Odysseia (e.g. Book V, l. 204):

Di:ogenès	*La:ertiáde:*	*polumé:xan'*
Zeus-born.VOC	Laertes. patronymic.VOC	many.deviced.VOC

Odussêu
Odysseus.VOC
'O Zeus-born son of Laertes, Odysseus, full of devices'

the vocative is retained (and Ukrainian even has a separate vocative plural form *panove* for the word *pan* 'gentleman'). However, in Slovak, Lower Sorbian, Polabian, Russian and Byelorussian the original vocative is now lost (except for a few fixed expressions, such as *bože moj!* 'my God!' from *bog* 'god'). (But see below for the Russian 'new vocative'.) This cross-linguistic patterning illustrates the unusual status of the vocative as an outlier, particularly the differential treatment between dialects or closely related languages. In Czech, the vocative maintains a complex morphology with a variety of allomorphs and various irregular forms, while in the mutually intelligible Slovak it has disappeared. Similarly Upper and Lower Sorbian are considered dialects of a single language yet they differ in the vocative, and the same is true of Polish, Cassubian and Polabian. Add to this the remarks made above about the survival of the vocative in Bulgarian, alone of all the cases, and it is clear that the vocative is in some sense 'outside' the system as a whole.

Formal (morphological) considerations in other languages also point to the peripheral character of the vocative. It is quite common to find forms labelled 'vocative' which depart from the normal nominal morphology of the language. Chukchee has a rich system of cases. Skorik (1961, p. 155f) distinguishes absolutive (nominative), instrumental, locative, ablative, dative-allative (really just an allative), orientative, comitative, associative and designative (equative). To these Dunn (1999, p. 97f) adds inessive, perlative, sublative and privative. (The first three of this last group are the recently grammaticalized cases mentioned above.) The cases are all formed by suffixation, with the exception of the comitative, associative and privative which are formed by circumfixes. In a separate part of his grammar, Skorik (1961, p. 306) describes a vocative form. In all instances, the vocative has final stress, which is otherwise not found with Chukchee nouns.[6] Nouns ending in a single consonant shift stress without further affixation: *Tégrəlqut* (proper name) ~ *Tegrəlqút!* Vowel-final nouns additionally suffix *-j*: *áte* 'Daddy', *atej!* Nouns whose stem ends in two consonants suffix (stressed) *-ój*: *Rowtən* (proper name, stem, *Rowt-*), *Rowtoj!* Nouns ending in the derivational suffix *-tegən* or *-gərgən* truncate the final syllable (a rare instance of subtractive morphology in the language): *Gərgoltágən* (proper name) *Gərgoltág!*, *Təŋagərgən* (proper name), *Təŋagər!* Plural forms of proper names are based on the plural absolutive, suffixed by *–nti*: *Rultənlínti* 'the Rultyn family'. The vocative is formed regularly by stress

[6] Final stress seems to be a popular way of forming vocatives cross linguistically, as is truncation, but it is not possible to draw firm conclusions here. For instance, Somali has a four-case system: absolutive, nominative, genitive and vocative. The nominative and genitive cases are distinguished from the absolutive mainly by tone/stress, while the vocative is formed by suffixation (Orwin, 1995, p. 12; Saeed 1999).

shift and *–j* suffixation: *Rultənlintíj!* Common nouns of the first declension, which lack the *–nti* plural absolutive suffix, just shift stress if the stem is vowel final: *nenéne-t* 'children' *nenenét!* If the stem is consonant final, the noun takes the vocative suffix *–ó-t* (where *–t* is presumably reflex of the absolutive plural suffix): *tumgət* 'friends' *tumgot! welwət* 'ravens' *welwot!*

Discussing the place of the vocative case in the case system generally, Skorik (1961, p. 309) makes the point that the vocative is always, so to speak, parasitic off the absolutive form. This is especially clear in second declension nouns which take a unique suffix in the absolutive plural (but not other plural case forms). This unique suffix forms the base of the vocative plural in these nouns. This is a good point, though it is not clear that this prevents us from analysing the vocative as a case form in Chukchee. After all, it is quite common for one case form to serve as the base of another case form. Thus, Estonian case forms are based on the partitive singular stem (Viitso, 1998), while in some Daghestan languages such as Archi oblique plural case forms are based on a stem formed from the ergative singular (Kibrik, 1998).

We mentioned that Russian had lost its original Common Slavic vocative case forms. However, during the 20th century, proper names in the *–a* declension (especially diminutives) have developed a 'new vocative', formed by truncating the *-a* ending of the nominative singular: *Mama* 'Mummy', *Mam!*, *Papa* 'Daddy' *Pap!*, *Maša*, *Maš! Mar´ina*, *Mar´in!* There are phonological constraints on what kind of clusters can remain (Yadroff, 1996), for instance, we can not form **Natašenk* from *Natašenka*. There are two phonological peculiarities about the new vocative. First, it involves subtractive morphology. Second, this subtraction often leaves behind a voiced obstruent: *Ser´oža*, *Ser´ož!*, *Ser´oga*, *Ser´og! D´ad´a Van´a* 'Uncle Vanya' *d´ad´ Van´!* Voiced obstruents are unpronounceable in final position in Russian. However, in forms such as *Ser´ož*, *Ser´og*, *D´ad´* we fail to find devoicing. This is very peculiar, since word final devoicing is very characteristic of the Russian accent and is generally very hard to suppress. It may well be that the process is not one of full vowel deletion but very strong reduction to a 'supershort' vowel, which cannot normally be heard. In that case, we might be able to argue that the final consonant is 'protected' by an (inaudible) final vowel. But however we analyse the phonology this is a highly marked phonotactic construction.

Finally, it is worth mentioning vocative markers which do not fit into the declensional paradigm of the noun. A simple example is provided by the Arabic vocative particle *ya*, which is found at the beginning of the noun phrase. (Classical Arabic has a three-case system of nominative-accusative-genitive realized entirely by suffixes.). In Greek, we have the interesting situation of a phrase initial particle, *o*, combined with an inflecting vocative case form *Petros* 'Peter' *o Petre*! In both these cases it seems highly counterintuitive to regard the particle as a vocative case marker, though it is not at all clear what should prevent us from making that analytical decision.

What conclusions should we draw from vocative case morphology? First, we should note that some languages fully integrate vocative case morphology into the morphological case system. For instance, it would be completely arbitrary to deny that the Czech morphological case system has a vocative case on a par with the other six cases. On the other hand, we often find that a vocative form exhibits rather different morphology to the rest of the case system (Chukchee, the Russian new vocative). However, it would be perverse to claim that such vocatives are not part of the inflectional system. The second point is that the function of the vocative is not integrated in any clear sense into the morphosyntax of the language. Rather, the vocative has a discourse level function (though one that has received very scant attention from students of discourse). In this respect the vocative does not fit into the characterization of case as signalling a dependency on a dependent noun. What dependency relation is being expressed by a vocative? (And normally the vocative fails to enter into agreement or concord systems, with the curious exception of Bulgarian.)

Bulgarian apart then, there is no real reason to posit a syntactic attribute-value pair CASE VOCATIVE. Instead, we must assume that the vocative signals directly whatever discourse-theoretic function corresponds to the pre-theoretical notion of addressing. The vocative, therefore, provides a simple illustration of the need to separate the morphological case system from a syntactic system of case.

4.3. Outliers

A traditional puzzle in English linguistics is the role of the 'Saxon genitive'. Is this really a genitive case? It is so treated in the recent Cambridge Grammar of the English Language (CGEL, Huddleston & Pullum, 2002), and one can argue that the corresponding construction in Mainland Scandinavian is equally a type of case. Of course, this is not an example of true inflection, but rather edge inflection, but this is simply a matter of formal realization. If we were to exclude the Saxon genitive on the grounds that it is not realized by true affixation we would presumably have to ask questions about languages such as Turkish, in which the case endings (and other suffixes) are subject to Gruppeninflexion or 'suspended affixation' as in example (6).

Swahili, like other Bantu languages, lacks a case system. However, all nouns with the appropriate semantics can take an adverbial or locative suffix *-ni* to form a word which can be used as a predicate together with one of a set of locational/directional copulas. The precise location or direction of motion can be specified by the verb. Examples are given in (7) (Ashton, 1944, p. 18f):

(7) a. *Ki-su* *ki-ko* *wapi?*
CL-knife CL-COP where
'Where is the knife?'

b. *Ki-ko* *nyumba-ni*
CL-COP house-LOC
'It is at the house'

c. *Ki-po* *meza-ni*
CL-COP table-LOC
'It is on the table'

d. *Ki-mo* *sanduka-ni*
CL-COP table-LOC
'It is in the box'

The obvious question arising from such examples is 'what is to stop us from treating these as instances of case morphology?' But if these are types of cases then they constitute a counter-example to the usual typologies of case, under which case systems first distinguish subject case from other types.

5. 'Case' in Japanese

Japanese is a language which is often described as having a case system. In this section, we review some of the properties of this system and its interaction with other morphosyntactic constructions, and come to the conclusion that there is no justification for adopting traditional case labels for any aspect of the language.

Japanese has a number of elements conventionally described as case markers: *ga* 'nominative', *o* 'accusative', *ni* 'dative', *no* 'genitive'. These elements look somewhat like noun suffixes, in that they invariably occur immediately to the right of the head noun. However, Japanese noun phrases are strictly head final, and Japanese also has a set of elements generally thought of as postpositions, some common examples being (Tsujimura, 1996, p. 133): *uti de* 'at home', *gakkoo e* 'to school', *tomodati to* 'with (a) friend', *gozi made* 'until 5 o'clock', *yama kara* 'from (the) mountain'. Blake (1994, p. 10) refers to the Japanese case markers as 'analytic case markers' and explicitly describes them as postpositions. Tsujimura (1996, p. 134) refers to them as 'case particles'. Others refer to them as 'grammatical cases' as opposed to 'semantic cases' (see Ono, 1996, p. 82 citing Sugioka). The particles are best thought of as phrasal affixes because they cannot be repeated on coordinated phrases:

(8) a. *[Taroo to Hanako] ga* *kita*
Taroo AND Hanako NOM came
'Taroo and Hanako came'

b. **[Taroo ga to Hanako ga]* *kita*
Taroo NOM AND Hanako NOM came

In order to remain as neutral as possible about structure and function, and in particular to avoid prejudicing judgements about whether the particles constitute some kind of syntactic head, we shall avoid the term 'postposition' where the markers have case-like functions. We shall refer to *ga/o/ni/no* as 'grammatical case particles' and the other group as 'semantic case particles'.

In addition to these case markers, there is a set of particles with discourse or adverbial functions, which also appear immediately after noun phrases. We shall follow Ono (1996, p. 82) in referring to these as 'adverbial particles', though they fall into two or three distinct subgroups, and the term 'adverbial' erroneously suggests that the particles themselves are immediate constituents of verb phrases or clauses, when in fact they are attached to noun phrases. The adverbial particles include *wa* 'topic', *mo* 'emphasis', *nado* 'such as', *dake* 'only', *made* 'even', *bakari* 'only', *sae* 'even', *demo* 'even, or something', *sika* 'only (followed by the negative form of verb)'.

Disentangling these formatives from each other is not straightforward. The problem lies in the fact that many, though not quite all, of the case particles co-occur with each other (again, the consequence of incomplete grammaticalization). Moreover, the different particles show different privileges of co-occurrence with the adverbial particles, but not in ways that neatly distinguish grammatical from semantic uses. The grammatical case particles can be elided, especially in the spoken language (so-called 'case particle drop'). Moreover, in some instances, they are replaced by one or other of the adverbial particles. The behaviour is summarized by Ono (1996, p. 96f) and an exhaustive set of examples is given in the appendix to his paper (Ono, 1996, p. 103f). We first present some illustrative examples involving the nominative and accusative particles. In (9), we see a basic intransitive in which the nominative particle can be dropped (10a) or replaced by an adverbial particle (10b–f):

(9) *Gaizin ga* *kita n dattte*
foreigner NOM come I heard
'I heard that foreigners had come'

(10) a. *Gaizin Ø* *kita n dattte*
foreigner come I heard
'I heard that foreigners had come'

b. *Gaizin wa* *kita n dattte*
foreigner TOP come I heard
'I heard that foreigners had come'

c. *Gaizin mo* *kita n dattte*
foreigner EMPH come I heard
'I heard that *foreigners* had come'

d. *Gaizin sae* *kita n dattte*
foreigner EVEN come I heard
'I heard that only foreigners had come'

e. *Gaizin dake* *kita n dattte*
foreigner ONLY come I heard
'I heard that only foreigners had come'

f. *Gaizin demo* *kita n dattte*
foreigner EVEN come I heard
'I heard that even foreigners had come'

In the next set of examples we see that the nominative particle can appear outside some adverbial particles (11a,b) but not others (11c–e):

(11) a. *Gaizin sae ga* *kita n datte*
foreigner EVEN NOM come I heard
'I heard that only foreigners had come'

b. *Gaizin dake ga* *kita n dattte*
foreigner ONLY NOM come I heard
'I heard that only foreigners had come'

c. **Gaizin demo ga* *kita n dattte*
foreigner EVEN NOM come I heard

d. **Gaizin wa ga* *kita n dattte*
foreigner TOP NOM come I heard

e. **Gaizin demo ga* *kita n dattte*
foreigner EVEN NOM come I heard

In examples (12) we see the accusative particle and the colloquial equivalent with dropped particle:

(12) a. *Hanako ga* *zizyoden o* *kaita n datte*
Hanako NOM autobiography ACC wrote I heard

b. *Hanako ga* *zizyoden Ø* *kaita n datte*
Hanako NOM autobiography wrote I heard
'I heard that Hanako had written her autobiography'

In examples (13a) we see that the accusative particle can be replaced by a variety of adverbial particles, and that it can appear after some (13b,c) although not others (13d):

(13) a. *Hanako ga* *zizyoden* *wa/mo/sae/dake/demo*
Hanako nom autobiography TOP/EMPH/EVEN/ONLY/EVEN
kaita n datte
wrote I heard
'I heard that Hanako had written her autobiography (even, only, ...)'

b. *Hanako ga* *zizyoden dake/made* *o* *kaita n datte*
Hanako NOM autobiography ONLY/EVEN ACC wrote I heard
'I heard that Hanako had written her autobiography (even, only, ...)'

c. *?Hanako ga* *zizyoden sae* *o* *kaita n datte*
Hanako NOM autobiography EVEN ACC wrote I heard
'I heard that Hanako had written only her autobiography'

d. **Hanako ga* *zizyoden wa/mo/demo* *o*
Hanako NOM autobiography TOP/EMPH/EVEN ACC
kaita n datte
wrote I heard

Finally, the accusative particle can appear before some (14a,b) but not other (14c) adverbial particles:

(14) a *Hanako ga* *zizyoden o* *mo* *kaita n datte*
Hanako NOM autobiography ACC EMPH wrote I heard
'I heard that Hanako had written her *autobiography*'

b. *?Hanako ga* *zizyoden o* *sae/demo/dake/made*
Hanako nom autobiography acc EVEN/EMPH/ONLY/EVEN
kaita n datte
wrote I heard

c. **Hanako ga* *zizyoden o* *wa* *kaita n datte*
Hanako NOM autobiography ACC TOP wrote I heard

The 'dative' particle *ni* is particularly puzzling. This particle can appear in a considerable variety of different constructions. Broadly speaking, in some constructions, it has the properties of a case particle while in other constructions it has the properties of a postposition (Sadakane and Koizumi, 1995; Ohara, 2000). Two important uses of *ni* are as a marker of an indirect object, as a goal or allative marker and as a locative marker. These functions are illustrated in (15, 16):

(15)

Taroo ga	*Hanako ni*	*eigo o*	*osieta*
Taroo NOM	Hanako DAT	English ACC	taught

'Taroo taught Hanako English'

(16) a.

Hanako ga	*Huzisan ni*	*nobotta*
Hanako NOM	Mt Fuji GOAL	climb

'Hanako climbed Mt. Fuji'

b.

Hanako ga	*Pari ni*	*itta*
Hanako NOM	Paris ALLAT	went

'Hanako went to Paris'

c.

Taroo wa	*Nagasaki ni*	*sunda koto ga aru*
Taroo TOP	Nagasaki LOC	live nominal

'Taroo lived in Nagasaki'

Now, as a 'semantic' particle with a locational/directional meaning *ni* behaves much like *ga*, in that it can be dropped or replaced by an adverbial particle. Like *o* (and to some extent *ga*) it can follow certain adverbial particles. Thus, compare (11), (13b,c) with (17):

(17) a.

Hanako ga	*Huzisan nado ni*	*nobotta*
Hanako NOM	Mt. Fuji and so on GOAL	climbed

'Hanako climbed Mt. Fuji and other mountains'

b.

Taroo wa	*Nagasaki dake ni*	*sunde ita*
Taroo TOP	Nagasaki ONLY LOC	lived

'Taroo only lived in Nagasaki'

c.

Hanako ga	*Pari dake ni*	*itta*
Hanako NOM	Paris ONLY ALLAT	went

'Hanako only went to Paris'

However, in its purely grammatical function of indirect object marker, *ni* fails to pattern fully like the other grammatical cases, in that it cannot be dropped or replaced with most of the adverbial particles. However, it can follow some of the adverbials and can be followed by some (18):

(18) a.

**Taroo ga*	*Hanako Ø*	*eigo o*	*osieta n datte*
Taroo NOM	Hanako	English ACC	taught I heard

b.	**Taroo ga*	*Hanako mo*	*eigo o*	*osieta n datte*
	Taroo NOM	Hanako EMPH	English ACC	taught I heard
c.	*Taroo ga*	*Hanako dake ni*	*eigo o*	*osieta n datte*
	Taroo NOM	Hanako ONLY DAT	English ACC	taught I heard
	'I heard that Taroo taught only Hanako English'			
d.	*Taroo ga*	*Hanako ni wa*	*eigo o*	*osieta n datte*
	Taroo NOM	Hanako DAT TOP	English ACC	taught I heard
	'I heard that Taroo taught Hanako English'			

To summarize:

1. With the exception of dative *ni*, the grammatical case particles, including locational/directional *ni*, can be dropped or replaced by adverbial particles;
2. The nominative particle *ga* (and the genitive particle *no*) cannot be followed by any adverbial particle;
3. In all functions, *ni* can be followed by adverbial particles and *o* can be followed by most adverbial particles except for *wa* (though only *mo* gives completely acceptable constructions);
4. The grammatical particles (including dative *ni*) can follow certain adverbial particles (notably *nado*, *dake*, *made*, *bakari* and to some extent *sae*, *sura* 'even', *koso* 'emphasis').

Now we should compare the grammatical particles with the semantic case particles. For reasons of space we shall just present a bare summary of the behaviour of *kara* 'from':

1. *kara* cannot be dropped or replaced by adverbial particles;
2. *kara* can be followed by any adverbial particle;
3. *kara* can follow certain adverbial particles (notably *nado*, *dake*, *made*, *bakari* and to some extent *sae*, *sura* 'even', *koso* 'emphasis').

Finally, we present some additional contrastive facts.

No grammatical case particle can be followed by any other grammatical case particle. However, the semantic case particle *kara* can be followed by genitive *no*:

(19) *haha kara no tegami*
mum KARA GEN letter
'a letter from Mum'

The semantic particle *kara* can be followed by the nominative particle *ga*:

(20)	*Sakkaa wa*	*koohan kara ga*	*omosiroi*
	soccer TOP	second half KARA NOM	exciting
	'The soccer match is getting exciting from the second half'		

Clearly, this very complex patterning fails to delineate in a straightforward fashion a set of 'purely' grammatical case markers and a set of 'purely' semantic case markers. Particularly puzzling is the behaviour of *ni* which patterns more closely with the grammatical markers when used as a semantic case than it does when used as a grammatical case.

To what extent can we speak of a case system here? In a sense this question is impossible to answer without a detailed syntax of Japanese, but the question can be reformulated in the following way: is it necessary for Japanese syntax to refer to a syntactic attribute [CASE: {NOM, ACC, ...}]? From the evidence presented in this section it would seem unlikely. Syntactic principles need to place the phrasal affixes *ga*, *o* and so on with respect to noun phrase edges and with respect to adverbial particles, but it seems that abstract case feature labels would be entirely otiose. Moreover, the realization of such labels as phrasal affixes is made more complicated by the fact that the adverbial particles can entirely replace them. In order to handle that phenomenon, we surely need a set of realization principles which say that when a noun phrase is topicalized, focussed, emphasized and so on the appropriate adverbial particle will or may take precedence over any grammatical particle. This will require us to subcategorize the particles in various ways, but that is clearly required for the non-case adverbial particles just as much as it is for the grammatical and semantic particles.

In this connection it is worth considering the arguments of Shibatani (1977) concerning the nature of the case particles. Shibatani argues that we need to distinguish grammatical relations such as subject and object from surface case features. He points out that only subjects can be the antecedent of a *zibun* reflexive, and only subjects can trigger subject honorification. Given these criteria, however, a variety of *ga*-marked nominals cannot be analysed as subjects (Shibatani, 1977, p. 794). In (21), we illustrate this for subject honorification:

(21) **Yamada sensei-ga okusan-ga owakai*
Yamada teacher-NOM wife-NOM young.SUBJ.HON
intended: 'It is Professor Yamada whose wife is young'

On the other hand, there are other constructions that seem to be sensitive to 'surface case', most notably quantifier floating. Quantifier floating refers to a construction in which a quantifier is detached from its position to the left of the nominal head of the nominal phrase and placed after the NP. A *ga*-marked nominal permits quantifier floating, while a nominal marked with any other particle does not permit floating, even if the grammatical relation borne by the nominal is the same as that of a corresponding *ga*-nominal. Thus, the ungrammatical example (22c) is identical to the grammatical example (22b) except that the (non-subject) noun phrase is marked by *no* and not *ga*:

(22) a. *Sorerano* *3-nin-no* *sensei-no/ga* *okusan-ga*
those 3-CL-GEN teacher-GEN/NOM wife-NOM
wakai
young
'Those three teachers' wives are young/It is those three teachers whose wives are young'

b. *Sorerano* *sensei-ga* *3-nin* *okusan-ga*
those teacher-NOM 3-CL wife-NOM
wakai
young
'Those three teachers' wives are young/It is those three teachers whose wives are young'

c. **Sorerano* *sensei-no* *3-nin* *okusan-ga*
those teacher-GEN 3-CL wife-NOM
wakai
young

Shibatani (1977, p. 800) similarly shows that a nominal, which passes subject tests but which is marked by *ni* fails to license quantifier floating. Shibatani therefore argues that some notion of 'surface case' has to be appealed to.

More recent accounts of quantifier floating, however, treat it in terms of constituent structure differences (Miyagawa, 1989) or with a more refined interpretation of the relevant grammatical relations (Ono, 1992; Ohara, 2000). Moreover, as shown by Sadakane and Koizumi (1995) nominals marked by the dative *ni* can license floating, but only when *ni* is a clitic and not a postposition. We conclude from the observations that the real generalization is that quantifier floating is possible when the nominal has a marker with clitic-like properties (i.e. what students of Japanese grammar generally call 'case particle') but not when the nominal has a marker with the properties of a postposition. Crucially, there is no need to set up any abstract feature CASE with values such as NOM, ACC, GEN, DAT, and so on.

6. Conclusions

By taking Beard's Criterion seriously, we have been led to ask the question of when and under what circumstances the grammar of a language needs to appeal to an attribute 'case', either in the morphology or in the syntax. Morphological and syntactic case are distinct properties and, in principle, can be found independently of each other. In the morphology we need an attribute of morphological case to generalize over inflectional classes. In the syntax, we need an attribute of syntactic case in order to state certain types of agreement, multiple case marking and some forms of government. On the formal side we will tend to include an inflectional category in the case paradigm if it competes with case markers. In Indo-European it is common for a noun to have a morphological vocative form expressed by a suffix in the position of a case marker. However, it is far from clear whether the concept 'syntactic vocative case' makes any sense for the majority of languages, since the vocative has discourse structure functions rather than grammatical functions proper. On the syntactic side we are inclined to posit a case when syntactic principles appeal to a specific set of forms. However, unless the syntax requires appeal to a property of noun phrases (as opposed to just the lexical noun heads of those phrases) then we are not entitled to gratuitously assume a case feature in the syntax. This is the conclusion from our survey of Japanese. We would also argue that essentially the same reasoning applies to the so-called case particles of Indo-Aryan languages such as Hindi/Urdu.[7]

What does this say about typological generalizations built on the notion of case? Here it is necessary to recall that these generalizations are usually about grammatical functions of various kinds. If a language has just one morphosyntactic way of distinguishing grammatical functions, then we expect the split to occur between subjects and non-subjects. Further to that non-subjects are split into direct and indirect objects, and so on. But notice that typologies of this sort are likely to be question-begging unless we have a clear notion of morphological case to work with, independent of our typology. However, when we look more closely at grammatical systems, we find that the typologies are not as clear as we would like. English distinguishes subjects from non-subjects by word order and (marginally) by verb agreement, but the only nominal case form is the genitive (marked by edge inflection). The fact is that case marking, whether canonical or non-canonical, is just one way of distinguishing grammatical functions and the fact that it tends to respect our universal hierarchies is therefore a reflection of the properties of grammatical functions not of case systems. Essentially, the same hierarchies can be found for predicate agreement (though agreement and case do not necessarily match with each other in languages which have both).

If this reasoning is correct, why is it that grammatical descriptions so frequently appeal to case labels and implicit or explicit case features? The intuitions captured by such descriptions are real and the phenomena which are labelled are real. However, we propose that the motivation is best thought of as

---

[7] However, in the case of Hindi/Urdu, one could certainly argue that the so-called 'oblique form' of nouns, which historically was a case form, remains a case form (contra Masica, 1991, p. 237), since it triggers agreement on adjectives (cf. the rather contradictory remarks in Masica, 1991, p. 250, where he states that adjectives inflect for 'case', meaning the direct/oblique form distinction, not his 'Layer II' markers).

a type of metagrammatical description. By this we mean that it is a generalization over types of grammatical phenomena found cross-linguistically, and which in some sense can be grouped together. However, this is not to say that the labels and categories of that metagrammatical description have any role to play in grammars themselves. The grammars of Japanese or Turkish simply do not require an attribute of case either at the level of morphology or syntax. To be sure, it is convenient in some respects to think of these languages as having a case system in which subjects are expressed by a nominative, possessors by a genitive and so on, by analogy with Latin or Finnish. But the grammar of Japanese needs only work in terms of the particles *ga/o/no* and so on and a rather complex set of rules of placement, while Turkish just needs to appeal to form features identifying bare nouns and nouns suffixed with *-in*, *-de* and so on. An analogy can be drawn with the inflection/derivation distinction. It can be very useful to distinguish between morphology that provides forms of a lexeme, and morphology that is used for increasing the stock of lexemes. However, it is very difficult to motivate any grammatical feature within the grammar of a language which takes the values {Inflectional, Derivational}. These categories of description are simply lacking in grammars themselves. Putting it in another way, the inflection/derivation distinction is a metagrammatical distinction that can only be drawn by grammarians by considering the interaction of all the principles of morphology, syntax, semantics and lexical relatedness in the language. It is not a distinction that is made by learners and speakers of the language.

Given this metagrammatical perspective on case, it becomes less puzzling to see so many non-canonical types of case marking, whether at the level of form or function. Grammars grammaticalize all sorts of relationships using a great variety of morphosyntactic devices. The mere fact that we can identify a metagrammatical category of 'case' does not entail that languages will necessarily respect that category in their own grammars.

Acknowledgements

Spencer gratefully acknowledges the support of a British Academy/Leverhulme Trust Senior Fellowship 2002–2003. The chapter has benefitted from very helpful comments from two anonymous reviewers.

References

Ackerman, F., & Moore, J. (2001). *Proto-properties and grammatical encoding: A correspondence theory of argument selection.* Stanford University: Center for the Study of Language and Information.

Andrews, J. R. (1975). *Classical Nahuatl*. Austin: University of Texas Press.
Ashton, E. (1944). *Swahili grammar*. London: Longmans.
Baerman, M., Brown, D., & Corbett, G. G. (2005). *The syntax-morphology inter face. A study in syncretism*. Cambridge: Cambridge university press.
Beard, R. (1995). *Lexeme morpheme base morphology*. Stony Brook, NY: SUNY Press.
Blake, B. (1994). *Case*. Cambridge: Cambridge University Press.
Chomsky, N. (1981). *Lecturers on government and binding*. Dordrecht: Foris.
De Bray, R. G. W. (1980). *Guide to the Slavonic languages*. (In three volumes). Columbus, Ohio: Slavica Publishers.
Dixon, R. M. W. (1980). *The languages of Australian*. Cambridge: Cambridge University Press.
Dunn, M. (1999). *A grammar of Chukchi*. Unpublished PhD dissertation. Canberra: Australian National University.
Durrell, M. (1979). Some problems in the morphology of the German noun phrase. *Transactions of the Philological Society*, 66–88.
Evans, N. (1995). *A grammar of Kayardild*. Berlin: Mouton de Gruyter.
Fennel, T. G. (1975). Is there an instrumental case in Latvian? *Journal of Baltic Studies*, *6*, 41–48.
Fraser, N. M., & Corbett, G. G. (1995). Gender, animacy, and declensional class assignment: A unified account for russian. In: G. Booij & J. van Marle (Eds), *Yearbook of morphology 1994* (pp. 123–150). Dordrecht: Kluwer.
Friedman, V. A. (1993). Macedonian. In: B. Comrie & G. G. Corbett (Eds), *The Slavonic languages* (pp. 249–305). London: Routledge.
Goddard, C. (1982). Case systems and case marking in Australian languages: a new interpretation. *Australian Journal of Linguistics 2*, 167–196.
Greenough, J. R., Kittredge, G. L., Howard, A. A., & D'Ooge, B. L. (Eds). (1983). *Allen and Greenough's new Latin grammar*. New Rochelle: Aristide D. Caratzas Publisher.
Huddleston, R., & Pullum, G. K. (2002). *The Cambridge grammar of the English language*. Cambridge: Cambridge University Press.
Hudson, R. (1995). Does English really have case? *Journal of Linguistics*, *31*, 375–392.
Karlsson, F. (1987). *Finnish grammar* (2nd ed., translated by A. Chesterman). Helsinki: Werner Sörderström Osakeyhti.
Kenesei, I., Vágo, R. M., & Fenyvesi, A. (1998). *Hungarian*. London: Routledge.
Kibrik, A. E. (1998). Archi (Caucasian - Daghestanian). In: A. Spencer & A. Zwicky (Eds), *Handbook of morphology* (pp. 455–476). Oxford: Blackwell Publishers.
Lewis, G. L. (1967). *Turkish grammar*. Oxford: Clarendon Press.
Masica, C. P. (1991). *The Indo-Aryan languages*. Cambridge: Cambridge University Press.
Mathiassen, T. (1997). *A short grammar of Latvian*. Columbus, OH: Slavica Publishers, Inc.
Mel'čuk, I. (1986). Toward a definition of case. In: R. D. Brecht & J. S. Levine (Eds), *Case in Slavic* (pp. 35–85). Columbus, OH: Slavica Publishers, Inc.
Miyagawa, S. (1989). *Structure and case marking in Japanese*. San Diego: Academic Press.
Mohanan, T. (1994). *Argument structure in Hindi*. Stanford University: Center for the Study of Language and Information.

Ohara, M. (2000). *An analysis of verbal nouns in Japanese*. Unpublished doctoral dissertation, University of Essex.

Ono, K. (1992). *A study of syntactic and discourse phenomena in Japanese*. Sydney: Eramboo Press.

Ono, K. (1996). Syntactic behaviour of case and adverbial particles in Japanese. *Australian Journal of Linguistics*, *16*, 81–129.

Orwin, M. (1995). *Colloquial Somali*. London: Routledge.

Saeed, J. (1999). *Somali*. Amsterdam: John Benjamins.

Sadakane, K., & Koizumi, M. (1995). On the nature of the dative particle *ni* in Japanese. *Linguistics*, *33*, 5–33.

Scatton, E. (1984). *A reference grammar of modern Bulgarian*. Columbus: Slavica Publishers.

Schiffman, H. F. (1983). *A reference grammar of spoken Kannada*. Seattle: University of Washington Press.

Shibatani, M. (1977). Grammatical relations and surface cases. *Language*, *53*, 789–809.

Skorik, P. Ja. (1961). *Grammatika čukotskogo jazyka, tom 1* [A Grammar of Chukchee, Vol. 1. In Russian]. Leningrad: Nauka.

Spencer, A. (2005). Syntactic vs. morphological case: Implications for morphosyntax. In: L. Kulikov, A. Malchukov & P. de Swart (Eds), *Case, valency and transitivity*. Amsterdam: John Benjamins.

Tsujimura, N. (1996). *An introduction to Japanese linguistics*. Oxford: Blackwell Publishers.

Viitso, T.-R. (1998). Estonian. In: D. Abondolo (Ed.), *The Uralic languages* (pp. 115–148). London: Routledge.

Werner, H. (1997). *Die Kettische Sprache*. Wiesbaden: Harrassowitz Verlag.

Yadroff, M. (1996). Modern Russian vocatives: A case of subtractive morphology. *Journal of Slavic Linguistics*, *4*, 133–153.

Chapter 6

Case as Feature Checking and the Status of Predicate Initial Languages

Lisa deMena Travis

1. Introduction

The notion of Case is central to grammar. This is immediately clear to a learner of a second language which has case morphemes. Some grammatical notions such as categories (e.g. N and V) have, at least in the core examples, a link to extra-linguistic notions (e.g. thing and action). Case, however, appears much more closely tied to purely linguistic phenomena. Case alternations offer a glimpse into the inner mechanisms of language such as category distinctions (nouns vs. verbs), grammatical functions (subjects vs. objects vs. indirect objects), the status of arguments and adjuncts (e.g. Dative vs. Accusative DPs within German PPs), and aspectual distinctions (e.g. Accusative vs. Partitive Case in Finnish). Nominative Case is particularly interesting in this respect because it is closely connected to the syntactic notion of subject and independent of the more semantic notion of theta-role (e.g. Agent). Below we see Nominative Case on an Agent (1a), an Experiencer (1b), an Instrument (1c), a Theme (1d, 1e) and a Goal (1f).

(1) a. The children broke the gift
    b. The children loved the gift
    c. The key opened the door
    d. The gift broke
    e. The gift was given to the children
    f. The children were given a gift

Competition and Variation in Natural Languages: The Case for Case
Edited by M. Amberber and H. de Hoop

ISBN: 0-08-044651-5

In this chapter, I look at the role of Case, and in particular Nominative Case, in a specific theoretical context and then in a certain language family. The theoretical context is that of the Minimalist Program of Chomsky (1995, 2000, 2001) and its theoretical predecessors such as the Government Binding (GB) framework of Chomsky (1981). The language family which will be of interest is the Western Malayo-Polynesian (WMP) family and particularly the languages Malagasy and Tagalog. My goal is to outline one trajectory of thought that is suggested by the direction that the Minimalist Program has taken and see how subjects and Nominative Case fare in languages like Malagasy and Tagalog within this context. The only defense of this trajectory that will be offered is in the form of a language typology that appears to have promising results. Otherwise, the view of Case that is assumed in this paper is quite abstract in that it departs quite quickly from any overt realization of Case. First I give a brief history of the role of Case in the progress of Chomskian syntactic theory in Section 2, followed by an equally brief overview of some Western Malayo-Polynesian case systems in Section 3. Just as the status of subjects is controversial in these languages (see Schachter, 1976, 1996), the status of Nominative Case can be questioned. In Section 4, I raise further questions concerning the typology of syntactic movements as they exist in, for example, Chomsky (2001), and suggest that this typology of syntactic movements can be utilized to create a language typology. In Section 5, I situate WMP languages within this typology and present a view of Case in these languages that is consistent with its language type. I conclude with Section 6.

2. Case in the Chomskian Framework(s)

Case in Chomskian syntax has enjoyed a central role for close to 30 years, but its guises and the range of data that is meant to account for has changed substantially over time. Further, the details of Case assignment or Case checking has changed in ways that may seem trivial but they offer a radically different view of the relation. I outline some of these issues in more detail below.

2.1. Government and Binding (GB) Theory

Morphological case at first posed no real interest within the development of Chomskian theory beyond its link to theta-roles. It existed mainly as a surface effect of a syntax that was created by independent means. It was with the work of Vergnaud (cited in Chomsky, 1980, p. 24) that Case started its increasingly central role as a driving force in syntax. In this context, overt morphological case was only suggestive of a larger phenomenon that included abstract Case, i.e. non-overt instantiations of Case. At this point, (abstract) Case was proposed to be the motivation behind A-movement (e.g. passive and raising). The link between Case and syntactic movement was created through the Case Filter, which had to be satisfied by S-structure (i.e. in the overt syntax).[1]

(2) CASE FILTER (Chomsky, 1981, p. 49)
  *DP if DP has phonetic content and has no Case

In both of the examples below, movement of a DP to subject position is triggered by its need to be in a Case-marked position by S-structure. To account for the movement in (3a) it was assumed that passive verbs cannot assign Accusative Case to their objects. To account for (3b) it was assumed that [−tense] T, encoded by the infinitival marker *to*, cannot assign Case to its subject. In both examples, a DP which has not been assigned Case in its usual (merged) position moves to a subject position where it receives (abstract) Nominative Case from the [+tense] T.

(3) a. [The children]$_i$ will be [given] t_i new books.
  (+Case: will → The children; -Case: given → t_i)

  (vs. Someone will [give] [the children] new books.)
  (+Case: will → Someone; +Case: give → the children)

b. [The children]$_i$ will seem t_i [to] be eating all the cookies.
  (+Case: will → The children; -Case: to → t_i)

  (vs. It will seem that [the children] [are] eating all the cookies.)

At this point in the theory, Case designated a syntactic relation between a Case assigner (typically V, P, or +tense T) and a DP. Structurally, the core Case relation was between a head and its complement (usually where the head was V or P), but sometimes the relation was between a head and its specifier (e.g. the relationship between +tense T and the subject). These relations are sketched below.

[1] I have changed NP to DP (Determiner Phrase) in Chomsky's (1981) formulation.

(4) a. OBLIQUE b. ACCUSATIVE c. NOMINATIVE

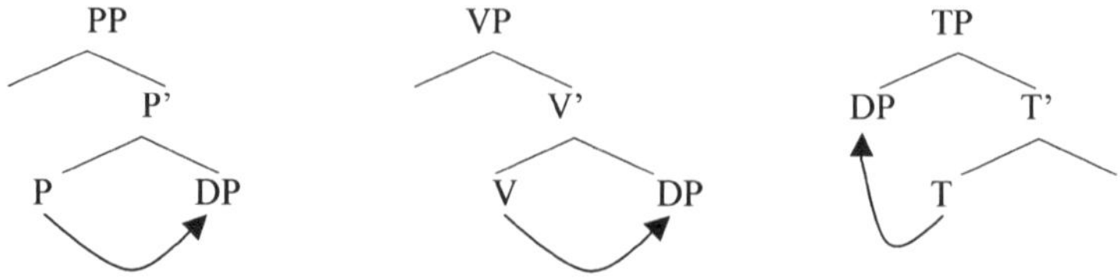

2.2. Nominative Case and Subjects

Clearly, in the theories outlined above, there is a link with the definition of subject (Spec, TP) and Nominative Case (assigned by +tense T to its Spec position).[2] Other theories that take grammatical subjects as primitives such as Relational Grammar also assume a tight connection. Case assignment and agreement are often used as one of the diagnostics of subjecthood. For example, we see in Rosen and Walli (1989, p. 4) that Case and verbal agreement are used to test for final 1 (subject) status of an element.

There are clear problems, however, with this link. The most celebrated cases are Quirky Case constructions (called inversion structures in Relational Grammar (see Harris, 1984a,b; Legendre, 1989; Perlmutter, 1984)) where we find in subject position a DP which bears a lexically determined Case. Such constructions often also have a Nominative DP within the VP. The problem, then, is two-fold. The subject does not have Nominative Case (nor does the verb agree with it) and the Nominative Case-marked DP (with which the verb agrees) is not the subject.[3]

(5) Quirky Case construction (from Sigurðsson, 1989, p. 201)[4]

a. *Mér* *áskotnuðust* *peningar*
1SG.DAT lucked-onto money.NOM

b. *Mér* *batnaði* *veikin*
1SG.DAT recovered(-from) disease.NOM

---

[2] In exceptional case-marking (ECM) constructions such as *Mary believed them to be liars,* a DP in the embedded subject position exceptionally receives case from the matrix V.

[3] Since verbal agreement is also used for a diagnostic for subjecthood in Relational Grammar, both diagnostics create problems for these constructions. In inversion constructions, the dative DP was considered a "working 1" (1→3) and the nominative was the final 1. The subject-like properties of the final 3 were assumed to be due to the original 1 status.

[4] Sigurðsson gives no translations here but I assume that (5a) means 'I lucked into money' and (5b) means 'I recovered (from a disease)'.

Whatever the analysis for Case assignment, it needs to be able to handle constructions where non-Nominative DPs function as subjects, and Nominative DPs can function as non-subjects. Below we will see that such constructions can be accounted for in the more recent views of the Minimalist Program. The link between Nominative Case and the structural subject position, however, must be loosened.

2.3. Early Minimalism

In early Minimalism, there was a shift in what was taken to be the core relation in Case assignment. This was prompted by the investigation of object shift in Scandinavian languages. Data such as the example given below, where objects appear in derived positions, suggest that objects as well as subjects undergo movement.

Icelandic (Holmberg, 1986, p. 166)[5]

(6) a. *Hvers vegna lasu stúdentarnir ekki allir <u>greinina</u>*
why read the students not all the article

b. *Hvers vegna lasu stúdentarnir <u>greinina</u> ekki allir*

The assumption was that, like subjects, objects move for reasons of abstract structural Case assignment.[6] With this development came a change in the view of what was the core relation in structural Case assignment. The typical configuration was no longer a head-complement relation but rather a head-specifier relation. In this version of Minimalism, Case was assigned in the following environments.

(7) a. OBLIQUE b. ACCUSATIVE c. NOMINATIVE

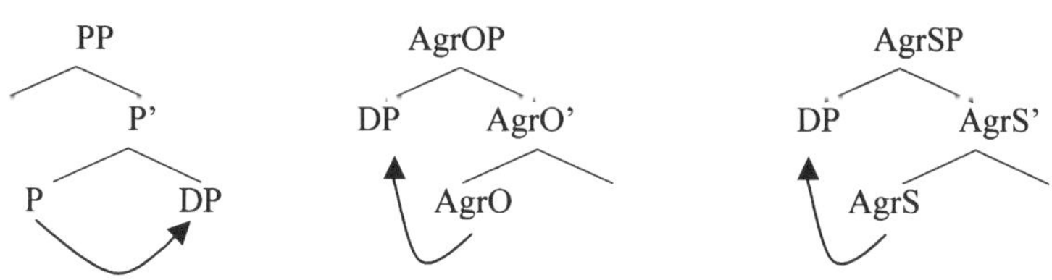

The shift in importance of the Spec, head relationship in Case assignment allows Case assignment to become part of a larger syntactic process. Now Case

[5] Holmberg gives no translation here but I assume that it means something like 'Why didn't all the students read the article'.

[6] Object shift differs from movement to subject position for Nominative case in several ways, the most obvious from the example given is that it appears to be optional. It has been shown that there is a link to movement/Case and interpretation (see e.g. Bobaljik, 1995; Diesing, 1996; de Hoop, 1996) which, perhaps, points to an additional difference. I leave this issue aside here.

assignment had much in common with other movement processes such as WH-movement. Both Case assignment and WH-movement involve movement of an element to a specifier position of a functional head to satisfy some mutual need of the moved element and the head itself. AgrO and AgrS need to assign Case and the subject and the object DP, according to the Case Filter, need to be assigned Case. A WH Comp needs to have a WH-XP in its specifier position, and the WH-XP has to be in a position of operator scope (i.e. Spec of a +WH COMP). These similarities prompted a further re-evaluation of Case. Case now can be seen as a particular example of a larger grammatical phenomenon where a head contains something (a feature) that must be matched (checked) by an element that has already been merged into the structure (i.e. is lower on the tree). We can compare the following structures (where CS=Case).

(8)

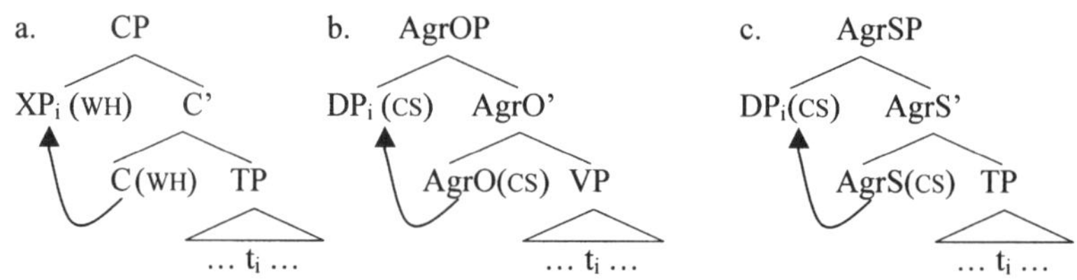

Now we take the next step, which is to say, at a very abstract level, all head-specifier relations are, in some sense, Case relations. This is the thought that is at the heart of what Marantz (1995, p. 365) calls the generalized form of the Case Filter — 'the requirement that morphological features get checked off before the interface levels'. While it might be too strong to state the generalization over the structures in (8) in these terms, it is useful to see the close connection in order to better understand the subsequent developments in this theoretical framework.

In the GB approach, it was a need for Case, dictated by the Case Filter, that triggered movement of the subject to the Spec, TP position. In the Minimalist Program, all movement is triggered by a need to check off uninterpretable features and the feature used to trigger movement of the subject DP was, in an early version of the theory, the feature D. It was posited that the functional category T housed two features, D and V. D was responsible for movement of the subject DP to the specifier position, and V was responsible for movement of the V (in languages like French where V movement is overt) to adjoin to the T. This is sketched out in the trees below with a relevant French sentence given in (10).[7]

[7] Double strikethrough (~~DP~~) indicates a copy that is left by movement but deleted at the phonological interface. Single strikethrough (~~D~~) indicates an uninterpretable feature that is deleted when in an appropriate checking configuration with a matching interpretable feature.

(9)

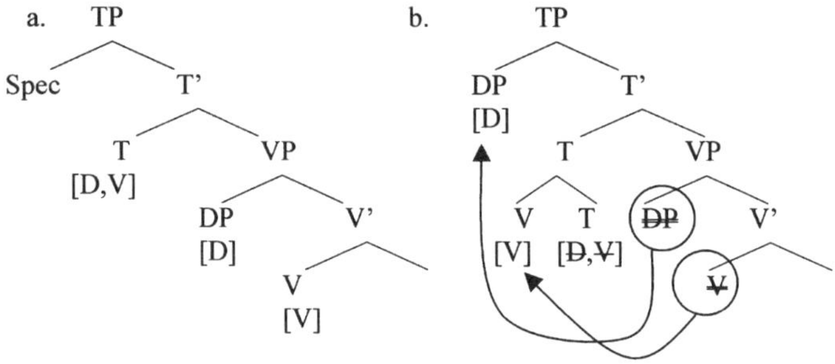

(10)

Les	*enfants*$_i$	*ne*	*mangent*$_k$	*pas*	t_i	t_k	*à*	*l'école*
DET	children	NEG	eat	NEG			at	DET-school

'The children do not eat at school'

The main difference between Government and Binding Theory and the early Minimalist Program is that in the latter Case is part of a larger system of feature checking. Note, however, that the link between the subject and Nominative Case is just as strong and that the Quirky Case Subject examples in (5) are just as problematic.

2.4. Later Minimalism

In later versions of the Minimalist Program, there are further shifts in the technology that can be viewed as a return to a view of Case that is closer to that taken in the GB system, or even to an earlier system where Case was not linked to movement at all. In this stage of the theory, MOVE is broken down into two parts. First, there is another relation, AGREE, which allows a pairing to be set up between the PROBE (the functional head which contains the relevant feature) and the GOAL (the closest element within the domain of the probe which has a matching feature). The AGREE relation is sufficient to check this feature. Only if there is another feature (often encoded as EPP[8]) is the additional mechanism of movement involved. Typically, the AGREE relation can encode a Case dependency between a Case-assigning head and an *in situ* DP. For example, in a *there*-construction in English, the D feature can enter into an AGREE relation with a DP internal to the VP. This AGREE relation accounts for the checking of the D feature (which amounts to Case assignment) and the agreement on the verb. The EPP feature is satisfied by the expletive *there*.

---

[8] The Extended Projection Principle (EPP) in Chomsky (1981) ensured that there was a subject in every clause. An extended version of this exists in the Minimalist Program, which ensures that there is an element in the relevant Spec position.

Similar technology can be used to account for the Quirky Case constructions discussed in Section 2.2. To take one of Sigurðsson's examples, we can capture the Case and agreement of the element within the VP through AGREE. A DP with inherent Case has moved to the Spec, TP checking the EPP feature in the T head.[9]

(11) a. There were two children in the classroom.

b.

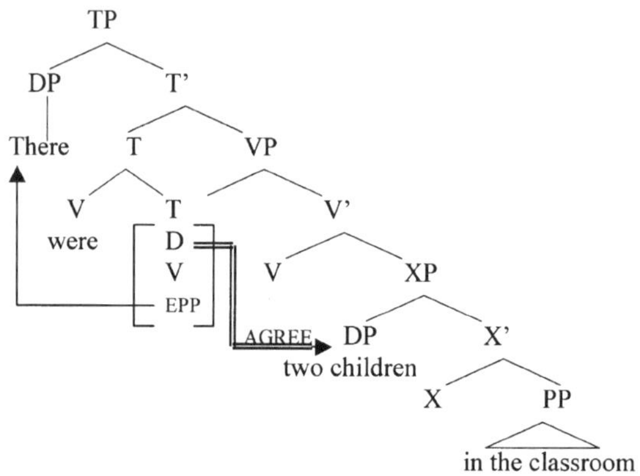

(12) a. *Mér* *áskotnuðust* *peningar*

1SG.DAT lucked-onto money.NOM

b.

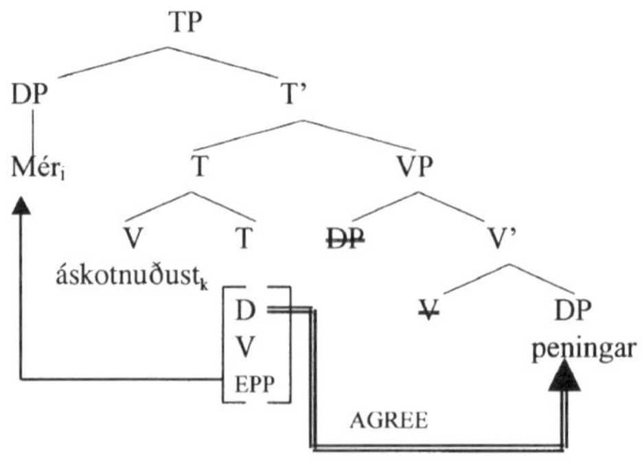

---

[9] I have based this analysis of Quirky Case subjects on Rackowski (2002, p. 16). Her goal, like mine, is to use these mechanisms to explain the syntax of Western Malayo-Polynesian languages. In her case, the focus was on Tagalog. Her work will become central in Section 3.2.

With this theoretical background, let us now look at Case in Western Malayo-Polynesian languages. I will suggest that the theoretical tools that we have seen lead to an analysis of these languages that forces a further reinterpretation of Case.

3. Case in Western Malayo-Polynesian Languages

Schachter (1976, 1996) famously raised the question of whether Philippine languages have a single nominal phrase that can be labeled SUBJECT. When looking at characteristics that are typically associated with subjects, these characteristics split over two candidates in these languages (and other related languages such as Malagasy) — the Topic (or pivot) DP and the Actor[10] (or external argument) DP. As we see in (13) below, morphology on the verb will designate the role of the DP that is behaving as the topic (pivot) of the clause and this DP will appear with the *ang* particle.[11] In (13a), an Actor Topic (AT) construction, the pivot is the Actor, *bata* 'child'. In (13b), a Theme Topic (TT) construction, the pivot is the Theme, *tela* 'cloth'. In (13c), a Location Topic (LT) construction, the pivot is the location, *palengke* 'market'. In (13d), a Benefactive Topic (BT) construction, the pivot is the Beneficiary, *Nanay* 'mother'.

(13) a.

*b-**um**-ili*	*ang*	*bata*	*ng*	*tela*	*sa*	*palengke*	*para*	*sa*	*Nanay*
AT-buy	ANG	child	NG	cloth	SA	market	P	SA	mother

'The child bought cloth at the market for mother'

b.

*b-in-ili-**Ø***	*ng*	*bata*	*ang*	*tela*	*sa*	*palengke*	*para*	*sa*	*Nanay*
PST-buy-TT	NG	child	ANG	cloth	SA	market	P	SA	mother

'The child bought the cloth at the market for mother'

c.

*b-in-ilih-**an***	*ng*	*bata*	*ng*	*tela*	*ang*	*palengke*	*para*	*sa*	*Nanay*
PST-buy-LT	NG	child	NG	cloth	ANG	market	P	SA	mother

'The child bought cloth at the market for mother'

d.

***i**-b-in-ili*	*ng*	*bata*	*ng*	*tela*	*sa*	*palengke*	*ang*	*Nanay*
BT-PST-buy	NG	child	NG	cloth	SA	market	ANG	mother

'The child bought (the) cloth at the market for mother'

---

[10] The term Actor is used instead of, for example, Agent since it is meant to cover other sorts of external arguments such as Experiencer (*The children respect their parents*) or Goals (*The child received the present*).

[11] I will call the *ang* DP the pivot of the clause following Pearson (2001). It has been referred to as Topic (Richards, 2000), Subject (Keenan, 1976; Kroeger, 1993; Paul, 2000; Travis & Williams, 1982) and Absolutive DP in an Ergative system (Aldridge, 2004; de Guzman, 1988; Gerdts, 1988; Maclachlan, 1996; Manning, 1996). I return to this issue below.

Schachter concludes, twice, that nothing is gained by identifying one of these DPs as the subject. In this paper I raise the possibility that Case in these languages should be viewed just as skeptically. My work is foreshadowed by claims made in Sells (1998) and Pearson (2001) that what appears to be Case is not tied to grammatical functions in the same way as in other well-studied languages.

3.1. Case Problems

There are reasons to believe that Case, as conceived of within the Chomskian framework, may work differently in Austronesian languages, the larger language family of WMP languages. I give some examples below. First, quite superficially, overt Case marking on non-pronominals in these languages is different from languages such as Latin in that Case in languages like Tagalog is realized as separate particles.[12] We can see the different cases in the Tagalog example in (13) above realized as the particles *ang*, *ng*, and *sa*. In a view where the morphology of the verb designates advancement to subject of various arguments, *ang* would mark Nominative Case.

Beyond the oddity of having Case realized on a separate particle, certain theory internal uses of Case create problems in these languages. I mention one here.[13] Above, in example (3b), we saw an instance of raising where a [–tense] embedded clause failed to check the Case of its subject. Because of the Case Filter (or feature checking) the embedded subject was forced to move to the matrix subject position. In a language like Malagasy there are no cases of infinitival verbs, but there are arguably cases of raising as shown below.[14]

[12] As pointed out by an anonymous reviewer, the realization of case as a particle is found in many other languages such as Polynesian languages, Indo-Iranian languages, etc. This is part of my point. It may be that the form that Case takes may tell us something about the nature of the syntactic relationships that are set up. So-called particles in head-final languages most often occur on the nominal head making their status less clear. The particle status in head-initial languages such as Tagalog may be clearer since material may occur between the particle and the nominal head as in *ng kanyang ama* 'CS his/her father'. This shows clearly that the case morphology is not crucially realized on the head.

[13] Control structures also have [+tense] Ts in them unlike in English (see Law, 1995). Because PRO is not supposed to be assigned Case (or is supposed to be assigned a different sort of Case as in null Case), the analysis of these structures also has to be reworked. A reviewer points out that this may be too simplistic and that it may be that certain matrix verbs select complements with overt tense that nevertheless do not assign case to their subjects. While an appropriate response requires a paper in itself, I maintain that this alternative account points in the same direction — Case is different in these languages.

[14] In writing the Malagasy sentences, I am following orthographic conventions which use apostrophes and hyphens. These are not meant to be indications of anything linguistic (i.e. phonological or morphological.).

(14) *No heverin-dRasoa ho nividy vary ve Rabe*
PST TT.believe-N-Rasoa *HO* PST.AT.buy rice Q Rabe
'Was Rabe believed by Rasoa to have bought rice?'

The positioning of the question particle *ve* before the DP *Rabe* shows that this DP is the subject of the main clause. It appears, however, that it has raised out of a tensed clause where it should have been assigned Nominative Case by the [+tense] T. Case, then, cannot be what is driving such movement. This is compounded in another Austronesian language, Niuean, where both objects and subjects of embedded clauses can move to the matrix subject position. Such constructions are discussed in Seiter (1978), Massam (1985, 2005), and Bejar and Massam (1999). The example below, from Seiter (1978, p. 321), shows that with a raising verb such as *maeke* 'be possible', there are three choices. All the embedded arguments may stay within the embedded clause (15a), the Ergative DP may raise to the matrix clause where it appears with Absolutive Case (15b), or the embedded Absolutive DP may raise (15c). Importantly, with these three choices, there is no change of verb form.

Niuean

(15) a. *To maeke ke lagomatai he ekekafo e tama ē*
FUT possible SBJ help ERG doctor ABS child this
'The doctor could help this child'

b. *To maeke e **ekekafo** ke lagomatai e tama ē*
FUT possible ABS doctor SBJ help ABS child this
'The doctor could help this child'

c. *To maeke e **tama ē** ke lagomatai he ekekafo*
FUT possible ABS child this SBJ help ERG doctor
'The doctor could help this child'

This type of raising looks much different from what we see in English where only the subject of the embedded clause may raise, and only when the embedded clause is non-finite.

3.2. Two Accounts

Next we look at two accounts of Tagalog which treat Case quite differently from the more traditional GB approach to Case. One account is situated in late Minimalism and the other in the framework of LFG. In both accounts, the morphology on the verb is not linked to normal voice morphology which would promote an argument to the Spec, TP position, but rather to the underlying Case of the *ang* DP. Further, in both accounts, *ang* is analyzed as a marker outside of the normal Case-marking system.

Rackowski (2002) views movement of the *ang* DP as being related to EPP and not to Case. One way of looking at the verbal morphology which, as we have seen in (13) above, indicates the role of the *ang* DP, is that the verbal morphology reflects the Case that is no longer visible on the *ang* DP. Rackowski (2002) expresses this technically through a feature on the V that realizes the Case value of the pivot DP. Generally a Case feature is realized within DP, but in her account of Tagalog, it can also be realized on the V. I sketch her account briefly.

First, all DPs receive Case in their merged positions through the appropriate AGREE relation. It is the highest argument that will enter into an AGREE relation with T and receive Nominative Case. Basically, then, Cases show underlying grammatical functions. Independently, an EPP feature on the Voice head will force an already Case-marked DP to the outer Spec position of VoiceP. Finally, a different feature on T which represents the unmarked Case value will copy whatever the Case value of the moved DP is. The tree below represents the structure for a Theme Topic construction as exemplified in (13b) above.[15]

(16)

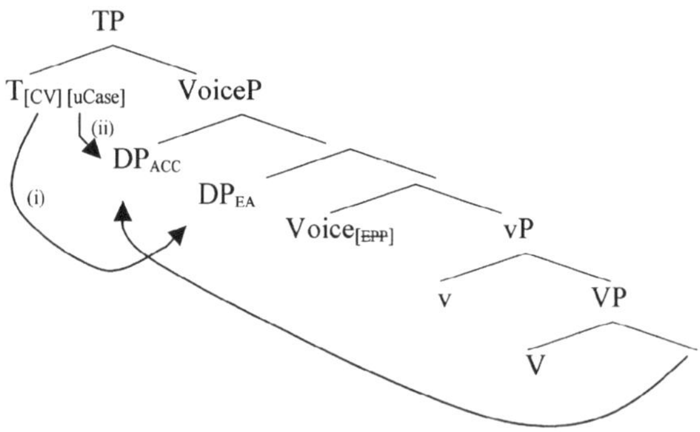

Note that the T enters into two relations.[16] As Rackowski (2002, p. 116) writes

> Multiple Agree (Richards, 1997; Chomsky, 2001) of the T head is necessary in order to account for the nominative case marking on the Agent. Multiple Agree means that the head may enter into more than one Agree relation with an argument in its domain if

---

[15] Tagalog has fairly free word order post-verbally so position on the tree does not necessarily coincide with the order of the constituents.

[16] I have reversed the (i) and (ii) given in Rackowski (2002, p. 116) so that they correlate with her discussion given below.

uninterpretable features are left on the head after the first instance of Agree. In the case of Tagalog, T performs two separate Agree relations: (i) checking case (nominative) on an argument and (ii) checking its own uninterpretable case feature by copying the case features of some argument.

In this account, Case marking on the DPs has nothing to do with surface relations and everything to do with underlying functions. Rackowski separates *ang*-marking from Nominative marking and claims that *ang* simply marks subjecthood. She saves the term Nominative for the particle that occurs with a non-subject Actor. In this analysis of Tagalog, then, Case has none of the triggering effects that we saw in the GB approach to syntax.

In Sells' (1998) account of Tagalog, while he calls *ang* the Nominative Case marker, he claims that Nominative Case is assigned to an A'-position.[17] Although his technology is quite different from that of Rackowski, he also links the verbal morphology with the underlying functions of the subject DP. More accurately, the *ang* DP is generated in an A'-position and is related, through the verbal morphology, to an empty subject pronominal which appears in the f-structure, a structure within the LFG model that encodes functional relations. His f-structure for a Theme Topic construction in (17)[18] is given in (18).

(17) *b-in-ili* *ng lalake* *ang* *isda* *sa tindahan*
PST-buy-TT *NG* man *ANG* fish *SA* store
'The man bought the fish in the store'

What is common to both Sells' and Rackowski's accounts is that the *ang* DP is not promoted to any position for reasons of Case. In fact, in these languages, Case has very little to do with syntactic phenomena.[19]

3.3. Case in Malagasy

Now let us turn to Malagasy. Malagasy is generally assumed to have three Cases — Nominative, Genitive, and Accusative. This is seen most clearly in the form of the pronouns as shown in the table in (19) with exemplifying sentences in (20).

---

[17] Richards (2000) also posits that the *ang* DP is in an A'-position within the Minimalist framework.
[18] I have changed the presentation of this example to be consistent with the other examples in this paper.
[19] Sells (1998, p. 125) points out that these languages do not have VP internal grammatical function changing operations, such as raising to object, double objects or applicatives, that feed movement to subject.

(18) F-structure

SUBJ	PRED	'pro'
	INDEX	i
OBJ	CASE	GEN
	PRED	'man'
	INDEX	j
PRED	'buy'	
NADJ	CASE	NOM
	PRED	'fish'
	INDEX	I
OBL	CASE	LOC
	PRED	'store'
	INDEX	k

(19) From Pearson (2001, p. 41).

	NOMINATIVE	GENITIVE	ACCUSATIVE
1SG	aho, izaho	-ko	ahy
1PL.EXCL	izahay	-nay	anay
1PL.INCL	isika	-ntsika	antsika
2SG	ianao	-nao	anao
2PL	ianareo	-nareo	anareo
3	izy	-ny	azy

(20) a. NOMINATIVE

Mahita *ny* *mpampianatra* *aho*.
PRES.AT.see DET house 1SG.NOM
'I see the teacher'

b. ACCUSATIVE

Mahita *ahy* *ny* *mpampianatra*
PRES.AT.see 1SG.ACC DET teacher
'The teacher sees me'

c. GENITIVE (for possession)
ny *tranoko*
DET house.1SG
'My house'

d. GENITIVE (for non-subject actors)
Hitako *ny* *mpampianatra*
TT.see.1SG DET teacher
'The teacher is seen by me'

e. GENITIVE (for objects of many prepositions)
alohako
before.1SG
'before me'

However, there is another way to view this three-way distinction. The Genitive Case is an enclitic form of the pronoun. This form may have less to do with Case and more to do with cliticization under certain structural conditions. Further, as shown by Pearson (2001, p. 43 clearly), the Nominative Case is a strong form of the pronoun that appears as the subject but also in other positions that are not in what might be called a 'tight' construction with a head. Another example given by Pearson is a co-ordinated structure. Normally a non-subject Agent will appear in the Genitive Case as in (21a). However, if the Agent is modified or part of a co-ordinated structure, it will appear in the Nominative form as in (21b).

(21) a. *Hitany* *tany* *an-tokotany* *iKoto*
TT.see.3 PST.there AN-garden I.Koto
'She/he/they saw Koto in the garden'

b. *Hitan'* *[izy* *sy* *ny* *zaza]* *tany* *an-tokotany* *iKoto*
TT.see [3SG and DET child PST.there AN-garden I.Koto
'She/he and the child saw Koto in the garden'

As a hypothesis, Case in these languages, if it exists, does not have the same syntactic function as in languages such as English and French. In the next section I explore this hypothesis in the context of the Minimalist Program and in the context of a language typology that I suggest the Minimalist Program leads us to.

4. Movement Typology and Language Typology

In other works (Travis, in press-a, in press-b), I have investigated some of the particular mechanisms of the Minimalist Program and questioned how they apply to

Western Malayo-Polynesian languages such as Malagasy. Chomsky (2001, p. 38) makes the following observation concerning the technology outlined in (9) above.

> It has always been taken for granted that the strong V-feature is satisfied by V-raising to T (French vs. English), not VP raising to SPEC-T; and the strong NOMINAL-feature by raising of the nominal to SPEC-T (EPP), not raising of its head to T. But the theoretical apparatus provides no obvious basis for this choice. The same is true of raising to C and D. In standard cases, T adjoins to C, and an XP (say, a WH-phrase) raises to SPEC-C, instead of the WH-head adjoining to C while TP raises to SPEC-C. And N raises to D, not NP to Spec-D.

He notes that there is an asymmetry in movement. The D feature triggers XP movement while the V feature triggers X^0 (head) movement. He explains this asymmetry away by suggesting that head movement is not part of the syntactic/computational system but rather part of the phonological component driven by a need to attach affixes. In his words,

> .... These conclusions ... follow naturally if overt V-to-T raising, T-to-C raising, and N-to-D raising are phonological properties, conditioned by the phonetically affixal character of the inflectional categories.

In my work, I argue that we do not want to rule out the opposite side of the asymmetry, i.e. languages which have VP-movement and D^o-movement (rather than DP-movement and V^0-movement). I argue that this other language type is, in fact, instantiated. I summarize some of my findings here.

4.1. VP-Movement

Much of the work on Malagasy (and related languages) suggests that there is iterative predicate movement both within the VP and outside of the VP. Pearson (1998) shows this VP internally where, when compared with other VO languages such as English and/or Icelandic, the word order of certain elements in Malagasy is unexpected. Below we see examples of the relative order of double objects, adverbs, and object shift. In examples (22) and (23) the word order of the Malagasy elements should be compared with the word order in English. In (24) and (25) the word order of Malagasy should be compared with that of Icelandic.[20]

[20] I have changed the glosses slightly to be consistent with other examples given in this paper.

(22) *Nanolotra* *ny dite* ***ny*** ***vahiny*** *ny* *zazavavy*
PST.AT.offer DET tea DET guest DET girl
'The girl offered **the guests** the tea'
(Pearson, 1998, (3))

(23) *Manasa* *tsara* *ny* *lambany* ***foana*** *i* *Ketaka*
PRES.AT.wash well DET clothes.3 always DET Ketaka
'Ketaka **always** washes his clothes well'
(Pearson, 1998, (7))

Icelandic

(24) *Á barnum drakk stúdentinn [* ***bjórinn$_i$*** *] stundum* t_i (Bobaljik 1995: 75)
in bar.the drank student.the beer.the sometimes
'In the bar, the student sometimes drank all the beer.'

Malagasy

(25) a. *Nijinja* ***vary*** *haingana ny mpamboly* (Pearson 1998: (9))
PST.AT.cut rice quickly DET farmer
'The farmer harvested rice quickly.'

b. *Nijinja* t_i *haingana [****ny vary$_i$****] ny mpamboly*
PST.AT.cut quickly DET rice DET farmer
'The farmer harvested the rice quickly.'

Pearson uses these data to argue for a language typology that divides VO languages into two types — direct and inverse. The word order arguments are summarized in the table in (26) below and more examples of each type of language are given in (27).[21]

(26) Two types of (VO) languages (Pearson, 1998)

	DIRECT	INVERSE
Double objects	IO>>DO	DO>>IO
Adverb order	Adv1>>Adv 2	Adv2>>Adv 1
Object shift	$DP_i \ldots t_i$	$t_i \ldots DP_i$

(27) DIRECT: English, French, Indonesian, Icelandic…
INVERSE: Malagasy, Tzotzil, Quiavini Zapotec, Palauan, ….

[21] x>>y is meant to indicate that x precedes y.

Further arguments have been given that the VOS word order of Malagasy comes about from predicate movement of the VP leftward across the subject (see e.g. Rackowski, 1998; Rackowski & Travis, 2000).[22]

It is important to point out that, if these word orders are accounted for through predicate fronting, this predicate fronting does not have the same effect as VP-fronting in a language like English (as in (28a)) or German (as in (28b)).

(28) a. The children said they would finish their homework
… and [$_{VP}$ finish their homework]$_i$ they did t$_i$

b. *[$_{VP}$ Gelesen das Buch] haben die Kinder.*
read.PP DET book have the children
'The children have read the book'

In the case of the English example in (28a) and the German example in (28b), there is exceptional movement of the VP for some discourse effect. The proposal for Malagasy is that the unmarked word order is achieved through (obligatory) predicate fronting. One way of thinking about the differences is to consider predicate fronting in a language like Malagasy to be similar to A-movement in that it moves elements within the inflectional (INFL) domain of the clause, while predicate fronting in languages like English and German is similar to A'-movement in that it moves elements into the peripheral (COMP) domain of the clause.[23] In this light, it does seem that this predicate movement is the other side of the asymmetry that Chomsky highlights. The VP (or any other maximal projection along the extended projection line of the V), will be the goal of the V feature probe. Other languages that have been argued to have similar predicate movement are Niuean (Massam & Smallwood, 1997; Massam, 1998a, b), Quiavini Zapotec (Lee, 2000), and Irish (Oda, 2002). Now we turn to the question of whether there are languages that have D^0-movement.

4.2. D^0-Movement

D^0-movement is a much harder phenomenon to find arguments for.[24] There are two cases in the literature of which I am aware that argue for D^0-movement. The first that

---

[22] Massam and Smallwood (1997), Massam (1998), and Oda (2002) have argued that predicate fronting in these languages moves the predicate to the subject position because the EPP feature is set to target predicates rather than DPs. The view I am espousing in this paper owes much to their work though I take a slightly different view.

[23] In Pearson's (2001) work, he likens predicate fronting to V^0-movement in a V2 language in that it is iterative and obligatory. My view of predicate fronting is closer to his than to Massam and Smallwood (1997), Massam (1998), and Oda (2002).

[24] There are many cases discussed of head movement within DPs, but what is of interest to us here are cases of head movement of D to a position in the extended projection of the verb. Pronominals clitics may also be argued to be D-movement but the trigger for their movement is presumably not to satisfy a D feature in T. More research needs to be done on this.

I discuss is that of Baker and Hale (1990). They propose D^0-movement for Breton to account for the complementary distribution of subject agreement and overt subjects. The relevant data are given below followed by the type of tree that would be created for (29) where, as indicated by the underscore, there is no overt subject.[25]

(29) a. *Bemdez e lenn ar vuglae eul levr*
every.day PRT read the kids DET book
'The kids read a book every day'

b. *Bemdez e lenn-ont__eul levr*
every.day PRT read-3PS DET book
'They read a book every day'

c. **Bemdez e lenn-ont ar vuglae eul levr*
every.day PRT read-3PS the kids DET book

(30)

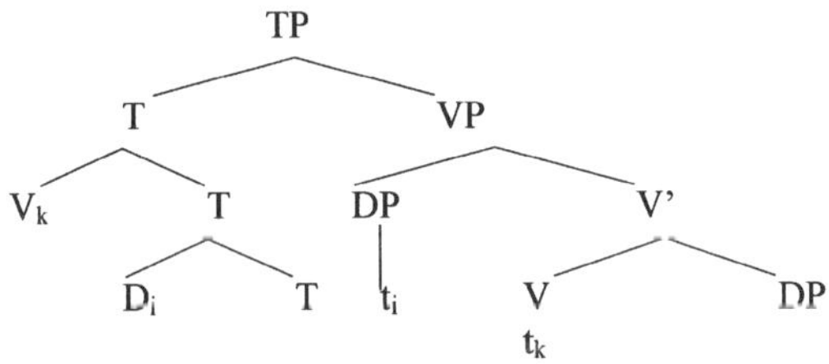

For reasons particular to their account, Baker and Hale argue that only V-initial languages have this type of D^0-movement and the other example that they give of a language with possible D^0-movement is Yatee Zapotec. While I do not use their account of the generalization, I take the correlation of D^0-movement and V-initial word order to be significant. I also take as important the fact that the languages they discuss, Breton and Yatee Zapotec, are closely related to two languages that have been argued to undergo predicate fronting, Irish (see Oda, 2002) and Quiavini Zapotec (see Lee, 2000). My hypothesis is that D^0-movement and VP-movement are linked through grammatical principles.

Now we turn to another case from the literature where D^0-movement has been posited. In Guilfoyle, Hung, and Travis (1992), we argued that the word order facts of a certain passive-like construction in Malay was best accounted for through D^0-movement. The relevant data are given below; (31) shows that Malay is generally SVO. However, as we can see in (32), when we get a bare verb form, certain elements may appear in a position between the subject and the verb. In

---

[25] The details of adjoined structure under T may not be as shown. This is for future research. See the tree in (33) for another possibility.

particular, pronouns and proper names may appear in this position (see (32a)) but not common nouns (see (32b)).

(31) *[Ali/Saya/Lelaki itu] [membaca buku itu dengan teliti]*
Ali / I / boy the AT-read book the with care
'Ali/I/the boy read the book carefully'

(32) a. √proper names / √pronouns
[Buku itu] ***Ali/kamu*** *baca dengan teliti.*
book the **Ali/you** read with care
'The book was read by Ali/you carefully'

b. *common nouns
[Buku itu]* ***lelaki itu* *baca dengan teliti.*
book the **man the** read with care
'The book was read by the man carefully'

Postal (1969) proposed that pronouns and proper names are Determiners. It appears, then, that in the bare verb form constructions in Malay, where the Theme appears in the subject position, the Agent may appear pre-verbally if it is exhaustively dominated by D^0. In our paper, we accounted for this by proposing a D^0-movement to a pre-verbal position. The relevant tree structure is given below.[26]

(33)

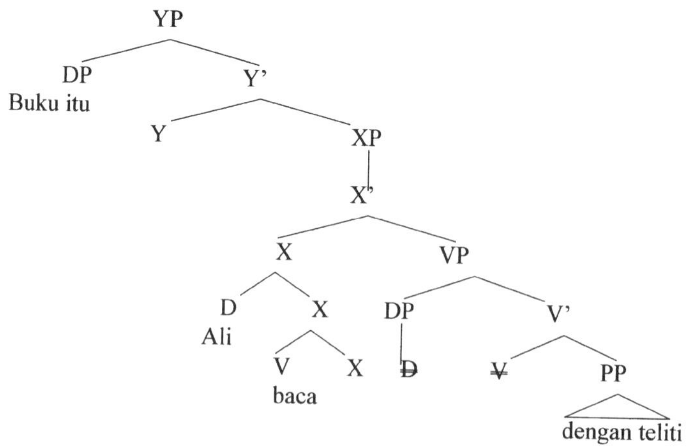

[26] I have labeled the projection above VP XP and the projection which houses the pivot YP to be deliberately vague. I have argued that the head to which the V and the D moves is something below T. In any case, I want to reserve the Spec, TP (or the highest functional projection within the inflectional domain) for the predicate fronting in these languages. As we will see shortly, I believe that there are reasons to argue for a covert predicate fronting analysis for Malay and Indonesian.

Before looking at Malay, the generalization seemed to be that D^0 movement languages, such as Breton and Yatee Zapotec, were also predicate fronting languages. Malay, clearly an SVO language, is a counterexample to this generalization. In spite of what is obvious, I will try to keep the generalization. My main empirical reason to do this stems from the way in which Malay flies in the face of another generalization. Oda (2002) argues that WH-clefting languages are also predicate fronting languages. He concentrates on Irish but also discusses Malagasy. Below we can see the WH-cleft formation in these two languages First we see an Irish example (from Oda, 2002, p. 75). The comparison between (34a) and (34b) shows that similarity of structure between the WH-structure and relative clause formation. The Malagasy case in (35) compares the WH-structure with a non-WH-cleft.[27]

(34) a. WH-cleft

Cé	*a^L*	*dhíol*	*an*	*domhan*
who	COMP	sold	the	world

'Who sold the world?'

b. relative clause

an	*fear*	*a^L*	*dhíol*	*an*	*domhan*
the	man	COMP	sold	the	world

'The man sold the world?'

(35) a. WH-cleft

Iza	*no*	*manasa*	*lamba?*
WH	*NO*	PRES.AT.wash	clothes

'Who is washing clothes?'

b. cleft

Rakoto	*no*	*manasa*	*lamba.*
Rakoto	*NO*	PRES.AT.wash	clothes

'It is Rakoto that is washing clothes'

Oda argues for a one-way correlation between being predicate initial and having WH-cleft structures, but a bidirectional correlation would be stronger correlation. The problem arises with languages, like Malay, that have WH-cleft structures but which are not predicate initial. An example of a WH-cleft construction in Malay is given below.[28]

---

[27] There are more arguments for a cleft analysis (or more accurately a pseudo-cleft analysis). See Oda (2002) and Paul (2001).

[28] The *yang* particle indicates a cleft structure.

(36) WH-Cleft

Siapa	*yang*	*membaca*	*buku*	*itu*
who	COMP	AT-read	book	the

'Who read the book?'

(from Guilfoyle, Hung, & Travis 1992)

However, there is the possibility that predicate fronting is done in the covert component.[29] There are languages that have covert WH-movement (e.g. Chinese) and covert head movement (English). I propose that Malay has covert predicate fronting. Only in the case of WH-constructions is the movement overt.[30]

The hypothesis now is that there are two types of language with respect to movement. The A languages move maximal projections to check D features and move heads to check V features. The B languages move heads to check the D features and move maximal projections to check the V features. This is outlined in the table below.

(37) Two types of language

	A LANGUAGES	B LANGUAGES
D^x movement	DP	D^0
V^x movement	V^0	VP

Before turning to the question of Case within these two systems, I argue that Malagasy also has overt D-movement.

4.3. Checking D-feature in Malagasy

In Malagasy, a predicate fronting language, the element that would move to the Spec position would be a predicate. In terms of features, the V feature in T is checked through movement of a predicate projection to Spec, TP. If D-movement

[29] An anonymous reviewer asks where the covert predicate would move to as it seems that the subject is already in the Spec, TP position (Spec, YP in the tree given in (33)). As we will see in section 5, I argue that subjects in predicate fronting languages are base-generated in positions outside the VP which do not interact with movement. It is also clear that the projection which acts as the landing site of D^0-movement is different from the projection which is the landing site of predicate movement (a projection that dominates YP). This appears quite different from languages like French where DP movement and V^0-movement appear to land within the same projection. I argue elsewhere (Travis, in press-a) that certain characteristics of XP vs. X^0-movement that we have taken for granted to be true in languages like French have to be re-evaluated when looking at similar rules in this other type of language. In that paper I look at (i) whether the relevant movements gather material or whether they 'excorporate' (as in Roberts, 1991), and (ii) whether they target the extended projection of the head or a specifier. The nature of landing sites also must be re-evaluated, but I assume that all B type languages would work similarly. I leave this for future work.

[30] Again, I make a hypothesis here that must be tested, i.e. the WH-cleft languages are B-type. This needs to be further tested on a range of languages.

were overt in Malagasy, we should see some indication that the D of the DP in Spec, VP would move.

I suggest that one of the confirmations of the proposed analysis of Malagasy as a B-type language is the effect of Genitive Case described in (20c) and (20d) above. The morpho-phonological effect of Genitive Case is called *N-bonding* by Keenan (2000) and it occurs when the external argument is not a subject. Below we see first an example where there is no external argument. The next three examples show that determiners, pronouns, and proper names all undergo the N-bonding process with the preceding V.

(38) a. *[Hita] ny trano*
TT.see the house
'The house is seen'

b. *[Hitan'ny lehilahy] ny trano*
TT.see-DET man the house
'The house is seen by the man'

c. *[Hitanao] ny trano*
TT.see-2SG the house
'The house is seen by you'

d. *[Hitan-dRabe] ny trano*
PST TT.see-Rabe the house
'The house was seen by Rabe'

This process is reminiscent of the facts described for Malay in (32) above where pronouns and proper names could appear before the bare verb.[31] I assume, therefore, that N-bonding is, in fact, the result of D-movement and it is this D-movement that checks the D feature in T.

The analysis, however, is not as straightforward as it is in Malay due to other processes that occur in Malagasy. First I assume that there is a functional head, E(vent), that is at the juncture of the event structure represented by the VP and the functional structure represented by various functional categories above the VP including T. In Malagasy, the V moves to this position as the highest head in the event structure domain for a reason other than the need to check a V feature.[32] The

[31] For some reason, if the head movement of the D is string vacuous as in Malagasy, the Determiner of a common noun can be included in the process. When the movement is not string vacuous as in Malay, only Ds that do not take a complement NP can undergo movement. This is a concern and I leave it for further study.

[32] Presumably we need this sort of V movement independently for a language like English which does not have V movement into the functional domain (e.g. across adverbs or negation, see Pollock, 1989) but which does have V movement within the VP shells proposed by Larson (1988).

D, however, would also move to this position as the lowest head in the functional domain because it has been targeted by the D feature.

(39)

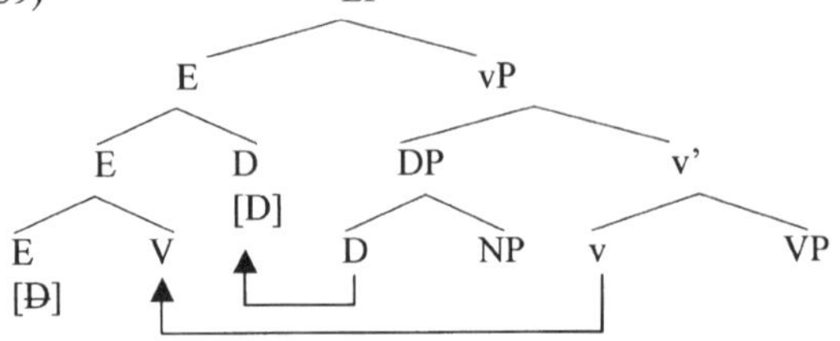

Further predicate fronting will front the whole EP to a higher Spec, XP. A sketch of the final structure is given in (40) below.

(40)

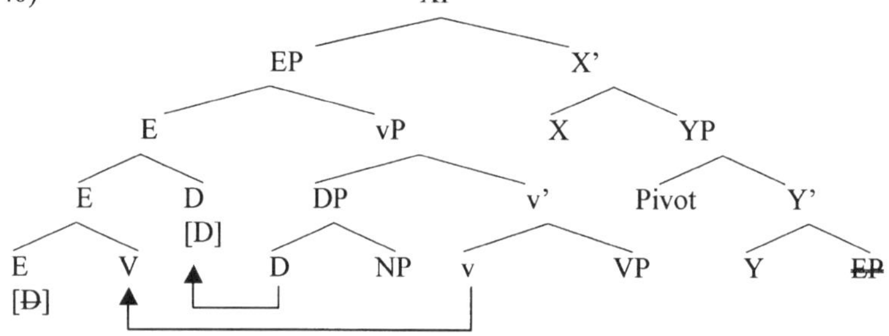

As with any new proposal, the claim that there is D^0-movement raises many questions.[33] I address some here but leave other for future research. One question has to do with several situations where there does not seem to be an element available to check the D feature. One case we saw above in (38a) where a non-subject Actor is not realized. I assume that this is a case similar to pro-drop. In one view of pro-drop, it must be the pro-DP itself which checks the D feature.[34] Presumably in these implicit Agent constructions, there also is a pro, perhaps in the D^0 position,

[33] I thank a reviewer for pointing these out to me. Some that I leave for further research are the following. Is there D^0-movement out of conjoined structures? Irish apparently can do this, which is a problem for me. Malagasy cannot in some cases such as (21b) above. How can there be head-movement out of a Spec position? I assume that this is what happens with Noun Incorporation in a Larson-type structure. This is discussed in Hale and Keyser (2002).

[34] This is not the view proposed by Alexiadou and Anagnostopoulou (1998), where it is the material in the agreement morphology on the verb that checks the D feature.

which is able to undergo the appropriate D^0-movement. Another instance where there appears to be no available DP to launch D^0-movement would be a construction where it is the external argument itself which is the pivot (i.e. outside the VP). In Guilfoyle, Hung, and Travis (1992), we proposed that Actor Topic morphology indicated that Case was not assigned to Spec, VP (i.e. to the external argument). In an updated version of this, the Actor Topic morphology would indicate, among other things, the lack of a D feature. This would also be true of constructions with non-verbal predicates such as (41) below where the predicate phrase is a PP.

(41) *Any an-tsena* *Rabe*
at ACC-market Rabe
'Rabe is at the market'

This construction, like the Actor Topic construction, realizes the external argument as the pivot and presumably there is no D feature in E.

This, however, leaves the sentence final subject/pivot of Malagasy unexplained in terms of feature checking since, by hypothesis, the D feature will trigger head movement not XP movement. But, in a way, this apparent problem reconfirms the suspicions of many other researchers that the subject of WMP languages is different from the more well-known Nominative subjects of languages such as English. In the next section I offer my view of WMP subjects and voice-marking.

5. Malagasy: Case and Features

Within the context of the language typology outlined above, we can re-evaluate the connection between being a subject and being in a Spec, TP position. We can also re-evaluate the connection between being a subject and having Nominative Case (where Case now is the checking of the D feature). If what I have argued for here is correct, i.e. that languages such as Malagasy move D^0s to check the D feature and move VPs to check the V feature, we are left in an odd position. The strongest claim would be that not only do D^0s move, but DPs do **not** move. In fact, DPs do not set up a relationship with a Case assigning head. In other words, if the D feature is going to target anything, it will be a head D^0. This raises questions as to what Case looks like in this sort of language.

Combining observations made by many other researchers, we can develop a view of subjects in Malagasy and Tagalog. Keenan (2000) notes that voice morphology in Malagasy simply denotes the role of the subject. In his account which he calls a predicate building approach to voice affixes, there is no movement but simply a relation that is set up between the subject and the predicate through the morphology. In Sells (1998), the verbal morphology in Tagalog designates an empty category pro in the f-structure that is co-indexed with the nominal adjunct *ang* DP. This empty category does not appear in the syntax. In both of these accounts, voice morphology works differently from active/passive distinctions set up in English.[35] Because of my own theoretical background and certain morphological facts of Tagalog that can be accounted for through a distribution of empty categories in syntax, I have argued that the empty categories are syntactically active (see Travis, 2000, 2002). The first conclusion, then, is that the subject is related to an empty category within the predicate via the morphology on the verb. The preliminary sketch is given below.

(42) a. $[\ V_{AT} \ldots ec_i \ldots\ ]$ $Agent_i$
b. $[\ V_{TT} \ldots ec_i \ldots\ ]$ $Theme_i$
c. $[\ V_{CT} \ldots ec_i \ldots\ ]$ $Oblique_i$

Pearson (2001) refines this relation between the pivot and the verbal morphology by positioning bits of the morphology in heads which are related to the empty categories in the Spec positions. This is set within the proposals of Sportiche (1996). Sportiche's structure for clitics is given in (43) and Pearson's development of this for Malagasy morphology is given in (44).

(43)

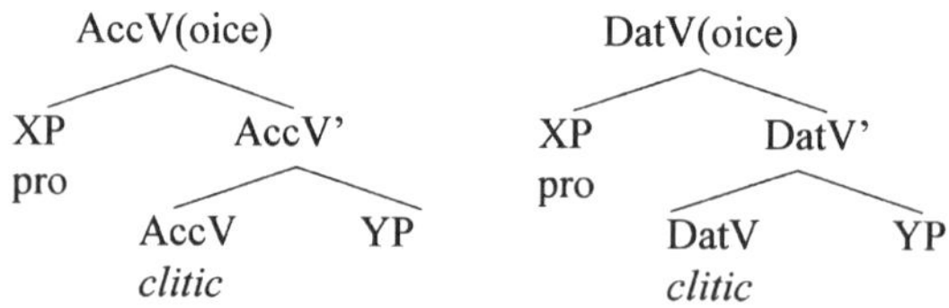

(44) a. ACTOR TOPIC b. THEME TOPIC

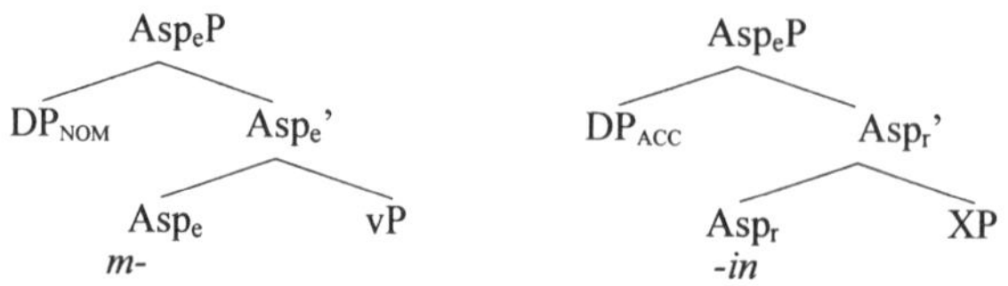

[35] Neither Keenan nor Sells works within frameworks which encode relations through movement. Both, however, still distinguish English-type voice morphology from WMP-type voice morphology. Sells (1998, p. 114) labels the former type M-Voice, and the latter P(hilippine)-Voice.

Taking Sportiche's terminology seriously, we can, in fact, view the voice morphology as a type of verbal clitic. It is a clitic in that it designates an empty DP category in its Specifier position. It is verbal in that it is realized on an extended projection of the verb. Also, taking Pearson's account still closer to Sportiche's, I propose that the empty category that is identified is pro rather than a trace as in Pearson's account. Putting this all together, we now have a construction where a predication relation is set up between a DP base-generated outside the VP and a verbal clitic that designates a grammatical role through its relation to an empty argument position in the phrase structure. When expressed this way, the WMP voice constructions are more similar to Clitic Left Dislocation (CLLD) structures in Italian (Cinque 1990) than to either passive constructions or topicalized constructions. Let us explore this possibility next.

We have seen arguments for why WMP pivots are not subjects in a theory-internal sense. There are also reasons to believe that they are not topics. Kroeger (1993) has argued that Tagalog pivots are not discourse or semantic topics, but there are syntactic facts as well that suggest they are not topics. Topics, being found in a high position within the CP domain, are often excluded from specific environments. Haegeman (2004) discusses this in her investigation of different types of adjunct clauses. She shows that topics cannot appear in central adverbial clauses (45) but can appear in peripheral adverbial clauses (46).[36]

(45) a. *Before **this book**, Mary read, John had already read it
(from Haegeman, 2004; credited to Maki, Kaiser, & Ochi, 1999, p. 4)
b. *If **some of these precautions** you take, you will pass the exams
(from Haegeman, 2004)

(46) a. We don't look to his paintings for common place truths, though **truths** they contain none the less
(from Haegeman, 2004; taken from *The Guardian*)
b. His face not many admired while **his character** still fewer felt they could praise
(from Haegeman, 2004; taken from Quirk et al.)

Malagasy pivots, as we see below, can appear in central adverbial clauses suggesting that they are not topics.[37]

(47) a. *Raha manasa lamba Rakoto, dia omeko sakafo izy*
if PRES.AT.wash clothes Rakoto, dia TT.give.1s food he
'If Rakoto washes clothes, then I give him food'

---

[36] See Shaer (1997), however, for criticisms of Haegeman's data and generalizations.
[37] Richards (2000) makes interesting comparisons between Icelandic topics and Tagalog pivots and he concludes that they are similar. The comparison of Icelandic topic constructions requires more research.

b. *Omeko* *sakafo* *Rakoto,* *raha* *manasa* *lamba* *izy*
TT.give.1s food Rakoto if PRES.AT.wash clothes he
'I give Rakoto food if he washes clothes'

Interestingly, CLLD constructions behave differently. As the example below shows, CLLD constructions may appear in central adverbial clauses.

(48) *Se* *queste* *cose* *Maria* *non* *le* *sa,* *non* *supererà* *l'esame*
if these things Maria NON them knows, NON will pass the exam
'If Maria doesn't know these things, she will not pass the exam'

Given the similarities in internal make-up and external distribution, I suggest that WMP clausal structure is closest to CLLD constructions in Indo-European languages. The basic Malagasy predicate structure, then, is as in (49) below.

(49) $[_{\text{PREDP}} \ldots \text{cl}_i \ldots]$ Pivot_i

The pivot is in a predication relation with the rest of the clause, a relation that is set up via the empty argument position identified by the verbal clitic. Given this, the question of subject and Nominative Case appropriately does not arise. If there is Case at all, it has none of the effects that are seen in other languages. It will never trigger DP movement. In fact, the claim is that these languages have no DP movement. WH-constructions are cleft constructions also created through these predication structures (see Travis, in press-b). Apparent voice alternations are simply different CLLD constructions. If Case is taken out of the grammatical system completely in these languages, the problem of doubly case-marked chains never arises (see Bejar and Massam, 1999).

6. Conclusion

Within Chomskian frameworks of the past 30 years, the role of abstract Case has been an important one in that it has been the trigger of A-movement, one of the central phenomena of Generative Grammar. Within the Minimalist Program, the role of Case has both expanded and diminished. Movement for reasons of Case first became the model for all movement as a process that is triggered by feature checking. With the introduction of the EPP feature, however, it has become less clear that Case itself triggers movement. Nominative Case within this framework is most closely related to the checking, by a DP, of a D feature in T. While movement to Spec, T, triggered by the EPP feature, is often combined with the checking of the D feature, there is nothing in principle that forces this.

Now there are two possibilities for subjects. They may be that element in Spec, T or the DP that checks the D feature. In Quirky Case constructions, we can see the division between the DP in the Spec, TP and the nominative DP within the VP. In this paper, however, I pursue a third possible relation – one that is more semantic. Subjects are the DPs that are predicated of where predication may be set up in a variety of ways that ensure that there is one open argument position (see e.g. Williams, 1980).

In Malagasy, the predication relation is created not through feature checking movement but through verbal morphology that licenses an empty argument. I set the discussion within a language typology that is created in a system which views movement as feature checking. Languages like English check the D and V features of the functional domain through DP and V movement, respectively, and languages like Malagasy check them through D and VP movement. In languages like Malagasy, then, where there is no DP movement, Case cannot be linked to DP movement. Nominative Case, if anything, is related to D-movement. We would expect, then, that subjects that appear outside of the VP must be merged in this external position. A stronger stand, however, would be to say that Case is not a syntactic force at all in these languages explaining a range of phenomena. It is not so much that this provides an answer for the behavior of these languages, as it expands the domain within which an answer might be found.

Acknowledgements

I appreciate comments from audiences at UC-Santa Cruz, GURT04, AFLAXI, and the financial support of SSHRCC 410-2001-1486 and FQRSC 2002-ER-75657.

References

Aldridge, E. (2004). *Ergativity and word order in Austronesian languages*. Unpublished Doctoral dissertation, Cornell University.

Alexiadou, A., & Anagnostopoulou, E. (1998). Parameterizing AGR: Word order, V-movement, and EPP-checking. *Natural Language and Linguistic Theory*, *16*, 491–539.

Baker, M., & Hale, K. (1990). Relativized minimality and pronoun incorporation. *Linguistic Inquiry*, *21*, 289–297.

Bejar, S., & Massam, D. (1999). Multiple case checking. *Syntax,* 2, 65–79.

Bobaljik, J. (1995). *Morphosyntax: The syntax of verbal inflection*. Unpublished Doctoral dissertation, MIT, Cambridge, MA.

Chomsky, N. (1980). On binding. *Linguistic Inquiry*, *11*, 1–46.

Chomsky, N. (1981). *Lectures on government and binding*. Dordrecht, the Netherlands: Foris Publications.

Chomsky, N. (1995). *The minimalist program*. Cambridge, MA: MIT Press.

Chomsky, N. (2000). Minimalist inquiries: The framework. In: R. Martin, D. Michaels & J. Uriagereka (Eds), *Step by step: Essays on minimalist syntax in honor of Howard Lasnik* (pp. 89–155). Cambridge, MA: MIT Press.

Chomsky, N. (2001). Derivation by phase. In: M. Kenstowicz (Ed.), *Ken Hale: A life in language* (pp. 1–52). Cambridge, MA: MIT Press.

Cinque, G. (1990). *Types of A'-dependencies*. Cambridge, MA: The MIT Press.

De Guzman, V. (1988). Ergative analysis for Philippine languages: An analysis. In: R. McGinn (Ed.), *Studies in Austronesian linguistics* (pp. 323–345). Athens, OH: Ohio University Center for International Studies.

de Hoop, H. (1996). *Case configuration and noun phrase interpretation*. New York: Garland Publishing.

Diesing, M. (1996). Semantic variables and object shift. In: H. Thrainsson, S. D. Epstein & S. Peter (Eds), *Studies in comparative germanic syntax, Vol. II* (pp. 66–84). Dordrecht, the Netherlands: Kluwer Academic Publishers.

Gerdts, D. (1988). Antipassives and causatives in Ilokano: Evidence for an ergative analysis. In: R. McGinn (Ed.), *Studies in Austronesian linguistics* (pp. 323–345). Athens, OH: Ohio University Center for International Studies.

Guilfoyle, E., Hung, H., & Travis, L. (1992). Spec of IP and spec of VP: Two subjects in Austronesian Languages. *Natural Language and Linguistic Theory*, *10*, 375–414.

Haegeman, L. (2004). Adverbial clauses, the left periphery and implications for the syntax of topics. Paper presented at Georgetown University Round Table (GURT). Georgetown University.

Hale, K., & Keyser, S. J. (2002). *Prolegomenon to a theory of argument structure*. Cambridge, MA: MIT Press.

Harris, A. C. (1984a). Case marking, verb agreement, and inversion in Udi. In: C. G. Rosen (Ed.), *Studies in relational grammar 2* (pp. 243–258). Chicago: The University of Chicago Press.

Harris, A. C. (1984b). Inversion as a rule of universal grammar: Georgian evidence. In: D. M. Perlmutter & C. G. Rosen (Eds), *Studies in relational grammar 2* (pp. 259–291). Chicago: The University of Chicago Press.

Holmberg, A. (1986). *Word order and syntactic features in the Scandinavian languages and English*. Stockholm: University of Stockholm.

Keenan, E. (2000). Morphology is structure: A Malagasy test case. In: I. Paul, V. Phillips & L. Travis (Eds), *Formal issues in Austronesian linguistics* (pp. 27–47). Dordrecht, the Netherlands: Kluwer Academic Press.

Keenan, E. L. (1976). Remarkable subjects in Malagasy. In: C. Li (Ed.), *Subject and topic* (pp. 249–301). New York: Academic Press.

Kroeger, P. (1993). *Phrase structure and grammatical relations in Tagalog*. Stanford, CA: CSLI.

Larson, R. (1988). On the double object construction. *Linguistic Inquiry*, *19*, 335–392.

Law, P. (1995). Grammatical relations in Malagasy control structures. In: C. Burgess, K. Dzewirek & D. Gerdts (Eds), *Grammatical relations: Theoretical approaches to empirical issues* (pp. 271–290). Stanford, CA: CSLI.

Lee, F. (2000). VP Remnant Movement and VSO in Quiavini Zapotec. In: A. Carnie & E. Guilfoyle (Eds), *The syntax of verb initial languages* (pp. 143–162). Oxford: Oxford University Press.

Legendre, G. (1989). Inversion with certain French experiencer verbs. *Language, 65*, 752–782.

Maclachlan, A. (1996). *Aspects of ergativity in Tagalog*. Unpublished Doctoral dissertation, McGill, Montreal.

Maki, H., Kaier, L., & Ochi, M. (1999). Embedded topicalisation in English and Japanese. *Lingua, 109*, 1–14.

Manning, C. D. (1996). *Ergativity: Argument structure and grammatical relations*. Stanford, CA: CSLI Publications.

Marantz, A. (1995). The minimalist program. In: G. Webelhuth (Ed.), *Government and binding theory and the Minimalist Program* (pp. 349–382). Oxford, UK: Blackwell.

Massam, D. (1985). *Case theory and the projection principle*. Unpublished Doctoral dissertation, Massachusetts Institute of Technology, Cambridge, MA.

Massam, D. (1998a). Predicate fronting in Niuean. In: M. Pearson (Ed.), *AFLA III and IV*. UCLA: UCLA Working Papers in Linguistics.

Massam, D. (1998b). VSO is VOS: Aspects of Niuean order. In: A. Carnie & E. Guilfoyle (Eds), *The syntax of verb initial languages* (pp. 97–116). Oxford: Oxford University Press.

Massam, D. (2005). Lexical categories, lack of inflection, and predicate fronting in Niuean. In: A. Carnie, H. Harley & S. A. Dooley (Eds), *Verb first: Papers on the syntax of verb initial languages* (pp. 227–242). Amsterdam: John Benjamins.

Massam, D., & Smallwood, C. (1997). Essential features of predication in English and Niuean. In: K. Kusumoto (Ed.), *North East Linguistics Society 27* (pp. 363–372). McGill University: Graduate Linguistic Student Association, UMass.

Oda, K. (2002). *Wh- questions in V-initial languages*. Unpublished MA Forum Paper, University of Toronto, Toronto.

Paul, I. (2000). *Malagasy clause structure*. Unpublished Doctoral dissertation, McGill University, Montreal.

Paul, I. (2001). Concealed pseudo-clefts. *Lingua,* 707–727.

Pearson, M. (1998). Two types of VO languages. In: P. Svenonius (Ed.), *The derivation of VO and OV* (pp. 327–363). Amsterdam: John Benjamins.

Pearson, M. (2001). *The clause structure of Malagasy: A minimalist approach*. Unpublished Doctoral dissertation, UCLA, LA.

Perlmutter, D. M. (1984). Working 1s and inversion in Italian, Japanese, and Quechua. In: D. M. Perlmutter & C. G. Rosen (Eds), *Studies in relational grammar 2* (pp. 292–330). Chicago: The University of Chicago Press.

Pollock, J.-Y. (1989). Verb movement, UG, and the structure of IP. *Linguistic Inquiry, 20*, 365–424.

Postal, P. (1969). On so-called pronouns in English. In: D. Reibel & S. Schane (Eds), *Modern studies in English* (pp. 201–224). Englewood Cliffs, NJ: Prentice-Hall.

Rackowski, A. (1998). Malagasy adverbs. In: I. Paul (Ed.), *The structure of Malagasy, Vol. II* (pp. 11–33). Los Angeles: UCLA Occasional Papers in Linguistics.

Rackowski, A. (2002). *The structure of Tagalog: Specificity, voice, and the distribution of arguments*. Unpublished Doctoral dissertation, MIT, Cambridge, MA.

Rackowski, A., & Travis, L. (2000). V-initial languages: X or XP movement and adverbial placement. In: A. Carnie & E. Guilfoyle (Eds), *The syntax of V-initial languages* (pp. 114–141). Oxford: Oxford University Press.

Richards, N. (1997). *What moves where when in which language*. Unpublished Doctoral dissertation, MIT, Cambridge, MA.

Richards, N. (2000). Another look at Tagalog subjects. In: I. Paul, V. Phillips & L. Travis (Eds), *Formal issues in Austronesian linguistics* (pp. 105–116). Dordrecht, the Netherlands: Kluwer Academic Press.

Roberts, I. (1991). Excorporation and minimality. *Linguistic Inquiry, 22*, 209–218.

Rosen, C. G., & Wali, K. (1989). Twin passives, inversion and multistratalism in Marathi. *Natural Language and Linguistic Theory, 7*, 1–50.

Schachter, P. (1976). The subject in Philippine languages: Topic, actor, actor-topic, or none of the above? In: C. N. Li (Ed.), *Subject and topic* (pp. 491–518). New York: Academic Press, Inc.

Schachter, P. (1996). The subject in Tagalog: Still none of the above. *UCLA Occasional Papers in Linguistics, 15*, 1–61.

Seiter, W. J. (1978). Subject-direct object raising in Niuean. In: D. M. Perlmutter (Ed.), *Studies in relational grammar 1* (pp. 317–359). Chicago: The University of Chicago Press.

Sells, P. (1998). The functions of voice markers in Philippine languages. In: S. G. Lapointe, D. K. Brentari & P. M. Farrell (Eds), *Morphology and its relation to phonology and syntax* (pp. 111–137). Stanford, CA: CSLI Publications.

Shaer, B. (1997). Adverbials, functional structure and restrictiveness. Unpublished manuscript, ZAS, Berlin.

Sigurosson, H. A. (1989). *Verbal syntax and case in Icelandic in a comparative GB approach*. Lund: Lund University.

Sportiche, D. (1996). Clitic constructions. In: J. Rooryck & L. Zaring (Eds), *Phrase structure and the Lexicon* (pp. 213–276). Dordrecht, the Netherlands: Kluwer Academic Publishers.

Travis, L., & Williams, E. (1982). The externalization of arguments in Malayo-Polynesian languages. *The Linguistic Review, 2*, 57–78.

Travis, L. d. (2000). Event structure in syntax. In: C. Tenny & J. Pustejovsky (Eds), *Events as grammatical objects: The converging perspectives of lexical semantics and syntax* (pp. 145–185). Stanford, CA: CSLI Publications.

Travis, L. d. (2002). Agents and causes in Malagasy and Tagalog. In: *Austronesian Formal Linguistics Association*. MIT: MITWPL.

Travis, L. d. (in press-a). VP, D^0 movement languages. In: R. Zanuttini, H. Campos, E. Herburger & P. Portner (Eds), *Negation, tense and clausal architecture: Cross-linguistic investigations*. Washington, DC: Georgetown University Press.

Travis, L. d. (in press-b). Through the looking glass: Malagasy in Wonderland. In: H.-M. Gärtner, P. Law & J. Sabel (Eds), *Clause structure and adjuncts in Austronesian languages*. Berlin: Mouton de Gruyter.

Williams, E. (1980). Predication. *Linguistic Inquiry, 11*, 203–238.

Chapter 7

The Case of Basque: An Accusative Analysis for an Ergative System

Mario van de Visser

1. Introduction

Every natural language is thought to distinguish between intransitive and transitive predicates. The former take only one argument, called intransitive subject (S), whereas the latter take at least two arguments, a transitive subject (A) and a direct object (O). In a nominative/accusative system, S and A are treated alike, whereas O is treated differently. This is illustrated by the German sentences below:[1]

(1) a. *Ich* *geh-e*
1SG go.PRS-1SG
'I am going'

b. *Du* *geh-st*
2SG go.PRS-2SG
'You are going'

c. *Ich* *wasch-e* *dich*
1SG wash.PRS-1SG 2SG.ACC
'I am washing you'

d. *Du* *wäsch-st* *mich*
2SG wash.PRS-2SG 1SG.ACC
'You are washing me'

[1] The following abbreviations are used in the glosses: (A)=Transitive Subject; ACC=Accusative case; (IO)=Indirect Object; ERG=Ergative case; FEM=Feminine; FAM=Familiar; IMP=Imperfective aspect; MASC=Masculine; PL=Plural; PRS=Present tense; 1/2/3=1st, 2nd, 3rd person; SG=Singular.

Competition and Variation in Natural Languages: The Case for Case
Edited by M. Amberber and H. de Hoop

ISBN: 0-08-044651-5

The first person singular pronoun *ich* 'I' functions as an intransitive subject in (1a), and as a transitive subject in (1c). This is the Nominative case, which I have omitted in the glosses because of presentational reasons. Whenever this pronoun functions as a direct object, as in (1d), a different form is used: *mich* 'me'. This is the Accusative case. Agreement morphology follows this nominative/accusative pattern closely: the verb agrees with nominative arguments only.

Basque, on the other hand, shows a completely different pattern:

(2) a. *Ni* *ibil-tzen* *naiz*
1SG walk-IMP 1SG.PRS.be
'I am walking'

b. *Zu* *ibil-tzen* *zara*
2SG walk-IMP 2SG.PRS.be
'You are walking'

c. *Zu-k* *ni* *garbi-tzen* *nau-zu*
2SG-ERG 1SG wash-IMP 1SG.PRS.HAVE-2SG(A)
'You are washing me'

d. *Ni-k* *zu* *garbi-tzen* *zaitu-t*
1SG-ERG 2SG wash-IMP 2SG.PRS.have-1SG(A)
'I am washing you'

A comparison between (2a) and (2c) shows that the unmarked pronoun *ni* '1SG' may function as an intransitive subject (S) or as a direct object (O). This is typical of ergative systems: the unmarked case is complemented by a marked case *ni-k* '1SG-ERG' (cf. (2d)) for transitive subject(A)-function (Ergative). The unmarked case in such a system is generally called Absolutive in order to avoid confusion with the Nominative, which covers a different combination of syntactic functions. Again, this case has been omitted in the glosses. In Basque, verbal morphology too seems to operate on an absolutive/ergative basis: absolutive arguments trigger (predominantly) prefixal person/number marking on the predicate, whereas ergative arguments are (consistently) represented by suffixes. In (2), this inflectional morphology appears on an auxiliary, which is a common process in Basque syntax. In intransitive sentences, a form of *izan* 'be' is used, whereas in transitive clauses, the auxiliary is a form of **edun* 'have'. Fuller descriptions of these facts can be found in Saltarelli, Azkarate, Farwell, and Ortiz de Urbina (1988) and Hualde and Ortiz de Urbina (2003), among others.

Apart from morphology, Basque does not show any other sign of ergativity: its syntax is nominative/accusative oriented. Various people have shown that intransitive subjects and transitive subjects form a class that is treated differently from direct objects when clauses are combined (see, for instance, Rebuschi, 1989; Ortiz de Urbina, 1989; Hualde & Ortiz de Urbina, 2003 for plenty of evidence).[2] In other words, interclausal syntax is generally taken to operate on a nominative/accusative basis, thereby excluding any *syntactically ergative* phenomena (cf. Dixon, 1994). In this chapter, I will therefore argue that although the case system in Basque is ergative, this is not the result of a syntactic make-up that differs fundamentally from nominative/accusative languages like English. In fact, I will show that it is possible to explain ergativity of the type found in Basque by analyzing every transitive sentence as a (kind of) passive construction. My analysis follows Hoekstra (1986), Jaeggli (1986) and Baker et al. (1989) in assuming that the passive verb assigns an external theta-role to an element in the functional projection dominating the verbal projection. In languages like English, this element is an invariant morpheme, implying a transitive subject which is not specified for features like person and number. In languages like Basque, I propose, this element is an unmarked fully specified pronoun. In both cases, the transitive subject is licensed by incorporation into the predicate (cf. Baker & Hale, 1990). By adopting this proposal, we can account for the inflectional morphology on the verb, as well as for the ergative case pattern, without having to postulate a parameter forcing a language to be either ergative or nonergative. This is a desirable result since ergative languages constitute a minority among the world's languages and morphologically 'pure' ergativity, as we find it in Basque, is also rarely found.

In Section 2, I briefly discuss recent analyses of ergativity and show that most of them base themselves on some parameter which is stipulated in order to account for the fact that some languages are ergative and others are not. Section 3 contains my proposal for a universal treatment of case and agreement, which allows arguments to be licensed by either structural case or agreement. A second part of the proposal deals with passive constructions. This part is subsequently applied to Basque in Section 4. In Section 5, I conclude the chapter with suggestions for future research.

[2] Throughout this chapter, I will use the terms 'subject' and 'object' as universal labels referring to structural positions: the complement of a transitive verb is called 'object', whereas its specifier is called 'subject'. The argument of an intransitive verb will also be called 'subject'. I consider these labels to be accurate enough for the description of ergative languages like Basque, since I propose that the verbal projection in that language does not differ fundamentally from verbal projections in non-ergative languages.

2. Previous Analyses of Ergativity

In this section, I will present four important and fundamentally differing analyses of ergativity: Marantz (1981, 1984), Murasugi (1992), Bobaljik (1993) and Bittner & Hale (1996a, b). For reasons of space, I will restrict myself to the basic ideas of these proposals. For a more detailed overview of the treatment of ergativity in generative linguistics, the reader is referred to Johns (2000) and van de Visser (in preparation).

One of the earliest syntactic accounts of ergativity within the Government & Binding framework is Marantz (1981, 1984). According to this proposal, there are languages in which theta-roles are assigned in a reversed way:

(3) a. b.

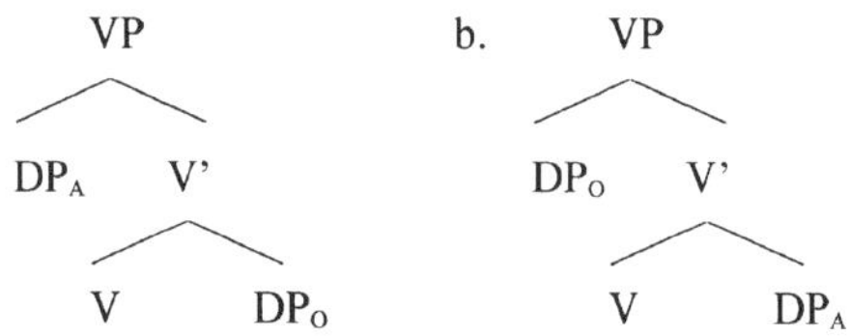

In languages like English, a transitive verb assigns its theme-role to its complement, which is the direct object, whereas the agent-role is assigned to the DP in the specifier (cf. (3a)).[3] In some ergative languages, this situation is reversed: the agent-role is assigned to the sister of the verb, whereas the theme-role is assigned to the specifier (3b). This has important syntactic consequences, since in such a configuration the object c-commands the subject. For instance, one would expect to encounter subject reflexive pronouns that are bound by object antecedents. As far as I know, there is no single ergative language where this is attested. However, there are languages like Dyirbal in which S and O behave alike (and differently from A) when it comes to clause combining (Dixon, 1972). This 'syntactically ergative' behaviour is found in only a handful of ergative languages (see Dixon, 1994, for an overview). The majority of ergative languages, Basque included, is only 'morphologically ergative', which means that VP looks like the one in (3a). For these languages, Marantz has to stipulate that ergative case and agreement differ from their counterparts in morphologically nominative/accusative systems. This does not really provide us with a reason why there should be such a difference.

[3] Marantz uses the labels 'Subj.' and 'Obj.' instead of DP_A and DP_O. I am avoiding Marantz' terminology here in order to keep the presentation consistent. This is also the case in the analyses to follow. Also, the notions 'theme-role' and 'agent-role' are approximations of what we usually call 'internal' and 'external' theta-role

A more recent analysis along the lines of the Minimalist Program is Marantz (1981, 1984) and Murasugi (1992). Murasugi assumes two functional projections where case and agreement features are checked. The lower one is called Tr(ansitivity), the higher one is T(ense). In nominative/accusative systems, Accusative case is checked in Spec,TrP, together with the phi-features of the object. Nominative case is checked in Spec,TP, as well as the phi-features of the subject. This results in 'crossing paths' (4a):

(4) Murasugi (1992)

a. nominative/accusative b. absolutive/ergative

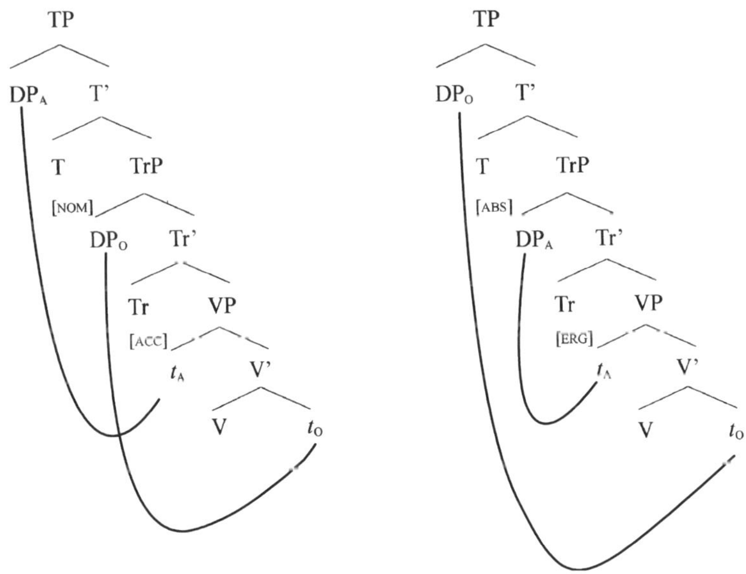

In ergative languages, Murasugi argues, the structural (Ergative) case checked by the subject is located in Tr, whereas the unmarked (Absolutive) case is checked by the object in Spec,TP. This yields 'nested paths' (4b). Again, the subject is c-commanded by the direct object (albeit after movement), making similar predictions as Marantz' theory. This means that for morphologically ergative languages like Basque, the structural absence of syntactic ergativity is not accounted for.

A slightly different approach by Bobaljik (1993) overcomes these problems. The two functional heads responsible for checking case and agreement are called Agrl and Agr2. In every language, case/agreement features of the (transitive) subject are checked in the higher Agr1P, whereas the object's features are checked in the lower Agr2P. The case features of a nominative/accusative system are therefore divided as in (5a) (cf. (4a)).

(5) Bobaljik (1993)

a. nominative/accusative

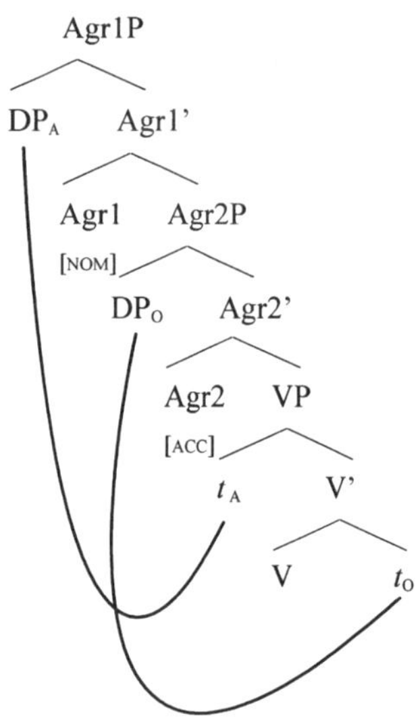

b. absolutive/ergative

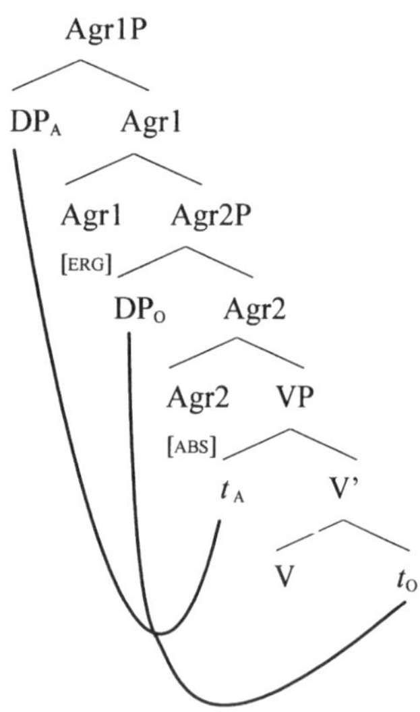

Absolutive/ergative case/agreement systems, on the other hand, are the reverse of Murasugi's (4b): Ergative is checked in the higher Agr1-projection, whereas Absolutive is checked in the lower Agr2-projection (cf. (5b)). This is a less elegant theoretical move, because it requires Universal Grammar (UG) to scatter marked and unmarked cases across functional heads, unlike Murasugi's theory. A further distinction between the two systems is made with respect to intransitive clauses. In nominative/accusative systems, Agr2 is inactive and Agrl is active; in absolutive/ergative systems it is the other way around. This stipulation is needed in order to support Bobaljik's Obligatory Case Parameter (based on Levin & Massam, 1985): the obligatory case in a nominative/accusative system is Nominative, in an absolutive/ergative system this is Absolutive. In order to prevent intransitive subjects from appearing in Ergative case in languages like Basque, Agr2 must be active in intransitive clauses. Although this analysis has been successfully applied to Basque by Laka (1993b), it still leaves us with a parameter that is necessarily stipulated in order to account for ergativity only.

We should wonder whether this is absolutely necessary. A striking fact about ergativity is that it seems to play a role in only a minority of the world's languages. According to Dixon (1994), some 25% of the languages that had been investigated in any detail at the time of writing can be said to be ergative. Moreover, most of these languages use the ergative pattern only in a specific area of their grammar. For instance, ergative Indo-European languages typically reserve it for past tense/ perfective aspect, employing an accusative pattern elsewhere. Several Pama-Nyungan (Australian) languages have an ergative case system for independent arguments (pronouns and full noun phrases), in addition to a nominative/accusative set of bound pronouns.[4] Often, the former function as optional doubles of the latter. Yet other Pama-Nyungan languages, as well as several native American and Tibeto-Burman languages, have a case-system which is ergative for third person arguments only, whereas first and second person arguments pattern nominative/accusatively (cf. Dixon, 1994). Even languages like Basque, whose morphology is ergative in a fairly consistent way, do not display syntactic ergativity like Dyirbal and a handful of other morphologically ergative languages, as noted above. In other words: nominative/accusative patterns are omnipresent, whereas ergative patterns appear sporadically, and when they do, they are often unable to replace the nominative/accusative pattern entirely. This suggests that *nominative/accusative* might be the *basic pattern in every single language*, dictated by UG. If this appears to be true, we need to be able to derive ergative morphology from an accusative structure. This is exactly what I will do in the next section. Before that, I will briefly discuss yet another influential account for ergativity: the analysis proposed by Bittner & Hale (1996a,b). The reason for this is that their proposal links the two patterns: ergative patterns may develop out of a construction that occurs independently in accusative languages (passive),

---

[4] The term 'bound pronouns' here refers to pronominal elements which behave like clitics. It has nothing to do with pronominal binding in a theoretical sense.

whereas accusative patterns develop out of a construction that occurs in ergative languages (antipassive). This makes more sense than the previous accounts, since it enables us to account for a range of split-ergative patterns. However, there is also an important drawback to this theory, as we will see shortly.

Bittner & Hale assume that Ergative and Accusative noun phrases (NP or DP) are headed by a functional Case projection (KP). The head of this projection, K, is empty at deep structure, and therefore these KPs are subject to the Empty Category Principle (ECP).[5] This means that they are to be Case-bound by V or I. When V is the antecedent, K is spelled out as Accusative; Ergative when the antecedent is I. Case binding is defined as follows:

(6) Case binding (Bittner & Hale, 1996a, p. 12):
Let α be a head that delimits a small clause, and let β be an argument. Then α Case-binds β, and β's head, iff
a. α locally c-commands β;
b. α governs a Case competitor for β.

In intransitive clauses, the sole argument of the verb is (universally) realized as a K-less NP or DP. This is because Case binding does not apply as long as requirement (6b) is not met. K-less noun phrases must be licensed under c-command and government by C, which is the sentential counterpart of K. This can be accomplished by head-movement of I to C.[6]

Transitive clauses come in two types, which correspond to the difference between ergative and accusative patterns. Remarkably enough, the ergative structure is less complex than the accusative one. Consider the trees in (7).[7]

[5] Only inherent cases have a K-head that is filled at deep structure, but these will not be discussed here.

[6] Another possibility is movement of the K-less noun phrase to Spec,IP. In ergative languages, direct objects are also K-less, and they are licensed in the same way as intransitive subjects. The language-specific choice between head movement and dependent movement distinguishes syntactically ergative languages from morphologically ergative ones. Basque is of the latter type, hence I will focus on the head movement analysis.

[7] These trees summarize the basics of Case licensing in Bittner & Hale's theory, and therefore they have been reduced for presentational reasons. For instance, agreement is not represented in (7). On Bittner & Hales's account, agreement can be expressed by C or I, and it does not require a specifier-head configuration (1996a: 17). Since my own proposal considers agreement to be able to license an argument (just like structural case), and, moreover, since it does not analyze every kind of person/number marking as agreement proper, discussing Bittner & Hale's approach to agreement is irrelevant.

(7) Bittner & Hale (1996a, b)

a. nominative/accusative   b. absolutive/ergative

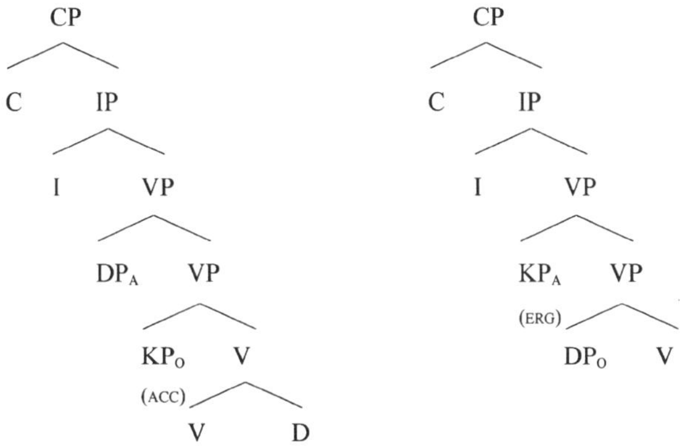

In both constructions, the subject is adjoined to VP, which — under Bittner & Hale's analysis — means that it is not governed by V. Hence, the direct object is the sole argument that is governed by the verb, so it is K-less. Like the intransitive subject, the direct object is licensed under c-command and government by C. The transitive subject is governed by I, which also governs the object. The latter counts as a Case competitor to the former, so the subject must be realized as a KP. K is spelled out as Ergative, because the Case binder is I. This situation is represented by the tree in (7b).

Accusative case is only found in clauses where V governs a Case competitor for the direct object. This situation arises when an element of category D incorporates into the verb. This is supposed to be the case when an antipassive morpheme is reanalyzed as a functional head or when pronominal object incorporation is reanalyzed as agreement (1996a, pp. 40,41). D counts as a Case competitor, and therefore the direct object is necessarily realized by a KP, where K spells out as Accusative. The transitive subject is K-less, since I does not govern another K-less noun phrase. Consequently, the subject is licensed by being c-commanded and governed by C.

Structurally, the ergative construction seems to be the most basic one since the verb does not have an incorporated D (cf. (7b)). In accusative constructions, the incorporated D is obligatorily present in order to be able to have Accusative objects (cf. (7a)). This D is either an old antipassive affix, or an incorporated object pronoun, which has been reanalyzed. Antipassives are only found in languages that have ergative case marking as well. Although there are examples of both ergative languages becoming accusative, and of accusative languages becoming ergative (cf. Dixon, 1994, pp. 187–203), it is not very likely that every accusative language has gone through a stage of ergativity. Within most familiar Indo-European languages, there does not seem to be explicit evidence in the morphology of the verb for the D-status of any functional head. And, moreover, object agreement is most often formally different from subject agreement, as the object agreement markers usually morphologically resemble independent pronouns with Accusative case. It is hard to account for these resemblances if the agreement marker is to be analyzed as a K-less D. I take this as a serious drawback of Bittner & Hale's theory.

The fact that ergative patterns are cross-linguistically rare is not at all reflected by their relative structural simplicity compared to accusative patterns. By itself, this is not necessarily an objection against an analysis along the lines of Bittner & Hale (1996a, b).[8] However, we should not be surprised if things would be the other way around: accusative patterns being more basic than ergative patterns. The proposal that I will present in the following sections is based on this assumption. It has in common with Bittner & Hale's analysis that ergative patterns may be based on a construction that is found in nonergative languages as well: the passive.[9]

3. Nominative and Accusative: Agreement and Case

In the generative framework, it is commonly assumed that DPs need case in order to be licensed. This means that cases like Accusative and Ergative are always complemented by cases like Nominative and Absolutive. The latter are typically unmarked. In languages with overt agreement, the predicate primarily agrees with a DP in the unmarked case: Nominative DPs trigger agreement if a language is nominative/accusative, and Absolutive DPs will do so if a language is absolutive/ergative. This correlation has led several people to consider Nominative and Absolutive DPs to be case less, an idea dating back to Jakobson (1936) (see also Bittner & Hale, 1996a,b; Neeleman & Weerman, 1999). It is an attractive idea

[8] One of the anonymous reviewers pointed this out to me, under reference to Newmeyer (1998).

[9] On Bittner & Hale's account, this happens when a lexical passive head (nominal or adjectival) is reanalyzed as a functional head I, which governs a K-less direct object and an empty-headed KP-subject (1996a, p. 33).

because it makes argument licensing more economical: a DP is licensed either by case or by agreement. The present proposal implements this analysis, and the structure in (8a) is adopted as a universal scheme, in principle applicable to every natural language.[10]

(8) a. active b. passive

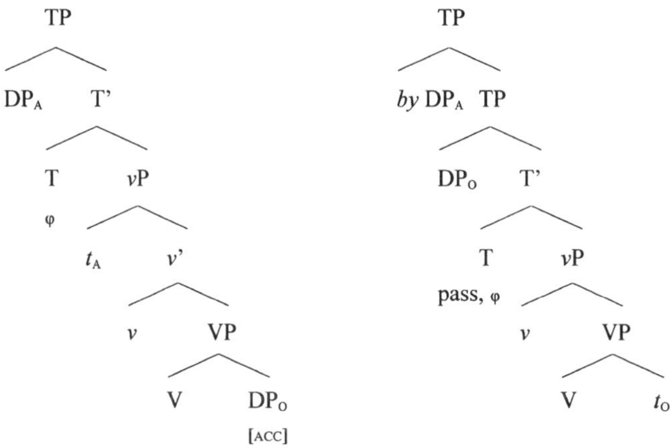

In *active* transitive sentences, the direct object is base-generated with a case feature (Accusative). I assume, contrary to most recent Minimalist views, that this feature is checked *in situ* by *v*.[11] The subject is base-generated without any case feature, but it moves to the specifier of T where it is licensed by agreement. Phi-features are checked in a Spec-Head configuration, (core) case features are not.

[10] I am well aware of the fact that this formulation might be too restrictive, as it excludes instances of agreement with case-marked direct objects, as well as instances where an unmarked argument that is licensed by agreement bears case. I consider this to be a topic for further research. Although Basque has intransitive (unergative) predicates which take an Ergative-marked subject, I do not consider this a problem for my analysis. Ortiz de Urbina (1989), as well as Bobaljik (1993) and Laka (1993b) convincingly analyze these constructions as syntacticaly transitive, which explains why the subject is overtly marked for case.

[11] The motivation for this assumption is presented in van de Visser (in preparation), but is irrelevant to the present topic.

Passive constructions differ from active ones with respect to the assignment of the external theta-role (cf. (8b)). I follow Baker *et al* (1989) in assuming that this role is assigned to the passive morphology in languages like English, which is expressed by the functional head T. From this, it follows that the direct object is the only argument that needs canonical licensing. We have a situation which is comparable to intransitive clauses, and hence agreement is chosen to license the direct object. Notice that according to this proposal, passive constructions are not only semantically transitive, but also syntactically. This point of view was first proposed by Jaeggli (1986) and elaborated in Baker *et al* (1989). The latter also argue for an alternative scenario, depicted in (9):

(9) passive (alternative scenario)

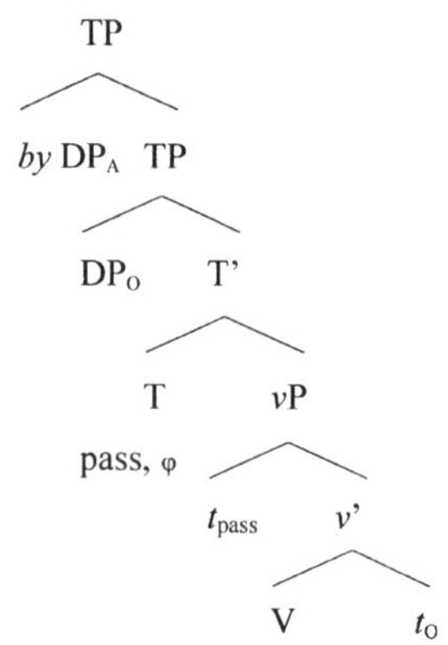

In (9), the passive morpheme is base-generated in the same position as the transitive subject of an active construction. From there, it incorporates into T. This analysis is required in order to account for languages like Lithuanian and Turkish where unaccusative predicates can be passivized. These predicates have a derived subject, which means that the passive morpheme receives an internal theta-role. Baker *et al* argue that in such constructions, the passive morpheme is base-generated in the complement of the verb, from where it moves to subject-position. Incorporation into T follows (1989, p. 232). My own analysis of Basque, to be presented in the next section, is based on the structure in (9).

Several people have claimed that ergative constructions may develop out of passive constructions diachronically (see Trask, 1979, or Dixon, 1994, for an overview).[12] The similarities between the two constructions are obvious: the direct object appears in the unmarked case and the verb typically agrees with it, whereas the subject appears together with an overt case marker (Ergative, Instrumental) or preposition (*by*). The fact that in many languages, the Ergative case form is homophonous with oblique cases like Instrumental and Locative, is particularly striking. For Indo-Aryan languages, which often have an ergative construction which is restricted to past tense or perfective aspect, it is well known that the (synthetic) past/perfective construction in the proto-language was gradually replaced by a (periphrastic) passive construction. The latter has become an ergative construction, which is found in the present-day languages. Within generative grammar, Hale (1970) suggests a development from passive to ergative in Australian languages, Chung (1978) argues for such a development in Polynesian (Austronesian). These authors assume that reanalysis of the passive needs to take place in order for this construction to become ergative, because most ergative constructions are syntactically accusative. Indeed, in most morphologically ergative languages, the patient argument of a transitive sentence is not the sentential subject, like in English passives. This is probably due to the difference between (8b) and (9). In English, the agent of a passive verb is always unspecified for person and number in the absence of a *by*-phrase. In Basque, on the contrary, the agent is specified for these features as a result of the structure in (9). Here, the external theta-role is assigned to an unmarked pronoun which incorporates into T, yielding specific information about the agent.

4. The Case of Basque

In Basque, intransitive verbs take one argument which is licensed by agreement, and not by case (but see footnote 10). In most clauses, main verbs are accompanied

[12] According to an anonymous reviewer, Basque has been analyzed as a language where every active sentence in fact is a passive construction as early as the 19th century. Ever since, various scholars have considered this proposal in detail, and discussed the problems it raises. The present proposal is of a different kind, since it does not predict the active sentence in Basque to display the characteristics of English passive clauses (see below).

by the intransitive auxiliary *izan* 'be'. The present tense paradigm of this verb is shown in (10a):[13]

(10) *Agreement in Basque*

a. intransitive auxiliary *izan* 'be'

ni	***n**aiz*	'I am'	*gu*	***g**ara*	'we are'
zu	***z**ara*	'you are'	*zuek*	***z**arete*	'you (pl) are'
hura	***d**a*	's/he/it is'	*haiek*	***d**ira*	'they are'

b. transitive auxiliary **edun* 'have'

har-k ni	***n**au*	's/he/it has me'
har-k zu	***z**aitu*	's/he/it has you'
har-k hura	***d**u*	's/he/it has him/her/it'

har-k	*gu*	***g**aitu*	's/he/it has us'
har-k	*zuek*	***z**aituzte*	's/he/it has you (pl)'
har-k	*haiek*	***d**itu*	's/he/it has them'

The paradigm in (10a) shows that agreement with intransitive objects is predominantly prefixal. In the case of first and second person, the first consonant of the pronoun is repeated in the verb. Third person pronouns are actually demonstratives, and no part of their morphology is copied in the auxiliary. Moreover, the agreement-prefix changes according to tense/aspect when the agreeing constituent is third person (Hualde & Ortiz de Urbina, 2003, p. 207). The paradigm in (10b) shows that direct objects in a transitive sentence trigger the same prefixes on the transitive auxiliary **edun* 'have'. Like intransitive subjects, they are unmarked for case, which in my theory implies that they are licensed by agreement (cf. (8a)).

Recall that the external theta-role of a passive verb is assigned to the passive morpheme in T (cf. (8b)). This is the case in languages like English in which unaccusatives cannot be passivized. However, as has been argued in the previous section, languages like Lithuanian and Turkish base-generate their passive morphology in argument-position (cf. (9)). In the case of Basque, I would like to

[13] Basque is a so-called *pro*-drop language, which means that pronominal subjects and (in)direct objects do not surface phonetically in most clauses (Laka & Uriagereka, 1987; Ortiz de Urbina, 1989). In this chapter, I am including these pronouns in every sentence in order to facilitate the presentation.

argue, this position may be filled by an unmarked pronoun. Like the passive morpheme, this pronoun cliticizes to T, which is reflected in the structure by incorporation into T. As is well known, this process ensures the licensing of the incorporated argument, leaving the agreement projection free to be used for the licensing of the direct object. The sentences in (11) show the complete paradigm of incorporated pronouns:

(11) *Incorporation of Basque transitive subjects*

transitive auxiliary **edun* 'have'

ni-k	*hura*	*du=t*	'I have him/her/it'
zu-k	*hura*	*du=zu*	'you have him/her/it'
har-k	*hura*	*du=∅*	's/he has him/her/it'
gu-k	*hura*	*du=gu*	'we have him/her/it'
zuek	*hura*	*du=zue*	'you (pl.) have him/her/it'
haiek	*hura*	*du=te*	'they have him/her/it'

The incorporated pronouns appear as suffixes on the transitive auxiliary. I have indicated this by an equals sign, in order to distinguish these forms from agreement affixes. If we would treat these suffixes as agreement markers, as most analyses have done so far, we would first of all have to assume that verbs can agree with more than one argument. This would be a departure from the structure in (8a), which assumes that arguments are licensed by either case or agreement.[14] Moreover, it would require a unique parameter stating that some languages treat direct objects and intransitive subjects morphologically alike and differently from transitive subjects, whereas others treat all subjects similarly (and differently from direct objects). Such a parameter would be designed to account for ergativity alone, whereas the present account makes use of a structure that is available in accusative languages anyway.

The idea that the person/number clitics are reduced pronouns is supported by their similarity to the independent pronouns. Especially the forms used for first person plural and second person singular/plural are almost identical to the pronoun. The vowel is often retained, which is not the case in the agreement

[14] Of course, this implies that every language showing person/number-marking for more than one argument incorporates pronouns. In van de Visser (in preparation), I argue that this might indeed be the case. The argument is based on Jelinek (1984) and Baker (1996, 2001, 2003), who argue that in certain nonconfigurational languages, every argument is necessarily realized pronominally.

paradigms in (10). The ergative case-marker, *-k*, does not appear in any of the forms. This supports the hypothesis that the incorporated pronouns are in the unmarked case.[15] The problematic forms are first person singular =*t* and third person plural =*te*. These do not resemble the independent pronouns *ni* '1SG' and *haiek* '3PL' in any transparent way. At present, I can only speculate about why these members of the paradigm are different.

One way to go about this would be to assume that an earlier form of the independent pronoun *ni* contained /t/.[16] This consonant is still present in the incorporated pronoun, but the independent form has lost it. If this is correct, the same would apply for a second person singular pronoun, the use of which is restricted to familiar addressees like siblings or close friends (Hualde & Ortiz de Urbina 2003, p. 151). I have omitted this pronoun, *hi*, from the paradigms given here, because *zu* is used more often. Like *ni-k* '1SG-ERG', *hi-k* '2SG.FAM-ERG' triggers a suffix on the transitive auxiliary which seems to be completely different from the independent pronoun: =*k* for masculine and =*n* for feminine reference. These markers differentiate for gender, something which is not found anywhere else in the grammar of Basque. If it is true that the incorporated pronouns show older stages of the independent pronoun, we predict that there have been two independent pronouns for second person singular informal: a masculine and a feminine variant. Note, that both the agreement prefixes for *ni* and *hi* do consist of the first consonant of these words: *n-* and *h-*. This might indicate that the new forms of the pronouns have been shaped after the agreement prefixes. So far, I have not been able to find similar claims or counterclaims in the literature, so nothing definitive can be said with respect to this issue.

As for third person plural (=*te*), a plausible assumption would be that this is merely a pluralizer, because /t/ is found in many places with precisely that function (compare, for instance, the agreeing forms *zara* '2SG.PRS.be' and

---

[15] Ura (2000) proposes an analysis of ergativity which has in common with my proposal that ergative languages license the transitive subject *in situ*. However, an important difference with the present account is the treatment of Ergative case. In Ura's analysis, this case is directly assigned to the transitive subject in its theta-position, whereas under the present analysis, Ergative is not considered to be a core case at all. The fact that the pronominal suffixes in Basque do not contain *–k* supports the latter idea, and this seems to be true for other languages with absolutive/ergative marking on the verb as well.

[16] But see (13b), where it is suggested that the original pronoun should have contained *-da*. Hualde and Ortiz de Urbina (2003, p. 213) argue that=*da*=, referring to a first person singular indirect object, is an allomorph of =*t*, and that the latter is the phonologically reduced form when=*da* appears word-finally.

zaitu '2SG.PRS.have(=3SG(A))' with *zarete* '2PL.PRS.be' and *zaituzte* '2PL.PRS.have(=3SG(A))' (cf. 10)). This would mean that the actual third person plural transitive subject form would be =*∅-te* '3-PL(A)'.

The table in (12) lists all the independent pronouns (unmarked form) together with the agreement prefixes and the transitive subject suffixes, as they would appear on verbs that occur without an auxiliary. There are no differences between these affixes and the ones listed in (10) and (11). Examples of verbs that (may) occur without an auxiliary are *etorri* 'come, arrive', *joan* 'go', *egon* 'stay', *ekarri* 'bring', *jakin* 'know', *eraman* 'take'. This class of verbs is relatively small, but it was bigger in earlier stages of the language (Laka, 1993a, p. 29).

(12) *Cross referencing subject and direct object on non-auxiliaries in Basque*

	agreement (ABS)	independent pronoun	transitive subject suffix
1SG	*n-*	*ni*	=*t*
2SG.FAM.MASC	*h-*	*hi*	=*k*
FEM	*h-*	*hi*	=*n*
2SG	*z-*	*zu*	=*zu*
3SG	*d-, z-*	*hura*	=*∅*
1PL	*g-*	*gu*	=*gu*
2PL	*z-*	*zuek*	=*zue*
3PL	*d-, z-*	*haiek*	=*∅-te*

Ditransitive verbs take an additional argument with Dative case. This argument is also cross referenced on the verb, as can be seen in the following examples:[17]

(13) a. *Ni-k* *zu-ri* *sagarr-a* *ema-ten*
1SG-ERG 2SG-DAT apple-DEF give-IMP
di=zu=t
3SG.PRS.have=2SG(IO)=1SG(A)
'I am giving you the apple'

[17] Intransitive verbs taking an (optional) dative argument also have indirect object suffixes. These will not be discussed further here, but I do not expect them to be of any problem to my analysis.

b. *Zu-k* *ne-ri* *sagarr-a* *ema-ten*
2SG-ERG 1SG-DAT apple-DEF give-IMP
di=da=zu
3SG.PRS.have=1SG(IO)=2SG(A)
'You are giving me the apple'

c. *Ni-k* *zaldi-a-ri* *sagarr-a* *ema-ten*
1SG-ERG horse-DEF-DAT apple-DEF give-IMP
di=o=t
3SG.PRS.have=3SG(IO)=1SG(A)
'I am giving the apple to the horse'

Independent pronouns and full noun phrases bear Dative case when they function as an indirect object. The auxiliary has a suffix cross referencing this argument, and this suffix always precedes the transitive subject clitic. Note that the vowel of the auxiliary has changed from /u/ to /i/. This is due to the fact that the verb is ditransitive. Hualde and Ortiz de Urbina (2003, p. 210) note that *-i-* is an allomorph of *–ki-*, which appears in other verbs that take an indirect object. Assuming that the structures in (8) and (9) are universal, we cannot allow verbs to agree with indirect objects. In the case of Basque, this means that in ditransitive clauses, the indirect object is also necessarily base-generated as a pronominal element which incorporates into the predicate. Evidence for this analysis comes from the fact that most of the indirect object suffixes are identical to the transitive subject suffixes:

(14) *Cross referencing indirect objects and transitive subjects*

	independent pronoun	indirect object suffix	transitive subject suffix
1SG	*ni*	*=t/=da*	*=t*
2SG.FAM.MASC	*hi*	*=k/=a=*	*=k*
FEM	*hi*	*=n/=na=*	*=n*
2SG	*zu*	*=zu*	*=zu*
3SG	*hura*	*=o*	*=Ø*
1PL	*gu*	*=gu*	*=gu*
2PL	*zuek*	*=zue*	*=zue*
3PL	*haiek*	*=e*	*=Ø-te*

The first three members of the paradigm for indirect objects have two allomorphs, depending on their being word-final or not. The indirect object clitic is word-final if it appears in a clause with a third person singular transitive subject or if it is attached to an intransitive verb. In both cases, it is not followed by any other (overt) marker. I assume with Hualde and Ortiz de Urbina (2003) that these form alternations are determined by phonology (cf. footnote 10). The indirect object form for third person singular, *-o*, differs from the transitive subject marker (empty). This can be explained by the very assumption that the transitive verb is a passive form. In English (and other nonergative languages), passive morphology implies that there is a subject, which is unspecified for person and number. In Basque it could well be that in the case of a third person singular subject, we still have a passive construction as in (9), with empty passive morphology. In absence of any first or second person suffix, or third person plural, the transitive subject is interpreted as third person singular. However, this does not tell us anything about indirect objects. If the incorporation process is standard for indirect objects, for whatever reason, there is no reason why third person singular should be excluded from that.[18]

The trees in (15) show the syntactic derivation of intransitive, transitive and ditransitive sentences in Basque. In an intransitive (unaccusative) sentence, the verb has a single argument which is moved to Spec,TP in order to become licensed (by agreement). The verb itself moves via the Aspectual projection (AspP) to the Auxiliary projection (AuxP), where it joins the auxiliary. The complex formed there moves to Tense, where it enters into a checking relationship with the verbal argument in Spec,TP. This becomes visible as agreement on the auxiliary (15a).

(15)a.

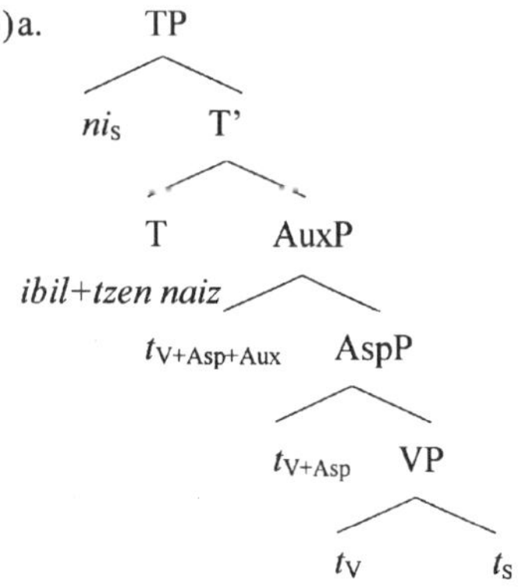

ni ibiltzen naiz 'I am walking'

[18] Note that the indirect object suffix does not show any resemblance with the independent pronoun, as is the case with several transitive subject clitics. I will leave this problem for future research.

In a transitive sentence, VP is extended to *v*P, the head of which is responsible for assigning the external theta-role to the transitive subject (A). Because the verb is passive in nature, this subject is obligatorily realized as a pronoun which incorporates into the predicate in order to be licensed (represented by underlining). This happens after head movement of the verb. The direct object is licensed by agreement again. Finally, a DP bearing oblique (Ergative) case may be adjoined. This DP is coreferential with the pronominal suffix, which ends up at the right of the auxiliary (15b).

(15)b. *zuk ni garbitzen nauzu* 'You are washing me'

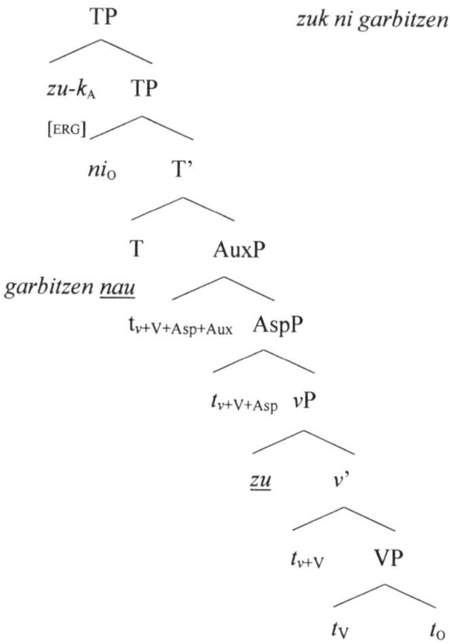

In ditransitive clauses, VP contains two argument positions, the higher one of which is occupied by the indirect object. This argument is obligatorily realized as a pronoun, and it is licensed in the same way as the subject, that is, by incorporation into the predicate. As usual, the direct object is licensed by agreement, and finally two DPs may be adjoined to TP: a Dative one and an Ergative one (15c).

(15)c. *zuk neri sagarra ematen didazu* 'You are giving the apple to me'

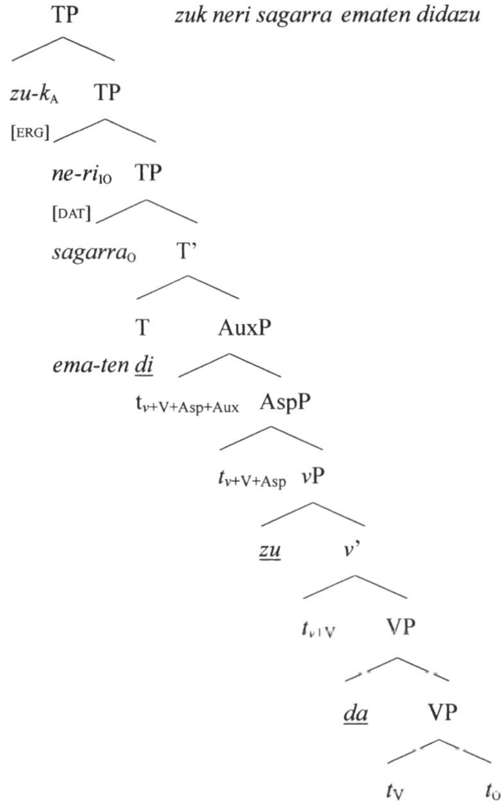

5. Conclusion and Perspectives

Basque differs from languages like English in that it lacks an active transitive construction where the direct object is licensed by structural Accusative case and the subject by agreement (cf. 6a). Instead, transitive verbs are always passive, which, in the case of Basque, means that the external theta-role is assigned to either an (empty) passive morpheme (cf. 9) or to an overt pronoun (cf. (15b,c)). This empty/overt element is licensed by incorporation: it appears as a suffix on the auxiliary or main verb. This suffix is optionally doubled by an adjunct-DP, which carries oblique case (Ergative). This optional DP is comparable to the English *by*-phrase. The direct object is licensed by agreement, just like the subject of an intransitive clause. This results in an ergative pattern, which is present both in the case system and in verbal morphology. Indirect objects are also realized by incorporating pronouns, which explains why the ditransitive auxiliary (or main verb) also carries a person/number marker cross referencing the indirect object. The fact that the DPs in transitive subject and indirect object function are adjuncts accounts for the fact that constituent order in Basque is highly flexible, although the language should not be considered to be strictly non configurational.[19]

The analysis presented here is inspired by the paradigm of prefixes and suffixes: the latter are more closely related to the independent pronouns than the former, which follows from my proposal which treats them as incorporated pronouns instead of agreement markers. However, there are a few exceptions, which need to be accounted for (first person singular, second person singular familiar).

I have argued that the main advantage of my proposal over other accounts of ergativity is that it relates ergative patterns to a construction that occurs independently in nonergative languages: the passive. No parameter is needed that stipulates a characteristic of syntactic structure that is only present in languages with ergativity. But we might still wonder why Basque should lack the active construction at all. The answer to this question is to be found in languages that have a passive construction which is obligatory under certain circumstances. Many native languages of North America cannot form active clauses when the patient outranks the agent on a person hierarchy. Instead, a passive construction has to be used. Something similar is the case in languages with inverse markers. It is imaginable that a passive or inverse construction becomes widely used, eventually taking over the function of the active or direct construction. I assume that such a development has taken place in Basque.

The idea that arguments can be realized by obligatory (clitic) pronouns is not new at all. The Pronominal Argument Hypothesis (Jelinek, 1984) argues that this is the case for every single argument in non configurational languages like Warlpiri.[20] This has repercussions for independent pronouns and full DPs doubling the pronominal arguments: they are adjuncts, and as such they are supposed to be referential (cf. Baker, 1996, 2001, 2003). This is because they are in an A'-position, from which quantifier-variable binding is impossible. For Basque, this would mean that transitive subjects and indirect objects are always

---

[19] See Elordieta (2001) for an overview of the debate on non configurationality in Basque, and for evidence against such an analysis.

[20] In the case of Warlpiri, ergative case marking applies to the adjuncts that double the pronominal arguments. In van de Visser (in preparation), I argue that ergativity is to be found among two types of languages: the Warlpiri type and the Basque type.

referential. Various tests with quantifiers have indicated that this is actually not true (Arantzazu Elordieta, p.c.). However, the incorporated pronouns that are bound by a quantified DP are necessarily third person, and we have seen that these do not correspond to the independent pronouns in a transparent way. Future research should point out what the exact status of these elements is.

The proposal presented here is expected to apply to every language in which person/number markers on the verb show an ergative pattern. The analysis should therefore carry over to Northwest Caucasian and Mayan languages, among others. Within the former, Abkhaz and Abaza provide an interesting case. Independent pronouns in these languages show a gender distinction for third person plural, which is also found in the 'ergative' paradigm (the incorporated pronouns), but not in the 'absolutive' paradigm (the agreement markers). Unlike Basque, transitive subject DPs are not overtly marked with Ergative case, but this might be a coincidence, as this case is overtly present in Kabardian and Adyghe, the sister languages of Abkhaz (Smeets, 1984). In Mayan languages, where there is no case marking, there is much correspondence between the 'absolutive' paradigm and the independent pronouns, whereas the 'ergative' paradigm is often quite different (Robertson, 1980; Penny Brown, p.c.). At first, this seems to be a serious counterexample against my proposal, but there is a plausible explanation for this. It appears to be possible to analyze the independent pronouns as containing a neutral base which is inflected for person and number. This is an instance of agreement, and we do not expect anything else but the 'absolutive' paradigm there.

Acknowledgements

I would like to thank Norbert Corver and Andrej Malchukov for reading and commenting on the draft. Two anonymous reviewers have encouraged me to be more explicit about my ideas. Many thanks are due to these persons. The help of my principal informants Asier Alcazar and Sonia Ortiz de Arri is also greatly acknowledged. The German examples are due to Jenny Audring. All errors are of course my own responsibility.

References

Baker, M. C. (1996). *The polysynthesis parameter*. Oxford: Oxford University Press.

Baker, M. C. (2001). The natures of nonconfigurationality. In: M. Baltin & C. Collins (Eds), *Handbook of contemporary syntactic theory* (pp. 407–438). Oxford: Blackwell.

Baker, M. C. (2003). Agreement, dislocation, and partial configurationality. In: A. Carnie, H. Harley & M. Willie, (Eds), *Formal approaches to function in grammar: In honor of Eloise Jelinek* (pp. 107–132). Amsterdam: John Benjamins.

Baker, M. C., & Hale, K. (1990). Relativized minimality and pronoun incorporation. *Linguistic Inquiry*, *21*, 289–297.

Baker, M. C., Johnson, K., & Roberts, I. (1989). Passive arguments raised. *Linguistic Inquiry, 20*, 219–251.

Bittner, M., & Hale, K. (1996a). The structural determination of case and agreement. *Linguistic Inquiry, 27*, 1–68.

Bittner, M., & Hale, K. (1996b). Ergativity: Toward a theory of a heterogeneous class. *Linguistic Inquiry*, *27*, 531–604.

Bobaljik, J. D. (1993). On ergativity and ergative unergatives. In: J. D. Bobaljik & C. Phillips (Eds), *Papers on case and agreement II* (pp. 45–88). Cambridge, MA: MIT Press.

Chung, S. (1978). *Case marking & grammatical relations in polynesian*. Austin/London: University of Texas Press.

Dixon, R. M. W. (1972). *The dyirbal language of North Queensland*. Cambrigde: Cambridge University Press.

Dixon, R. M. W. (1994). *Ergativity*. Cambridge: Cambridge University Press.

Elordieta, A. (2001). *Verb movement and constituent permutation in Basque*. Utrecht: LOT.

Hale, K. (1970). The passive and ergative in language change: The Australian case. In: S. A. Wurm, D. C. Laycock & C. L. Voorhoeve (Eds), *Pacific linguistic studies in honour of Arthur Capell* (pp. 757–781). Canberra: Linguistic Circle of Canberra.

Hoekstra, T. A. (1986). Passives and participles. In: F. Beukema & A. Hulk (Eds), *Linguistics in the Netherlands 1986 (2)* (pp. 95–104). Dordrecht: Foris.

Hualde, J. I., & Ortiz de Urbina, J. (2003). *A grammar of Basque*. Berlin/New York: Mouton de Gruyter.

Jaeggli, O. (1986). Passive. *Linguistic Inquiry*, *17*, 587–622.

Jakobson, R. (1936). Beitrag zur Allgemeinen Kasuslehre. In: *Selective writings 2: Word and language* (pp. 23–71). The Hague: Mouton, 23–71.

Jelinek, E. (1984). Empty categories, case and configurationality. *Natural Language & Linguistic Theory, 2*, 39–76.

Johns, A. (2000). Ergativity: A perspective on recent work. In: L. Cheng & R. Sijbesma (Eds), *The first 'Glot International' state-of-the-article book: The latest in linguistics* (pp. 47–73). Berlin: Mouton de Gruyter.

Laka, I. (1993a). The structure of inflection: A case study in X^0 syntax. In: J. I. Hualde & J. Ortiz de Urbina (Eds), *Generative studies in Basque linguistics* (pp. 21–70). Amsterdam/Philadelphia: John Benjamins.

Laka, I. (1993b). Unergatives that assign ergative, unaccusatives that assign accusative. In: J. D. Bobaljik & C. Phillips (Eds), *Papers on case and agreement I* (pp. 149–172). Cambridge, MA: MIT Press.

Laka, I., & Uriagereka, J. (1987). Barriers for Basque and viceversa. In: J. McDonough & B. Plunkett (Eds), *Proceedings of NELS 17* (*Vol. 2,* pp. 394–408). Amherst, MA: GLSA.

Levin, J., & Massam, D. (1985). Surface ergativity: Case/theta relations reexamined. In: S. Berman (Ed.), *Proceedings of NELS 15* (pp. 286–301). Amherst MA: GLSA.

Marantz, A. (1981). *On the nature of grammatical relations*. Unpublished doctoral dissertation, MIT, Cambridge, MA.

Marantz, A. (1984). *On the nature of grammatical relations*. Cambridge, MA: MIT Press.

Murasugi, K. G. (1992). *Crossing and nested paths: NP movement in accusative and ergative languages*. Unpublished doctoral dissertation, MIT, Cambridge, MA.

Neeleman, A., & Weerman, F. (1999). *Flexible syntax: A theory of case and arguments*. Dordrecht: Foris.

Newmeyer, F. J. (1998). *Language form and language function*. Cambridge, MA: MIT Press.

Ortiz de Urbina, J. (1989). *Parameters in the grammar of Basque*. Dordrecht: Foris.

Rebuschi, G. (1989). Is there a VP in Basque? In: L. Maràcz & P. Muysken (Eds), *Configurationality: The typology of asymmetries* (pp. 85–116). Dordrecht: Foris.

Robertson, J. S. (1980). *The structure of pronoun incorporation in the Mayan verbal complex*. New York/London: Garland.

Saltarelli, M., Azkarate, M., Farwell, D., & Ortiz de Urbina, J. (1988). *Basque*. New York: Croom Helm.

Smeets, H. J. (1984). *Studies in West Circassian phonology and morphology*. Leiden: The Hakuchi Press.

Trask, R. L. (1979). On the origins of ergativity. In F. Plank (Ed.), *Ergativity: Towards a theory of grammatical relations* (pp. 385–404). London. Academic Press.

Ura, H. (2000). *Checking theory and grammatical functions in universal grammar*. New York: Oxford University Press.

van de Visser, M. A. (in preparation). *The marked status of ergativity*. Dissertation, Utrecht University, Utrecht.

Chapter 8

Noun Phrase Resolution: The Correlation between Case and Ambiguity

Peter de Swart

1. Introduction

Upon hearing a sentence like *the man sees the woman*, we as hearers have to find the right interpretation for this sentence. One important step in this process is to find out which argument is the subject of the sentence and which one the object. For English this seems to be a trivial task most of the time. However, this is not the case for every language. In a recent paper de Hoop and Lamers (to appear) argue that the process of assigning the right grammatical function to the right argument and hence to avoid ambiguity of grammatical relations is guided by a universal constraint DISTINGUISHABILITY: The two arguments of a transitive relation should be distinguishable. In their view the distinguishability of arguments can be assessed on at least five different grounds: (i) morphological (case) marking; (ii) prominence of the arguments: the subject is higher in prominence than the object; (iii) word order: e.g. for English subject precedes object; (iv) verbal information: argument selection and argument structure; and (v) agreement.

In this chapter I investigate how ambiguities of grammatical function can arise and more importantly how they can be avoided. The most natural domain of investigation is transitive constructions with two arguments with equal features concerning person, animacy, definiteness, and the like. By illustration, if we have a situation comprising a seeing event with a man and a woman involved, both participants are equally qualified to be the subject of the sentence. The situation

Competition and Variation in Natural Languages: The Case for Case
Edited by M. Amberber and H. de Hoop

ISBN: 0-08-044651-5

changes when we replace 'the man' with 'a book.' Now, only 'the woman' can make a good subject, because 'the book' due to its animacy feature is simply ruled out as the subject of a seeing event. Therefore, in the first case an ambiguity of grammatical function might be at stake, whereas in the second case this is ruled out by the semantics. In this study I limit myself to the first situations: those cases in which the semantics does not determine a fixed interpretation. Of course since we are dealing with natural language we are not only talking about meaning but also about form. Even if the semantics allows for an ambiguous interpretation of a sentence, it is still possible that only one such interpretation is in accordance with the form of the utterance while the other is simply ruled out. That is, if two arguments are not distinguishable on the basis of their semantics they might be distinguishable on basis of some morphological (or morphosyntactic) feature (in line with the DISTINGUISHABILITY constraint)

In this respect, it is most interesting to contrast languages with morphological case with languages lacking morphological case. In the first type of languages the grammatical function can be simply read off the noun phrase in most cases. For instance, in a language with an accusative case system, the nominatively marked argument is generally interpreted as the subject whereas the accusative marked one is considered the object. In the second type of languages the grammatical function of an argument is not marked on the argument itself but often by means of agreement markers on the verb. Since the seminal paper by Nichols (1986) the first type of languages is often referred to as dependent-marking and the second as head-marking. Consider the examples in (1) and (2) (taken from Nichols, 1986) to illustrate this distinction. The same proposition is expressed first in Abkhaz, a head-marking language, and subsequently in the dependent-marking language Chechen.[1]

(1)	*a-xàc'a*	*a-pħ°əs*	*a-š°q°'ə̀*	*Ø-lə̀-y-te-yt'*
	the-man	the-woman	the-book	it-to.her-he-gave-FIN

'The man gave the book to the woman'

(2)	*cu*	*stag-a*	*zudčun-na*	*kni:ga*	*jelira*
	DEM.OBL	person-ERG	woman-DAT	book.ABS	gave

'The man gave the book to the woman'

[1] The abbreviations used in this chapter are: 3=third person; ABS=absolutive; ACC=accusative; AT=agent topic; AUX=auxiliary; CP=completive aspect; DAT=dative; DECL=declarative ending; DEM=demonstrative; DET=determiner; ERG=ergative; FIN=finite; IMP=imperfect; INC=incompletive aspect; IND=indicative; INTR=instransitive; INV=inverse; NOM=nominative; OBJ=object; OBL=oblique; OBV=obviative; PAST=past tense; PL=plural; PRES=present; Sg=singular; Su=subject; TR=transitive

The two sentences (1) and (2) are similar with respect to the syntactic relations that are assigned to the arguments. In both sentences 'man' is the subject, 'woman' the indirect object, and 'book' the direct object. Even so in both sentences the predicate is the head of the construction and the nouns are its dependents. The two languages, nevertheless, diverge in the way these relations are formally realized in the construction. In (1) the grammatical relations are marked on the predicate, the head of the construction, by means of affixes. If we were only to look at the nouns in the Abkhaz sentence we would not be able to figure out which noun is the subject, which one the direct object, and which one the indirect object. This is, on the other hand, very well possible in the Chechen construction in (2) in which these grammatical relations are marked on the nouns, i.e. the dependents, by means of case markers. By looking at the ergative marked noun *staga* we know that we have to assign this argument the function of subject independently of the functions of the other nouns or the type of verb present in the sentence.

In this chapter I investigate the correlation between the presence of morphological case on the one hand and the danger of ambiguity of grammatical relations on the other hand. I will show that in languages with morphological case danger of ambiguity of grammatical function is virtually absent due to the fact that in such languages the grammatical function is encoded in the noun phrase. Conversely, in languages lacking morphological case we expect potential ambiguity of grammatical relations to be lurking in constructions containing two 3rd person participants (as discussed above). This ambiguity is due to the nature of the language because of which grammatical function cannot be read off the argument. Therefore, we expect languages of the last type to reach for other means to avoid this potential ambiguity.

The chapter is structured in the following way: In the examination of structures with two 3rd person arguments I start with constructions containing one pronominal argument and one nominal argument (Section 2). In some languages we find the so-called *one nominal interpretation* constraint which states that in the case of a transitive sentence containing two 3rd person arguments of which only one is a full noun phrase (NP), this NP has to be interpreted as the direct object of the sentence. I argue that this constraint exists in order to reduce potential ambiguity of grammatical relations and furthermore that it is the way it is due to principles that structure discourse. After this section I turn to transitive constructions containing two 3rd person arguments, which are both expressed as full NPs. In Section 3 we first look at languages lacking morphological case. As stated above these languages suffer a greater risk of ambiguity of grammatical function and we will see that such languages reach for other means to avoid such ambiguities. But also in languages that do employ morphological case we can find situations in which ambiguity of grammatical relations is at stake (Section 4). A well-known example of this is provided by languages that exhibit *word order freezing*: in the absence of differentiated morphological case the variation in word order is reduced to one strict ordering. Furthermore, it is argued that in some languages the phenomenon of *differential object marking* can be explained in terms of ambiguity avoidance.

2. One Nominal Interpretation and Preferred Argument Structure

In this section I discuss a constraint on the interpretation of transitive constructions with two 3rd person arguments of which only one is a full noun phrase. This constraint is, for instance, found in the Salish languages spoken on the Northwest coast of North America. It is often referred to as the *one nominal interpretation* (ONI) constraint and was formulated by Gerdts (1988) in the following way:

(3) *One nominal interpretation:*
In the absence of marking for other persons, a single 3rd person nominal is interpreted as the absolutive.[2] (Gerdts, 1988, p. 59)

The example from Lushootseed — a Coast Salish language — in (4) illustrates this constraint nicely (example taken from Hess, 1973).

(4) *ʔu-č'áxʷa-t* *ti* *č'ač'as*
CP-club-TR DET boy
'He clubbed the boy' **not**: 'The boy clubbed him'

In (4) we see a transitive predicate with two 3rd person arguments, both expressed with a zero morpheme on the verb, and one full NP. The only feasible interpretation is one in which the nominal is assigned the role of the undergoer of the clubbing event, i.e. *ti č'ač'as* is the direct object. An interpretation in which the nominal is assigned the grammatical function of subject is ruled out.

This constraint on nominal interpretation has such a strong effect that even in pragmatically odd situations it cannot be overruled (cf. Gerdts, 1988):

(5) !!*niʔ* *q'ʷəl-ət-əs* *ɬə* *sɬeniʔ*
AUX bake-TR-3.Sg.Su DET woman
'He baked the woman' **not**: 'The woman baked it'

[2] Gerdts (1988) analyzes Halkomelem as a split ergative language in which 3rd person transitive subjects (A) are marked with ergative *-əs* and transitive object (P) and intransitive subject (S) are unmarked (=absolutive). We can thus reformulate the constraint as stating that the single NP in transitive sentences with two 3rd person arguments has to be interpreted as the direct object (P).

The Salish languages are not the only languages that have an ONI restriction that assigns the object function to the single NP. The same restriction has been reported for Navajo (Woolford, 1986) and Chamorro (Chung, 1998). An example from the former language is given in (6).

(6) *at'ééd* *yizts'qs*
girl 3.kissed.3
'He kissed the girl' **not**: 'The girl kissed him'

In this section I take up the question why this constraint assigns to the full NP the object function and not the function of subject. I argue that an explanation can be found in the principles that influence the way discourse is organized. In general we can say that topical arguments tend to go in subject position and have zero morphological realization. This means that the full NP is not likely to be associated with the function of subject and thus has to be interpreted as the object. We can state that ONI is the grammaticalization of what we find most in discourse.

First of all we can view the ONI restriction as a device to reduce potential ambiguity of grammatical relations. In head-marked constructions like the one in (5) the verb tells us that there are two 3rd person arguments involved in the situation denoted by the verb. However, there is nothing in the sentence which tells us to which 3rd person argument the full NP has to be linked. In principle, it is thus possible to link this 3rd person nominal either to the subject or the object function. Without a further rule like ONI such constructions are ambiguous with respect to the grammatical functions involved and everytime a hearer encounters such a construction he has to figure out what function has to be assigned to the nominal. However, for a hearer it is much easier to assign a fixed function to the single nominal. That is, to know that whenever you come across a sentence containing only one nominal you have to interpret this NP as the object. It is in some sense very efficient to have a rule that determines the interpretations of single NP constructions in a (standard) straightforward way.

The second question then is: if we want to have a fixed interpretation for sentences containing a single nominal, how are we going to decide which interpretation we standardly assign? Put differently, what reason do we have to link this single NP to the function of direct object? In the light of this second question I discuss briefly a study of Du Bois (1987) on the structuring of discourse in the Mayan language Sacapultec. In his analysis of several stories Du Bois (1987) distills two structural tendencies: (i) there are hardly any sentences with two full NPs; (ii) the full NPs that do appear are mostly found in the function of transitive direct object (P) or intransitive subject (S) but hardly ever in the function of transitive subjects (A). Du Bois (1987) uses the term Preferred Argument Structure to refer to these tendencies. Restated for transitive constructions we can say that full NPs almost always express direct objects. Complementing his structural analysis, Du Bois (1987) also investigated the pragmatic status of the constructions in his corpus. Interestingly he found pragmatic restrictions that mirror the structural tendencies he attested: (i) more than one new argument per clause is avoided; (ii) transitive subjects (As) that express new information are avoided. In sum transitive constructions are constrained by the fact that both maximally one full NP argument and maximally one new argument is allowed. Furthermore, the transitive subject is preferred to be neither a full NP nor one that expresses new information. We thus see that direct objects are preferably realized as full NP that express new information and transitive subjects as pronominal elements that express given information. This is in correspondence with the observations in Chafe (1976) about the relations between givenness, definiteness, and topicality on the one hand and grammatical function on the other. Table 1 shows cross-linguistic tendencies concerning the features of subjects and objects.

Thus 'prototypical' subjects of transitive sentences are definite and animate, they correspond to given information and are topic. Most typically they are realized as pronouns. For objects, the cases are less clear cut. We cannot straightforwardly state that 'prototypical' objects are the opposite of 'prototypical' subjects, i.e. indefinite, inanimate, new, and pronominal (see Næss, 2004 for discussion of the notion 'prototypical object'). We can only formulate the implicature that if an argument is indefinite, inanimate, new and/or a nominal it will most likely fulfill the role of direct object.

The correlation between topicality and reduced NP expression (zero >> agreement >> pronoun >> full NP) has also been discussed by Givón (1990) (see also the recent OT implementation by Bresnan, 2001). Following his term *referential iconicity* we can state that the following generalizations hold (where '↔' stands for 'correlates with'):

(7) a. A ↔ topic ↔ pronominal (zero, agreement, weak pronouns)
b. P ↔ new/comment ↔ full NP

Thus a topical argument correlates on the one hand with the function of subject and on the other hand with pronominal expression. New information, however,

Table 1: Properties of subjects and objects.

A (transitive subject)	**P (direct object)**
given	± new
pronominal	± nominal
topic	comment
+ definite	± definite
+ animate	± animate

correlates with the function of object on the one hand and full NP expression on the other. Given this referential iconicity principle it is not surprising that a restriction such as the ONI constraint comes about. When confronted with a construction containing one pronominal argument and one nominal argument it is most natural to interpret the pronominal element as the subject and the full NP as the object of the transitive verb. This interpretation constraint is simply the result of what we find most often in everyday language use. In other words ONI seems to be a grammaticalization of discourse principles. This grammaticalization assists us in avoiding the ambiguity of grammatical relations which would otherwise be at stake in these constructions.

3. NP Ambiguities in Caseless Languages

In the previous section I examined constructions containing only one full NP. In this section I turn to constructions with two full noun phrases as we find them in languages without morphological case. As stated in the introduction this type of languages is more sensitive to ambiguity of grammatical function due to the fact that the grammatical function of an argument is not expressed on the argument itself. Given that the function of the two 3rd person arguments is only marked on the predicate, and the observation that word order in most head-marking languages is relatively free, it is not surprising that it is not always immediately clear which NP fulfills which grammatical function. This is, for instance, what we find in the Mayan language Tzutujil (Dayley, 1985). In Tzutujil, as in most other Mayan languages (cf. England, 1991), word order is relatively free and can be altered under discourse considerations such as topicality and focus structure. Dayley (1985) reports that in Tzutujil transitive constructions with two NPs that share definiteness features are ungrammatical.

(8)	**X-uu-ch'ey*	*jun*	*ixoq*	*jun*	*aachi*
	INC-3.ERG-hit	DET	woman	DET	man
	'A man hit a woman'				

Duncan (2003), on the other hand, found in his research with his consultants that sentences with two NPs of equal definiteness do not result in ungrammaticality but rather in ambiguity:[3]

[3] Tzutujil Maya is spoken in villages surrounding lake Atitlán in Guatemala. Dayley's (1985) research is based on dialects spoken in villages to the north of lake Atitlán, whereas Duncan (2003) describes a variant spoken to the southwest of the lake.

(9)	*X-uu-tz'et*	*jun*	*tz'i'*	*jun*	*miix*
	INC-3.ERG-see	DET	dog	DET	cat

'A cat saw a dog' **or** 'A dog saw a cat'

Thus, within the same language, similar constructions are judged differently. Some speakers allow the ambiguity that arises from transitive constructions like the one in (8) and (9), whereas others clearly do not. What is important my analysis is to recognize that the ungrammaticality of (8) is fed by the danger of ambiguity that lurks in constructions with two NPs. This potential ambiguity itself is fed by the fact that in head-marking languages grammatical function cannot be read off the NP. In order to reduce potential ambiguities to a minimum and to take the success of communication to a maximum, restrictions as the ones in (8) come about. In the remainder of this section I discuss four types of strategies to deal with potential ambiguities found in the languages of the world: positional linking, argument hierarchies, exclusion of ambiguous constructions, and reliance on discourse.[4]

3.1. Positional Linking

The first strategy I discuss is what we can call *positional linking*. Consider the following example from Malagasy (Maclachlan, 2002):

(10)	*Mamono*	*ny*	*akoho*	*amin'ny*	*antsy*	*ny*	*mpamboly*
	AT.kill	DET	chicken	with'DET	knife	DET	farmer

'The farmer kills the chicken with a knife'

In Malagasy, the functions of subject and object are expressed by the position of the arguments. Final position corresponds to the function of subject and medial position preceding prepositional phrases is reserved for direct objects. Maclachlan (2002) identifies the subject position with nominative case and the object position with accusative. Whether or not this is felicitous, what matters to us is that due to this positional linking no ambiguity whatsoever arises. Even though Malagasy does not mark grammatical relations with overt morphology on the head or the dependents, hearers can still figure out which NP is the subject and which one the object.

3.2. Argument Hierarchies

Another strategy to reduce potential ambiguity is by setting up rigid *argument hierarchies* (see, for instance, Aissen, 1997). This is a strategy used by many languages

[4] I only discuss a subset of possible restrictions. See Nichols (1986) for a more complete list.

and most often applied to one of the dimensions of animacy and definiteness. Here I discuss, however, another type of hierarchical ranking. Namely, that found in languages that make a distinction between obviative and proximate morphology, such as found, for instance, in the Algonquian languages. As noted by Aissen (1997) "[t]his distinction is relevant only to 3rd person nominals, and only in contexts containing nominals which refer to more than one 3rd person participant."

Kutenai, a language isolate of British Columbia and adjacent areas of Montana and Idaho, is equipped with obviative morphology in combination with an inverse system. Consider the examples from Dryer (1994).

(11) *wukat-i* *pałkiy-s* *titqat'*
see-IND woman-OBV man
'The man sees the woman'

(12) *wukat-aps-i* *titqat'-s* *pałkiy*
see-INV-IND man-OBV woman
'The man sees the woman' (**or** 'The woman was seen by the man')

When we compare (11) and (12) we see two differences. First, in (11) the noun *woman* is marked with the obviative marker *–s,* whereas in (12) it is *man* which carries this marking. Secondly, the verb in (12) occurs with a so-called inverse marker (*-aps*), whereas in (11) this marker is absent and we say that the verb is in its direct form. Now let us turn to the interpretation. When we use the direct verb form the noun marked with the obviative marker is interpreted as the object of the clause (cf. (11)). However, when we use the inverse form of the verb the obviative noun is interpreted as the subject (cf. (12)). Due to these markings a hearer of a Kutenai sentence with two overt NPs can figure out which one is to be interpreted as the subject and which one as the object. Without such marking it would have been impossible. How does this relate to the setting up of hierarchies? We can establish for this language the hierarchy proximate > obviative. As Aissen (1997) puts it "If the direct form is used, then the higher-ranked participant [in the hierarchy] is linked to subject and the lower-ranked to object. The inverse form indicates the reverse — that the higher-ranked participant is object, and the lower-ranked subject."

Although obviation marking pertains not only to heads but also to dependents it is still clearly a strategy used by head-marking languages, together with direction marking, to resolve a potential ambiguity.

3.3. (Total) Exclusion of 3–3 Constructions

In the Salish languages of the Northwest coast of North America, sentences with two direct nominal arguments are almost entirely excluded from discourse.[5] Even in elicited speech such constructions are dispreferred, according to Jelinek (1996) because the order of the direct arguments is completely free. Instead speakers prefer a passive construction to express a proposition containing two direct nominal arguments.[6] Compare the Lummi — a Coast Salish language — example in (13) with the one in (14) below.

(13) *ʔkʷəniŋ-t-s* *sə* *sɬeniy'* *cə* *swey'qə'*
help-TR-3Sg.Su DET woman DET man
'The woman helped the man' **or** 'The man helped the woman'

(14) *kʷəniŋ-t-ŋ* *'ə* *sə* *sɬeniy'* *cə* *swey'qə'*
help-TR-INTR OBL DET woman DET man
'The man was helped by the woman' **not:** 'The woman was helped by the man'

Lushootseed, another Coast-Salish language, takes the restriction on active constructions with two direct nominal arguments even further as such sentences are deemed ungrammatical, cf. (15) from Hukari (1976).[7]

(15) **ʔu-kʷaxʷ-ad tə* *stubš* *cə* *sɬadəyʔ*
CP-help-TR DET man DET woman
'The man helped the woman'

By excluding the active constructions with two overt NPs and 'replace' them with a passive construction in which the agent argument is clearly signalled by an oblique marker, these languages find a way of getting round a possible ambiguity of grammatical relations. As with the Tzutujil examples cited at the beginning of this section we see again that closely related languages might differ slightly in how much ambiguity remains. Lushootseed clearly does not tolerate any ambiguity whereas Lummi has a preference for excluding it but not to the total extent.

[5] Kinkade (1983) argues that constructions with two direct nominal arguments are the result of English influence: "Probably no Salishan language allowed two direct complements aboriginally, but the restriction has been relaxed due to English language influence."

[6] No general agreement exists in the literature to characterize these constructions as passive (see, e.g. Gerdts, 1988; Thompson & Thompson, 1992; Thomason & Everett, 1993, for discussion). Formally they resemble passive to a large extent, i.e., the verb is detransitivized and the agent is marked as an oblique phrase. The patient is however not always cross-referenced with subject agreement but in many languages with transitive object agreement (see Wiltschko, 2001). Functionally these constructions do not always behave as an expected passive would (see Jacobs, 1994). Bearing these considerations in mind I will nevertheless use the term passive in this chapter to refer to constructions as the one in (14).

[7] The same restriction has been reported by Blake (1997) for the related Salish language Sliammon (Mainland Comox).

3.4. Leave it All to Discourse

So far we have seen a variety of strategies by which languages reduce the danger of ambiguity in double NP constructions. Languages differ as to which and how many types of constructions with two NPs are excluded but most of them do not go as far as Lushootseed (cf. (15) above) by simply excluding all such constructions. This means that some potential ambiguity may still arise and has to be resolved in some way or other. This disambiguation task will be taken up in most cases by discourse information (similar to that discussed in Section 2). Given considerations such as which element is discourse topic or which one is in focus, hearers can deduce the right grammatical function for the right argument.

An extreme case in point is found in the language Tohono O'odham spoken in Arizona and northern Mexico (Miyashita, Demers, & Ortiz, 2003). In this language two 3rd person arguments expressed by full NPs cannot be distinguished on the basis of agreement, case, word order or the like. Even intonation cannot help the hearer in distinguishing the two arguments. Thus the sentence in (16) is completely ambiguous.

(16)	*Huan*	*'o*	*g*	*Husi*	*ka:*
	Huan	3.Su.IMP	DET	Husi	hear

'Huan is hearing Husi' **or** 'Husi is hearing Huan'

That is, out of context both a subject initial and an object initial reading is possible. In order to find out which argument is the subject and which one the object we would have to rely totally on the context in which the sentence is uttered.

In this section we looked at transitive constructions with two 3rd person arguments in caseless languages. In these languages ambiguity of grammatical relations assignment to NPs is at stake due to the way these relations are formally realized. We saw that this type of languages reach for other means to resolve the possible ambiguities of grammatical function. In the next section I examine languages that do have case marking on their NPs. We will see that ambiguities of grammatical relations are absent in most case systems and discuss how such ambiguities are resolved when they do arise.

4. NP Ambiguities in Languages with Morphological Case: Freezing and Differential Object Marking

In the previous section we saw that in languages without case marking on nominal phrases, a potential ambiguity of grammatical relations is at stake. The question arises then if the same observation holds for languages that do have nominal case. Consider the following example from German.

(17) *Den Zaun* *hat* *der Mann* *zerbrochen*
[the fence].ACC has [the man].NOM broken
'The fence, the man broke'

In German, normally the subject precedes the object, but although in (17) this is not the case, no ambiguity arises due to the overt marking of subject and object with nominative and accusative case respectively.

In a recent paper, de Hoop and Lamers (to appear) argue that the process of assigning the right grammatical function to the right argument is guided by a universal constraint DISTINGUISHABILITY: The two arguments of a transitive relation should be distinguishable. In their view, the distinguishability of arguments can be assessed on at least five different grounds: (i) morphological (case) marking; (ii) prominence of the arguments: the subject is higher in prominence than the object; (iii) word order: for German subject precedes object; (iv) verbal information: argument selection and argument structure; and (v) agreement.

In the preceding section we saw what languages come up with when one or more of these clues for distinguishability are absent, i.e. case or word order. In the remainder of this section I examine two phenomena related to ambiguity avoidance in languages that do have morphological case: word order freezing and differential object marking. In the first case, we see what happens when the distinguishing factor of case is neutralized and in the second one we clearly see the relation between case marking and ambiguity avoidance.

4.1. Word Order Freezing

In our example in (17), case marking, prominence and verbal information all favor *der Mann* as the subject, whereas word order favors *den Zaun* to fulfill that function. Agreement is neutral in this sentence in that it is compatible with both arguments. In this example, case is the deciding factor in assigning the subject function to the nominative argument. Now compare the two sentences in (18) and (19) below:

(18) *Der Lehrer* *hat* *den Jungen* *geschlagen*
[the teacher].NOM has [the boy].ACC hit
'The teacher hit the boy'

(19) *Die Professorin* *hat* *die Studentin* *geschlagen*
[the professor].NOM/ACC has [the student].NOM/ACC hit
'The professor hit the student'

In (18) again case is the deciding factor supported by word order in linking the nominative marked argument to the subject function. None of the factors, prominence, agreement, and verbal information can make a distinction between the two arguments. Similar in (19). In this example, four of the five factors of DISTINGUISHABILITY do not favor one argument over the other. We have to rely solely on word order, in order to come to the interpretation of this sentence. Given that no additional contextual or intonational information intervenes, this sentence gets the interpretation such that 'the professor' is the agent and 'the student' the patient of the hitting event. The fact that word order determines the linking of arguments in situations in which case or other morphological information cannot help us, is observed in many languages with overt case marking on nominals and is known in the literature as *word order freezing*. The connection between ambiguity avoidance and word order freezing is well established in the linguistic literature, especially in the Slavic literature, initiated with the work by Jakobson on Russian word order (Jakobson, 1984).

Korean is also a language that exhibits word order freezing (Lee, 2001). In informal colloquial speech nominative and accusative case markers are often dropped in Korean. As a result such sentences are always interpreted as SOV, whereas different word orders are permitted when the case markers are present. Thus the following example in (20) can only have an interpretation in which *Mary* is chased by *a dog* and not the other way round.

(20) *Ku* *kay* *Mary* *ccoch-nun-ta*
that dog Mary chase-PRES-DECL
'That dog is chasing Mary' **not:** 'Mary is chasing that dog'

We clearly see that in cases where case marking as a ground for distinguishing grammatical function is neutralized due to case syncretism as in the German case or simply due to the drop of case markers as in colloquial Korean, we find that another clue of DISTINGUISHABILITY, i.e. word order, takes over and helps us in interpreting transitive constructions.

4.2. Differential Object Marking

A special case of interest in the study of the relation between ambiguity and case is formed by languages that exhibit so-called *differential object marking*. In such languages direct objects receive case depending on the semantic features of the object. An example of such a language is the Dravidian language Malayalam. In Malayalam animate objects receive accusative case whereas inanimate ones usually stay unmarked. Nevertheless, there are examples in which inanimate objects do receive accusative case. Consider the following two examples taken from Asher and Kumari (1997).

(21) *kappal* *tiramaalakaḷe* *bheediccu*
ship wave.PL.ACC split.PAST
'The ship broke through the waves'

(22) *tiramaalakaḷ* *kappaline* *bheediccu*
wave.PL.NOM ship.ACC split.PAST
'The waves split the ship'

In De Swart (2003, to appear) it is argued that in Malayalam, accusative case is not dependent on the dimension of animacy per se but rather on a principle that I call *Minimal Semantic Distinctness*: The two arguments of a transitive relation must be minimally semantically distinct and if not, they must be marked in a morphologically distinct way. In this view, (differential) case marking is thus clearly viewed as a mechanism to avoid a potential ambiguity of grammatical relations. This view is strongly supported by the following example from Malayalam.[8]

(23) *tiiyyə* *kuṭil* *naʃippiccu*
fire.NOM hut.NOM destroy-PAST
'Fire destroyed the hut'

Again, we see a sentence with two inanimate arguments but this time accusative case on the object is absent. In this case our knowledge of the world tells us which argument is playing which role. We know that fire tends to destroy huts and that it is very unlikely that the act of destroying is directed at the fire. Since we can infer this from our world knowledge we do not have to mark one argument overtly as the object. Thus in cases in which we would normally not mark the object — since it is inanimate — we find accusative case only in those cases where we want to avoid a potential ambiguity. Consider the following quote from Asher and Kumari's grammar of Malayalam (Asher & Kumari, 1997, p. 204):

> When both subject and object are [-ANIM] and both are in the nominative, their function is determined by semantic features of the verb, since it cannot be determined by word order. However, in the presence of a [-ANIM] subject it is possible for a [-ANIM] object to be in the accusative case to resolve potential ambiguity.

In this respect, differential object marking in Malayalam has a function very similar to word order freezing. However, this is not to say that all DOM systems can be explained in terms of Minimal Semantic Distinctness. Indeed, it is often the case that differentiated object marking occurs independent of a potential danger of ambiguity. In many languages differential object marking is solely

[8] This is not to say that the so-called *distinguishing* function is the only function that can be ascribed to case marking. In addition to the distinguishing function we can also ascribe an *indexing* function to case marking (see, for instance, Malchukov, to appear; de Hoop & Narasimhan, this volume).

determined by object features such as animacy, definiteness, and specificity. An example of a language in which the marking of objects is totally dependent on the specificity feature of the object is Turkish. In this language specific objects get accusative case whereas non-specific ones stay unmarked (see Aissen, 2003 and references therein for more examples and discussion).

Overall, we can conclude that in languages with overt case marking, the chance of potential ambiguities of grammatical relations is very small. Case itself is most of the time a clear indication of grammatical function. Nevertheless, there are cases in which it fails to help us and in those cases stringent restrictions on the order of constituents helps the hearer to determine the right interpretation. In caseless languages, on the other hand, ambiguities are excluded to a lesser extent and a large number of ambiguities remain that have to be resolved by relying on discourse information.[9] A language that comes closest to a sort of freezing effect is Malagasy in which grammatical relations are determined by position in the clause.

5. Conclusion

In this chapter I discussed the restrictions that languages place on transitive constructions with one or more noun phrases. In particular I examined the exclusion of double NP constructions and the obligatory interpretation of the NP in single NP constructions. We saw that languages that do not mark grammatical functions explicitly on their noun phrases have a greater chance of ambiguity of grammatical relations. Due to this danger of potential ambiguity they place restrictions on constructions with noun phrases in order to increase the success of communication. However, not every language uses the same restriction. As we saw, variation exists not only between languages but also within one single language. For some speakers it is fine to utter a sentence in which the grammatical relations are not clearly indicated. By doing so the speaker puts the burden of interpretation with the hearer, who has to deduce the right grammatical function for each argument from other information such as the previous discourse. We find a trade off between the effort the speaker has to make in order to produce a sentence on the one hand and effort the hearer has to make to come to the right interpretation on the other.

In languages that have morphological case the danger of ambiguity of grammatical relations is less likely to occur. Morphological case in itself is a means of ambiguity avoidance: through its case marking the function of a noun phrase is

[9] In order to fully substantiate these conclusions more research has to be done to reveal the rules and functions of word order both in languages with and languages without morphological case.

recoverable from the NP itself. This does, however, not mean that ambiguity of grammatical relations never occurs in languages with morphological case. If, for instance, these case markers disappear in casual speech or because of case syncretism, languages tend to place a strong restriction on the possible word order. Usually only one such possible order remains in order to secure one single interpretation. That is, if the major clue for the deduction of grammatical relations disappears, the speaker gives the hearer a new way of figuring out the right interpretation of the sentence.

The relation between morphological case and ambiguity is probably most clear from languages with differential object marking. In those languages case marking is employed when a potential ambiguity of grammatical relations is at stake. In the case we can infer from our knowledge of the world which participant is the subject and which one the object, it is not necessary to employ case marking. In these languages a trade off seems to exist between ambiguity avoidance on the one hand and the use of morphological marking on the other (cf. Aissen, 2003). If speakers do not have to use case marking because there is no danger of ambiguity of grammatical relations, they will indeed refrain from using it.

Acknowledgements

Part of the research for this chapter was carried out during a visit at the Linguistics Department of the University of British Columbia in the fall and winter of 2003/2004. This stay was financed by a travel grant from the VSB fund for which I am very grateful. Further financial support came from the Netherlands Organisation of Scientific Research (NWO) (grant 220-70-003, principal investigator Helen de Hoop). I am very grateful to Helen de Hoop for the discussion that led to this chapter and to Andrej Malchukov for his comments on an earlier version. Furthermore, many thanks go to the two anonymous reviewers for their constructive criticism. Finally, I would like to thank Mengistu Amberber and Sander Lestrade for their help during the final preparation of this chapter. All errors remain with the author.

References

Aissen, J. (1997). The syntax of obviation. *Language, 73* (4), 705–750.

Aissen, J. (2003). Differential object marking: Iconicity vs. economy. *Natural Language and Linguistic Theory, 21*, 435–483.

Asher, R. E., & Kumari, T. C. (1997). *Malayalam*. London/New York: Routledge.

Blake, S. (1997). Another look at passives in Sliammon (Salish). *International Conference on Salish and Neighboring Languages, 32*, 86–143.

Bresnan, J. (2001). The emergence of the unmarked pronoun. In: G. Legendre, J. Grimshaw & S. Vikner (Eds), *Optimality-theoretic syntax* (pp. 113–142). Cambridge, MA: The MIT Press.

Chafe, W. L. (1976). Givenness, contrastiveness, definiteness, subjects, topics, and point of view. In: C. N. Li (Ed.), *Subject and topic* (pp. 25–56). New York/San Francisco/London: Academic Press.

Chung, S. (1998). *The design of agreement: Evidence from Chamorro*. Chicago/London: The University of Chicago Press.

Dayley, J. P. (1985). *Tzutujil grammar, University of California Publications in Linguistics,* Vol. 107 Berkeley/Los Angeles/London: University of California Press.

Dryer, M. S. (1994). The discourse function of the Kutenai inverse. In: T. Givón (Ed.), *Voice and inversion* (pp. 65–99). Amsterdam/Philadelphia: John Benjamins Publishing Company.

Du Bois, J. W. (1987). The discourse basis of ergativity. *Language, 63*(4), 805–855.

Duncan, L. (2003). The syntactic structure of Tzu'utujil Maya. In: M. Butt & T. H. King (Eds), *Proceedings of the LFG03 conference* (pp. 164–183). Stanford: CSLI Publications.

England, N. C. (1991). Changes in basic word order in Mayan languages. *International Journal of American Linguistics*, *57*, 446–486.

Gerdts, D. B. (1988). *Object and absolutive in Halkomelem Salish*. New York: Garland Publishing Inc.

Givón, T. (1990). *Syntax: A functional-typological introduction* (Vol. II). Amsterdam/Philedelphia: John Benjamins Publishing Company.

Hess, T. M. (1973). Agent in a Coast Salish language. *International Journal of American Linguistics, 39*, 89–94.

de Hoop, H., & Lamers, M. (to appear). Incremental distinguishability of subject and object. In: L. Kulikov, A. L. Malchukov & P. de Swart (Eds), *Case, valency, and transitivity*. Amsterdam/Philadelphia: John Benjamins Publishing Company.

de Hoop, H., & Narasimhan, B. (this volume). Differential case marking in Hindi.

Hukari, T. E. (1976). Person in a Coast Salish language. *International Journal of American Linguistics, 42* (4), 305–318.

Jacobs, P. (1994). The inverse in Squamish. In: T. Givón (Ed.), *Voice and inversion* (pp. 121–145). Amsterdam/Philadelphia: John Benjamins Publishing Company.

Jakobson, R. (1958/1984). Morphological observations on Slavic declension (the structure of Russian case forms). In: L. R. Waugh & M. Halle (Eds), *Roman Jakobson. Russian and Slavic grammar: Studies 1931–1981* (pp. 105–133). Berlin: Mouton de Gruyter.

Jelinek, E. (1996). Definiteness and second position clitics in Straits Salish. In: A. L. Halper & A. M. Zwicky (Eds), *Approaching second: Second position clitic and related phenomena* (pp. 271–297). Stanford: CSLI Publications.

Kinkade, M. D. (1983). Salish evidence against the universality of 'noun' and 'verb'. *Lingua, 60*, 25–40.

Lee, H. (2001). Markedness and word order freezing. In: P. Sells (Ed.), *Formal and empirical issues in optimality — theoretic syntax*. Stanford: CSLI Publications.

Maclachlan, A. (2002). Optimality and three western Austronesian case systems. In: M. Amberber & P. Collins (Eds), *Language universals and variation* (pp. 155–184). Westport/London: Praeger.

Malchukov, A. L. (to appear). Transitivity parameters and transitivity alternations: Constraining co-variation. In: L. Kulikov, A. L. Malchukov & P. de Swart (Eds), *Case, valency, and transitivity*. Amsterdam/Philadelphia: John Benjamins Publishing Company.

Miyashita, M., Demers, R., & Ortiz, D. (2003). Grammatical relations in Tohono O'odham. In: S. Karimi (Ed.), *Word order and scrambling*. Oxford: Blackwell Publishers.

Næss, Å. (2004). What markedness marks: The markedness problem with direct objects. *Lingua, 114* (9–10), 1186–1212.

Nichols, J. (1986). Head-marking and dependent-marking grammar. *Language, 62*(1), 56–119.

de Swart, P. (2003). The case mirror. Master's thesis, University of Nijmegen. Available at: http://www.ru.nl/pionier/downloads/thecasemirror.pdf.

de Swart, P. (to appear). Case markedness. In: L. Kulikov, A. L. Malchukov & P. de Swart (Eds), *Case, valency, and transitivity*. Amsterdam/Philadelphia: John Benjamins Publishing Company.

Thomason, S. G., & Everett, D. (1993). Transitivity in Flathead. *International conference on Salish and neighboring languages, 28*, 317–344.

Thompson, L. C., & Thompson, M. T. (1992). *The Thompson Language*. University of Montona Occasional Papers in Linguistics 8.

Wiltschko, M. (2001). Passive in Halkomelem and Squamish Salish. *International conference on Salish and neighboring languages, 36*, 319–345.

Woolford, E. (1986). The distribution of empty nodes in Navajo: A mapping approach. *Linguistic Inquiry, 17*(2), 301–330.

Chapter 9

Changes in Case Marking in NP: From Old English to Middle English

Cynthia L. Allen

1. Introduction[1]

Two major questions that have been discussed in the extensive recent literature concerning case marking from a synchronic point of view are: (1) What types of case-marking systems are found in human languages, and how can we explain the apparent non-existence of some conceivable types of case-marking systems? (2) How can we capture the attested typological differences between languages that have fairly rich case morphology and those with no or very little morphological case? It has long been known that case marking in case-rich languages essentially performs the functions of constituent order or adpositions in languages which lack case morphology, but the equivalence between rich case marking and flexible constituent order is not exact,[2] a fact which causes difficulties, both for attempts to explain case systems from either a purely functional or purely formal perspective.

Associated with these synchronic questions is one very broad diachronic question: how do case-marking systems change? This question includes sub-questions such as: (3) What causes and what constrains the changes that can take place

---

[1] Research on this chapter was partially funded by Australian Research Council Discovery Grant DP0208153.

[2] For a discussion, see Haeberli (2001).

Competition and Variation in Natural Languages: The Case for Case
Edited by M. Amberber and H. de Hoop

ISBN: 0-08-044651-5

between one stage and another in a case-marking system? (4) How exactly are these changes implemented? What is the role of variation here?

These intertwined questions have excited a great deal of interest recently, particularly among linguists seeking to give formalist explanations to questions 1 and 2. It has been proposed that certain types of changes are not possible (or are unlikely) because they would involve stages with humanly impossible (or very marked) grammars. For example, Weerman and de Wit (1999) and Calabrese (1998) suggest that the combination of cases that a language may have is constrained by a hierarchy of cases which says that it will be either impossible or unusual for a language to have case X when it does not also have case Y.[3] Weerman and de Wit make a very specific and strong proposal: a language learner is unable to acquire a language that does not conform to the hierarchy, and the diachronic spin-off is that if the overt evidence for case Y disappears for some reason, case X will also disappear despite the overt evidence for it in the language-learner's data.

The question of implementation is particularly pertinent to formal accounts of the relationship between case marking and constituent order in which adjacency, adpositions, and morphological case all perform the function of 'licensing' NP/DPs (Chomsky, 1995 etc.). It is easier to propose a formal account of the two endpoints in a system which has shifted types — the earlier system with rich case marking and the system with greatly impoverished case marking — than to account for the intermediate stages where a language may not seem to be strongly of one 'type' or another. Some of the formal accounts which have been proposed also do not address the question of variation in the implementation of the changes. Furthermore, the question of implementation must remain largely guesswork when we are dealing with a language where there is not good documentation of the intermediate stages.

English is a good language for a case study of the implementation of the loss of a case-marking system, because it has shifted from a language with a substantial amount of case morphology to a language with very little within its written history. The reduction of case marking in English has naturally been the subject of a great deal of attention, but much remains to be done, particularly now that electronic corpora such as the *Penn-Helsinki Parsed Corpus of Middle English 2* (*PPCME2,* Kroch & Taylor, 1999) have made the task of data-gathering so much

[3] Unfortunately for such attempts, there is no agreement on what the hierarchy should be. While Weerman and de Wit (1999, p. 1182) propose that the dative case is higher on the hierarchy than the genitive, Calabrese (1998, p. 85) suggests the opposite ordering. Blake (2001, p. 159) considers that while case systems do seem to be typically built up in a particular sequence, it is impossible to establish implicational relationships in the hierarchy of cases (which he assumes has genitive higher than dative).

easier. There is a long tradition, going back to Kellner (1892, §483 and *passim*), of assuming that fixed word order 'stepped in' to make up for lost inflection. Recent works within the generative framework on the interaction of the fixing of word order and the loss of the accusative/dative distinction, such as the corpus-based investigations of McFadden (2002) and Polo (2002) among others, are essentially within this tradition. On the other hand, Jespersen (1894, §75) concludes that 'grammatical simplification' (more commonly referred to as 'deflexion' today) was only possible, once word order had already become relatively fixed.

Important changes involving case morphology within the NP/DP — the loss of case agreement morphology and the loss of what I will refer to as the 'post-head' genitive — have been the subject of some important claims about syntactic change (see below) without an adequate empirical base. In this chapter, I will look at these changes from Old English (OE, c. 700–1100) to Early Middle English (EME, c. 1100–1275), with a view to seeing what this case study can contribute to what we know about how a case-marking system can change. Using the results of my own investigation of case marking in the EME texts[4] I will argue that an explanation which is widely held, particularly by formally oriented linguists, is not adequate, and will argue for an explanation that probabilistic processing strategies are the key to understanding the long 'typological shift' to a language, which relies primarily on word order to calculate grammatical relations.

2. Case in OE

Old English was a language in which four cases, viz. nominative, accusative, genitive, and dative, were productively distinguished by morphology on nouns and their modifiers. There was already a good deal of syncretism in case morphology in OE; for example, the largest class of masculine nouns and all neuter nouns lacked a distinction between the nominative and accusative forms, while many singular feminine nouns had a distinct form for the nominative but had only one form for the other cases. OE was similar to Modern German in having a 'weak' declension of adjectives (essentially, adjectives preceded by a demonstrative or possessive) and a 'strong' one (essentially, the declension used elsewhere, but this is a simplification; see Mitchell (1985, §102 for a discussion). The weak declension

[4] For information about the texts used in the investigation, see the Appendix. The citations used in examples are from the *PPCME2,* unless otherwise specified; see the information supplied with the *PPCME2* for details of these texts and the editions used. The examples taken from the *PPCME2* can be distinguished from the others by the form of their citation, which is the form used in that corpus: they are surrounded by parentheses and begin with 'CM', e.g. (CMVICES1). I have supplied the references for the examples taken from my own reading.

had become highly syncretistic by OE; the suffix *-an* was used in this declension for 14 of the 24 combinations of gender, case, and number. The case of the noun they modified was therefore shown more surely by a determiner or quantifier rather than by the adjective, which is also the case in German. As noted by Hawkins (1994, p. 404), processing considerations cast these left-most elements of the NP/DP in a particularly important role: an inflected element here will establish the possible cases of the NP/DP.

Even when a noun was modified, however, its case was not always clearly identified. Saitz (1955, p. 83) found that fewer than half the pairs of subjects and objects headed by nouns in OE were distinctively inflected for case. This meant that OE speakers had to rely on other means of encoding grammatical relations when the case morphology was not sufficient. Saitz (p. 104) notes that in cases of ambiguity between the nominative and the accusative, the subject was nearly always placed before the object. But it was not only ambiguous morphology that caused this preference for subject–object (SO) order, which was dominant even when the case marking was quite clear. We can assume that the pragmatic considerations which cause subjects to be typically the most topical arguments in a sentence had led to an unmarked word order already in the pre-OE stage, and that this facilitated syncretism. By OE, there are some indisputably unmarked orders, although a more marked word order could be used when case marking made the grammatical relations clear. Prepositions were also an important part of OE grammar, meaning that OE was considerably advanced along the path towards less dependence on case marking to signal grammatical relationships.

One place in which there was considerable freedom of word order was in the positioning of the genitive NP/DP with respect to its head. The genitive phrase, which could be either a complement or a modifier, could be placed either before its head, which could be a noun, an adjective, or a quantifier, or after it. (1) and (2), both from the first series of Ælfric's *Catholic Homilies* (edited by Clemoes, 1997), represent the pre-head and post-head genitives, respectively. These examples also illustrate the fact that the genitive case was much more freely used for 'objective' genitives in OE than it is today:[5]

(1) *þurh ðines feondes lufe*
through thy(G) enemy(G) love
'through love for thy enemy'
ÆCHI III.176

[5] I use the following abbreviations in glosses: G=genitive, D=dative, U=uninflected. Note that I do not use 'U' for a zero morpheme in OE, but only for the later (EME) stage in which a form which would historically be inflected is lacking any inflection.

(2) *for þære micclan lufe godes*
for the(D) great(D) love(D) God(G)
'for the great love of God'
ÆCHI XIX.131

While from a grammatical perspective, either position was possible, the pre-head position was favoured in some situations while post-head position was dominant for others. For example, post-position was favoured for partitive genitives containing a noun, although the preposed position was also possible, while post-position is only rarely found when both the genitive and the head were single nouns. It should also be noted that there was a general shift away from post-head genitives within the OE period. This was first documented in the pioneering work by Thomas (1931). Crisma (to appear), who argues against the idea that case levelling was responsible for the loss of post-head genitives, gives further data. By late OE, the adpositional alternative to the post-head morphological genitive, viz. the '*of*-genitive,' was already used occasionally for partitive expressions; for some statistics, see Crisma (*ibid.*). The frequency of pre-head genitives also increased in this period. The shift towards pre-head genitives makes sense under the view that OE was a language that was relying more heavily on word order than on case. The post-head genitive had the advantage over the pre-head genitive of showing the case of the larger NP/DP immediately, since this NP/DP typically began with an inflected element such as a determiner, but had the disadvantage that in processing an NP/DP, a hearer could not 'close off' the phrase as soon as the head noun (N) was reached. As word order became a more reliable method of calculating grammatical relations in OE, early identification of case became less important, and the advantage of the post-head genitive became less.

It is important to note, however, that although there was a general decline in post-head genitives, the decline was not uniform across all types of genitives. In particular, while partitive genitive phrases containing a noun[6] are found fairly freely in the pre-head position in the *Orosius* of the late ninth century, McLagan (2004, p. 61) found post-position in 256 examples, nearly 99% of the nominal partitives found with quantifier heads in Ælfric's second series of *Catholic Homilies,* written about a century later.[7] Clearly, some types of post-head genitives were still flourishing in the late-OE period, even though the overall frequency of genitives in this position had declined. We turn now to the question of why these post-head genitives disappeared.

[6] The pre-head position was very strongly favoured in all periods for partitives consisting only of a genitive pronoun, although some post-head examples are found.

[7] It should be noted that McLagan did not include *of*-genitives (which were not very commonly used by Ælfric) in her statistics; the percentage is a comparison with the pre-head construction.

3. Loss of the Post-Head Genitives: Two Hypotheses

3.1. Ease of Processing

Ever since Thomas (1931), it has been recognized that there is a general correlation between the loss of agreement morphology and the replacement of the post-head morphological genitive by an *of*-genitive; generally speaking, the texts of EME which have the best-preserved case agreement also have the most post-head genitives (but see Section 5.1).

Thomas observes that deteriorating agreement cannot have been the explanation for the decreasing frequency of post-head genitives during the OE period, when case agreement morphology for determiners and the strong adjectives was still quite healthy. He concludes that loss of inflection could not have been the only reason why the 'post-positive' genitive was disfavoured in ME, but adds:

> Loss of inflection, however, in the definite article and strong adjective is an additional cause for the continued preference of the pre-positive genitives over the post-positive during the twelfth and thirteenth centuries, and this loss of inflection is one of the two main causes (the other being the use of the periphrastic construction) of the ultimate disappearance of the post-positive inflected genitive.
>
> (Thomas, 1931: 107)

The view that agreement inflection was decisive in the eventual loss of the post-head genitive has been adopted by recent writers such as Rosenbach (2002, p. 210). An advantage of this theory is that the small number of post-head genitives in the EME texts, e.g. (3), conform to Thomas's (p. 107) generalisation that we do not at any period find post-head genitives which are not inflected for genitive[8] case:

(3) *Hwa is wyrhte þære synne?*
Who is author the(G) sin(G)
'Who is the author of sin?'
CMKENTHO, 140.167

[8] This statement includes inflection which is not specifically genitive; as in (3); *þære synne* could also be a dative form. We include genitive pronouns here also.

The explanation for this generalisation has to do with ease of processing. Overt morphological marking of the genitive case was crucial in identifying an NP/DP which followed an N as a genitive phrase. There is an asymmetry here because with the pre-head genitives, the genitive N was always directly adjacent to the head. This means that pre-head genitives would usually have a genitive inflection directly before the head N, making them easy to process. Even in the instance of an ambiguously marked noun, the presence of another noun following immediately, with no determiner, would be enough to favour a parsing of that noun belonging to the same NP/DP as the preceding noun, rather than the beginning of a separate NP/DP (such as some sort of object).

With post-head genitives, however, the possessor N was not usually adjacent to the head N; even in OE, post-head genitives consisting only of the possessor N were unusual, and increasingly in EME, the post-head genitive was reserved for NP/DPs consisting of more than one constituent (in most texts). The post-head genitives imposed a greater burden of processing even in OE, but this was not a serious problem when agreement inflection gave the tip-off that a genitive N was coming. Once this inflection was lost, however, it would be easy for a hearer to misparse the genitive phrase as the beginning of a separate NP/DP. Speakers can reasonably be expected to avoid constructions that will cause parsing difficulties for the hearer. I will refer to the idea that the ease of processing pre-head genitives played a crucial role in the loss of the post-head genitive (and the fixing of word order generally) as the Ease of Processing hypothesis.

3.2. *Early Reanalysis*

The second hypothesis is that the genitive inflection was reanalysed as something else in EME, and the genitive disappeared as a morphological case. As pointed out by Jespersen (1894, §247–248; 1942, §17), the possessive marker of Present Day English (PDE) does not behave like a normal inflection, in that it is attached to the end of a 'nominal group' (hence the term 'group genitive') rather than to the possessor N. An early example of the group genitive is given in (4):

(4) *þe kyng of Fraunces men*
'the king of France's men'
CMPOLYCH,VIII, 349.380 (1387)

Janda (1980, 1981) suggests that the group genitive arose because changes took place in EME which made it impossible to maintain the genitive as a morphological case. Janda suggests that the old genitive inflection had to be reanalysed as something else[9] when the distinction between the dative and accusative cases collapsed in EME. He comments (Janda, 1981, p. 65):

> And it seems that no (other) language opposes a genitive case-inflection to a general case-inflection that conflates all of the possible cases (nominative, accusative, etc.)

Janda further comments (*ibid.*, p. 245) that the genitive was the most marked case inflection of OE, according to Greenberg's (1966) criteria, and that markedness predicts that more unmarked cases should never be lost before the more marked cases.

Janda's explanation is adopted by Lightfoot (1999, pp. 117–125), who refers to the 'new, caseless' system of EME, and Weerman and de Wit (1999) propose a variant of this explanation in terms of their hierarchy of cases. Although their paper is mainly about the loss of the genitive in Dutch, they extend their proposed explanation to English and argue that the genitive could no longer be maintained as a morphological case in English once the evidence for an accusative/dative distinction disappeared from the grammar. As Weerman and de Wit point out, an advantage of such an approach is that it explains why genitive case should have disappeared in both Dutch and English in spite of the fact that it was not subject to the same sort of phonological erosion (see below) as the accusative and dative cases were in these languages.

I will use the term 'Early Reanalysis' for this widely held hypothesis that a reanalysis of the genitive inflection as a clitic was forced in EME. In the next section, I will focus particularly on the idea that the collapse of the accusative/dative distinction made it impossible to maintain the genitive as a morphological case.

4. Evaluating the Hypotheses: Case in EME

In this section, we will look in some detail at deflexion in the NP/DP in EME and evaluate our two hypotheses against the attested facts. We can distinguish three major changes to case morphology which took place between c.1000 and c.1300 in most dialects of English:

(5) a. Disappearance of the dative/accusative distinction, both in the pronominal and nominal systems.

---

[9] Space does not permit a discussion of the role that the so-called '*his* genitive' (e.g. *the king his daughter*) is supposed to have played in this development, but see Janda (1980, 1981) and Allen (2003).

b. Disappearance of the distinction between the nominative and the object (i.e. non-genitive) cases in nominal morphology.
c. Disappearance of case agreement morphology on modifiers (determiners, quantifiers, and adjectives).

There are also three major changes involving genitives in ME:

(6) a. The post-head genitive disappears.
b. Genitive-marked objects of verbs disappear.
c. The genitive case becomes more limited in its use (e.g. it is no longer used with the complements of adjectives or to express partitive relationships).

An appealing feature of the Early Reanalysis hypothesis is that it offers the prospect of a unified explanation for all the changes of (6), as will be discussed in section 4.3. Before we can consider the interaction of the changes of (5) and (6), however, we must look in more detail at the loss of case morphology in EME.

4.1. Deflexion in EME: General

It is generally assumed that a combination of analogy and phonological erosion was largely responsible for the disappearance of nominal case morphology (see Lass, 1992, pp. 103–106 and *passim*). A view which is frequently presented is that phonological changes were primary, as they obliterated some important distinctions in suffixes. One particularly important phonological change which is always cited as important in deflexion in English is the reduction of vowels to schwa in unstressed syllables. Since some suffixes were distinguished only by vowel quality, they collapsed together. Final *m* and *n* merged, and in combination with the vowel reduction, the result was that, for example, *-um,* which (in the nominal system) represented only dative plural (of any gender) was now indistinct from the reflex of *-an,* a suffix which was already used for many combinations of features in some declensional classes (see Section 2). With the loss of the final nasal, the syncretism of forms was so great that the remaining suffixes were of very limited usefulness in distinguishing case and other features.

However, recent scholarship has tended to take the view, expressed by Lass (1992, pp. 104–105), that the phonological changes *supported* analogical changes, rather than being the driving force in the loss of case distinctions. As Lass notes, there was a good deal of analogical change even in OE; the extension of the genitive *-(e)s* to nouns which formerly inflected in a different way for the genitive is one example of many analogical developments.

At any rate, phonological reduction did not simply apply blindly. For example, it appears that weak nouns and adjectives were subject to the loss of final *-n* in all dialects, while in some dialects verbal forms were more resistant to this change. This is likely to be because the *-an* in the weak nominal and adjectival declensions was already of little help in marking distinctions.

Kitson (1992, p. 82) emphasises that the change of English from a synthetic language to an analytic one was the 'prime mover' of deflexion, rather than the result of it. That is, it was the increasing reliance on word order and prepositions that made it possible for the rather redundant case inflections to become non-functional, rather than the non-functionality of the case morphology that made fixed word order necessary. Lass (1992, p. 105) points out that 'mere phonological erosion does not explain how case and gender as concordial categories disappeared.' The disappearance of case 'concord' or agreement on modifiers of nouns is more plausibly attributed to a greater reliance on word order than to phonological change.

Whatever the causes of deflexion in EME, by 1250, an overt distinction between the dative and accusative cases had entirely disappeared except in some southern dialects, and the marking of the distinction was optional even in those dialects retaining the category distinction. Let us now look in a bit more detail at how the changes of (5) proceeded, paying particular attention to evidence for the continued existence of the genitive case when other case categories had disappeared and to the interaction of (5c) with the category changes of (5a) and (5b).

4.2. The Inflection-Rich Dialects

The texts that we have from the southern areas of England show that the OE case categories were well maintained into the 13th century. In MS Cotton Vespasian D. xiv, from the mid-12th century, for example, we find that agreement inflection is used pretty much as it was in OE in terms of category distinctions, although there is a good deal of syncretism of forms. In this text, there is still full inflection of the definite determiner for the four cases of OE, and adjectives still agreed with the nouns they modified. However, the processes mentioned in the preceding section had already rendered the adjectival morphology of very little use in distinguishing the cases, and by the beginning of the 13th century, agreement for case was marked only optionally on adjectives. It is interesting to note, however, that there was some tendency to use the more distinctive 'strong' genitive ending *-(e)s* on adjectives which would have required the (non-distinctive) weak ending in OE to modify genitive nouns.

By the beginning of the thirteenth century, we find more deterioration in the clarity of the case marking, and indeclinable *þe* 'the' is widely used in variation with forms which are inflected for the dative or accusative case in the *Vices and*

Virtues (*V&V*[10], southeastern, c. 1200), for example. However, a determiner modifying a noun in the genitive case was nearly always inflected for the genitive in this text.[11] There is therefore no doubt as to the continued existence of the genitive case in these 'inflection-rich' dialects. The variation which is found in a text such as the *V&V* looks chaotic at first and has often been attributed to 'confusion' and Millar (2000, p. 176) does not consider that the functional associations which are left residing in the particular forms of the determiners in *V&V* should be regarded as representing accusative and dative case. However, when we look closely at the use of forms in this text, we find that the variation is in fact highly structured. Importantly, we find variation in the third person masculine pronoun between the old dative form *him* and the old accusative form *hine* for the direct objects of verbs which took only *hine* in OE. In contrast, we find only *him* as the indirect object in this text and also as the object of some verbs which formerly governed only dative objects, such as *helpen* 'help'. This indicates a situation in which a clear category distinction between dative and accusative case remained, but in which *hine* represented only accusative case, while *him* represented either accusative or dative case.[12]

This encroachment of one form into the functional territory of another is a recurring pattern in the ME period. First we have variation, but only in one direction; *him* can be used instead of *hine*, but *hine* cannot be used instead of *him*, for example. Eventually *hine* stops being used altogether and when this happens with all the dative and accusative forms we have the complete disappearance of a category distinction, rather than the earlier syncretism of forms. A parallel exists in the variation of some quantifiers used for either mass or count nouns in colloquial English today: many speakers will use *less*, historically a quantifier for mass nouns, with count nouns in some contexts, as in *less men came to the party than I expected*. This 'confusion' of *less* with the specifically count quantifier *fewer* does not indicate a loss of the count/mass distinction for these speakers, however, who will say *I ate less pizza than you* but never **I ate fewer pizza than you*. Rather, *less* can be seen as a quantifier of comparison which is compatible with

[10] This text is designated as (CMVICES) in the *PPCME2*.

[11] The apparent discrepancy with Thomas' (1931, p. 92) finding of a high percentage of uninflected definite articles in genitive constructions in this text is due to the fact that Thomas was including *of*-genitives in his figures.

[12] Another change which is evident in the *V&V* as well as other texts is the rise of a single case, usually the reflex of a dative form, used for the objects of prepositions. The existence of a 'prepositional' case makes EME different from OE, where different prepositions governed accusative or dative case. However, the fact that prepositions no longer governed different cases in *V&V* simply means that these case categories were used in a modified way; it does not mean that they did not exist. See Allen (1995, pp. 185–195) for more details concerning the case-marking system in *V&V*.

either mass or count nouns, while *fewer* is compatible only with count nouns. Thus the count/mass distinction is somewhat obscured by the extension of *less* to count contexts, but has not yet disappeared. Similarly, the dative/accusative distinction had not disappeared in the inflection-rich dialects despite the extension of *him*.

4.3. The Inflection-Poor Dialects

In the more northerly dialects of EME, case agreement within the NP/DP was greatly reduced by the end of the 12th century. The term 'inflection-poor' is appropriate for these dialects, which did not maintain the four case categories of OE and in which agreement morphology is greatly reduced. In the *Ormulum* of c. 1180 or 1200,[13] for example, the definite determiner is always uninflected. It is particularly instructive to look at the variation found in the morphology of this text because it cannot be due to layers of copying, since we are in the unusually fortunate position here of having the author's autograph manuscript.

Adjectives in this text show no agreement for case, although they may still inflect for plural number and the weak/strong distinction still plays some role. What about the evidence for a genitive case in this text?

Examples such as (7) from this text have been adduced (e.g. by Kroch, 1997, p. 134), as evidence of the reanalysis of the possessive marker as a clitic in this text:

(7) *þe* *Laferrd* *Cristess* *are*
the(U) Lord(U) Christ(G) mercy
'the Lord Christ's mercy'
Orm Dedication 152

In this example (and many others in this text), the possessive marker is at the end of an NP in apposition with another, and this is an innovation; in OE, both parts of the appositional construction would have had genitive marking.

However, an other evidence from agreement within the NP/DP suggests that the genitive was retained as a case in Orm's dialect. Quantifiers[14] such as *an* 'a, one' and *nan* 'no' in the *Ormulum* are frequently inflected and indicate a three-way case distinction — subject/object/genitive — with the old dative and accusative cases merging as a single object case, as far as can be determined from agreement phenomena. Some quantifiers have a suffix *-ne* which is a reflex of the old accusative

[13] Examples are cited from Holt's (1878) edition.

[14] It is generally agreed that *an* had not developed into a true article by this time, and the term 'quantifier' is appropriate here because *an* patterned with quantifiers such as *nan* 'none' in retaining genitive inflection.

masculine singular suffix, but the same suffix is used for quantifiers modifying nouns which would have been dative in OE, rather than accusative, or which are the objects of prepositions. For example, at line 9197 we find *of ænne mann* 'of one man,' where *ænne,* historically an accusative form, is used with a preposition which governed the dative in OE instead of the 'correct' dative reflex *ane* (which is used in the same contexts as *ænne*). Case marking in the *Ormulum* does not appear 'confused' when a proper distinction is made between case categories and case forms, although there is considerable syncretism of forms; thus *ænne, ane,* and the unmarked *an* are all exponents of a single object case.[15] Agreement of a quantifier with a noun in the *Ormulum* is productive although optional and conditioned by such things as metre, rather than being a matter of a few fossilised forms. Such agreement with a genitive is common, although optional:

(8) a. *inn aness cullfress like*
in a(G) dove(G) form
'in the form of a dove'
Orm 10677

b. *inn an ærness like*
in an(U) eagle(G) form
'in the form of an eagle'
Orm 5867

In (8a), the quantifier is marked for genitive case, but in (8b), it is not. It is difficult to explain (8a) if we deny the existence of a category of genitive case in the *Ormulum.*

The evidence for the retention of genitive case is not incompatible with a weaker version of the Early Reanalysis hypothesis in which we assume that an early reanalysis did take place but assume that the newer analysis co-existed with the older one. We could take examples such as (7) as evidence that Orm had a clitic analysis as one possible analysis, and the examples with agreement for genitive case as evidence that Orm still had the genitive as a morphological category.[16] The idea of a dual analysis is plausible because it is certainly true that once there was widespread loss of agreement morphology, it would often be unclear

[15] This statement is based on my own examination of forms in the *Ormulum.* Unfortunately, the editor of this text listed forms in his glossary according to function, rather than according to form; the *ænne* of the example given here is listed as 'genitive,' apparently solely because genitive case, rather than a preposition, would have been used in OE.

[16] We can easily recast this variation in terms of 'competing grammars' for EME (as proposed by Kroch, 1989 and later works) and assume that Orm had one grammar with a genitive category, and another only with a clitic.

whether a possessive marker was to be analysed as an inflection or a clitic, since it would usually be marked only on the possessor N, which would furthermore usually be at the end of the NP, as in *þe kinges sune* 'the king's son.' We could assume that at the time when agreement morphology was essentially optional (in terms of grammar),[17] language-learners constructed grammars in which the possessive marker was either a suffix or a clitic. A problem with the assumption that the possessive marker was a full-fledged clitic already by the time of the *Ormulum,* however, is that we do not find true group genitives until nearly two centuries later; example (4), from 1387, is one of the earliest. For evidence that the absence of such examples for such a long period is unlikely to be simply a data gap, see Allen (2003). An attractive alternative to assuming a sharp distinction between inflections and clitics is to treat EME *-(e)s* as an edge-located inflection (cf. Zwicky, 1987) which was quite selective about its host: it must be a noun, and furthermore it must be the possessor N. It became more clitic-like in late ME when case agreement had been completely lost and the single genitive marker came to be used with all nouns.

Leaving aside the question of whether the possessive in (7) was a clitic, the important point here is that such examples do not prove that the genitive case had become unlearnable in EME. It is clear enough that Orm retained a genitive case category in some form, despite not having acquired case agreement morphology for determiners or adjectives or a distinction between dative and accusative objects.

In another important text of the 12th century, the Second Continuation of the *Peterborough Chronicle* (henceforth *PCII*), written c. 1154,[18] agreement for case is even more reduced than in the *Ormulum*. In this text, there is no case agreement at all between quantifiers, determiners, or adjectives and the nouns they modify, and there is no longer a subject/object case category distinction for nouns, although there is still a residual *-e* ending (a reflex of the old dative case) which is optionally used on the objects of prepositions. However, one type of agreement which is rather surprising in a text with such advanced deflexion as *PCII* is found with extraposed appositives to genitive nouns; for details, see Allen (2002).

Note that genitive-marked nouns of *PCII* do not behave differently in this text from inflection-rich texts in EME, as we might expect them to do if the changes of (5) had caused a reanalysis of the case marker as a clitic. That is, the genitive-marked possessor is directly adjacent to the possessum, and instead of a group

[17] Naturally, we expect to have variation according to style and register, but there is no reason to assume any variation of register in the *Ormulum* (which furthermore was written in a simple style for an illiterate audience). Variation here is likely to be controlled by such factors as metre.

[18] The *PPCME2* does not distinguish between the First Continuation (*PCI*) and *PCII,* but their very different morphology makes it important to treat them separately.

genitive we get examples like (9), just as in morphologically more conservative EME dialects:

(9) *& nam þe kinges suster of France to wife*
and took the king(G) sister of France as wife
'and married the king of France's sister'
PCII 1140.54

5. Morphological Genitives in EME

The preceding discussion of case morphology in EME has presented some evidence of a mainly morphological nature against the Early Reanalysis hypothesis. For a discussion of morphophonemic alternations which continue to give some evidence of a category of genitive case even in later ME, see Plank (1985) and Allen (2003). Let us now turn to our discussion of the syntactic changes involving genitive case in the EME period.

The Early Reanalysis hypothesis would get strong support if we found that the changes of (6) were simultaneous, all taking place as soon as the accusative/dative distinction disappeared. We have already seen how this hypothesis explains (6a). (6b–c) are also immediately explained: both these types of genitives are lexically selected, and if there is no genitive case, there can be no lexical selection for this case. Lightfoot (1999) indeed argues that the loss of inherent (lexically assigned) case was a function of the loss of morphological case, specifically the dative case. It should be noted, however, that the Early Reanalysis hypothesis will not be disconfirmed if we find that any of these changes happened *before* the loss of the accusative/dative distinction, although this would reduce the appeal of the hypothesis because of its reduced explanatory power. On the other hand, if we find the productive survival of any of the constructions of (6) after the collapse of the accusative/dative distinction, that will be very damaging to the hypothesis, which predicts that genitive should no longer be available in the grammar as a morphological case at that point.

The fact that the changes of (5) happened at different periods in different dialects affords an opportunity to test the relationship between deflexion and the changes of (6). By the Early Reanalysis hypothesis, we would expect that in a dialect where there is considerable deflexion in general there will also be evidence for the loss of the genitive as a morphological case in that dialect. If the changes of (6) were all directly due to the loss of the genitive as a case, we expect to find both post-head genitives and genitive objects in the inflection-rich dialects, as well as genitives which are lexically selected by quantifiers or adjectives. On the other hand, we expect that the inflection-poor texts should no longer have these constructions. In comparison, the Ease of Processing hypothesis makes no predictions that these changes should be simultaneous.

5.1. The Disappearance of Post-Head Genitives

When we look at the texts of the 12th century, we do indeed find a striking difference between the texts with rich case morphology and those with impoverished morphology where the use of post-head morphological genitives is concerned. Although post-head genitives are generally less frequent in the case-rich texts than they are in the texts of earlier periods, they are by no means rare, and they are used to express a range of genitive relationships, such as pure possession, partitive relations, etc. In contrast, the 12th century texts that show marked deflexion either have no post-head genitives at all or only have one or two examples, which are furthermore quite restricted as to type; they either express a partitive or some other part–whole relationship or can be treated as fixed expressions. For more details, see Allen (forthcoming). Thus, there appears to be quite a good correlation between deflexion and the disappearance of (productive) post-head genitives in this period. That is, we only have to see that deflexion is well advanced in a particular text to know that few, if any, post-head genitives will be found, while we can predict that there will be more than sporadic use of post-head genitives in a text in which deflexion is not far advanced.

At first, this would appear to support the hypothesis that the loss of case categories made the post-head genitive unavailable because the genitive had disappeared as a case category. When we look at the texts from the end of the 12th or beginning of the 13th century, however, we find that things are not this simple. Strikingly, by this time the correlation between rich inflection and post-head genitives has disappeared, because the post-head genitives have almost entirely disappeared even in the texts such as the *V&V* which show a clear retention of the dative/accusative category distinction. There are very few genitives of this sort in any text of this period, and furthermore they are very limited as to type; they are all lexically selected:

(10) *ne ðurh nan ðare þinge ðe...*
nor through none those(G) things(G) that
'nor through any of those things that...'
CMVICES1,29.314

(11) *and hwaðer ȝunker hes tobrecð*
and which you(G) it(G) breaks
'and whichever of you breaks it...'
CMVICES1,95.1131

Examples such as (10) and (11) were only possible in *V&V* when the post-head genitive began with a genitive-inflected word, such as *ðare*, as discussed above. Thus, it is no surprise that we do not find examples like (10) in the *Ormulum*, where genitive inflection was no longer possible for the determiner.

What is a surprise, however, by the Early Reanalysis hypothesis, that we do not find more post-head genitives in *V&V* and other inflection-rich dialects. There can be no question that the genitive was available as a case in this dialect for determiners, so we cannot attribute the near-loss of post-head genitives in this dialect to the loss of genitive case.

Thus, it seems clear that the demise of the post-head genitives was not a sudden change that resulted from a reanalysis. Furthermore, we still find some post-head genitives in texts which show no dative/accusative category distinction, such as the *Ormulum*:

(12) *Forr eȝȝþer here ȝede swa*
for either they(G) went so
'Because both of them (lit. 'either their') went like that'
Orm 119

Such examples (which were of course only possible when the genitive was a pronoun) seem unexpected if no genitive case was available to the author of the *Ormulum*. If Orm had a grammar so radically different from that of the author of *V&V*, it seems surprising that the post-head genitives should show such similar limitations (after allowances are made for the lack of inflected definite determiners in Orm's grammar).

5.2. The Disappearance of Genitive Objects

The next change to consider is (6b). Genitive objects pretty much disappear from the original prose texts by the end of the twelfth century,[19] and it is difficult to attribute this change directly to the phonological changes which are traditionally held responsible for the obliteration of most case distinctions, so some sort of explanation needs to be found in terms of the organisation of the grammar. It is reasonable enough to suggest that the loss of the genitive as a possible case is the culprit; the old genitive objects had to be expressed either by a prepositional phrase or simply as a bare object (both of which happened, with different verbs). However, there is a major problem for this idea: genitive objects seem to have suffered a gradual

[19] Some genitive objects are still found in thirteenth century texts which are copies of earlier texts.

decline at about the same time in the dialects in the inflection-rich dialects as in the inflection-poor ones.

Why did genitive objects disappear? It is likely that genitive case marking had a real function for objects at an earlier stage, similar to the indication of 'the limit of the referent's involvement in the content of the utterance' that Jakobson (1984[1936], p. 72) proposes for the genitive case in Russian. Typically, genitive objects were rather peripheral objects; for example, verbs such as *beniman* 'to deprive (someone of something)' typically had genitive case marking on the object of deprivation, the thing which is 'no longer there.' Highly transitive verbs did not govern genitive objects. However, by the OE stage, genitive and dative objects were in variation with some verbs; for example, the object of *helpan* 'help' could appear in either case, and no one has succeeded in giving an explanation that explains the variation really satisfactorily. It seems that although some sub-generalisations can be made, in OE some verbs simply selected the genitive objects, some selected genitive or dative, and some selected dative or accusative. In many instances, a verb which selected a genitive object could alternatively take a prepositional complement (e.g. *wlatian* 'to be nauseated at').

In the 12th century, some genitive-marked objects are found, but they have become unusual, even in the inflection-rich texts, and there is clear indication of replacement of genitive objects by either a prepositional phrase or an accusative-marked object in texts of all dialects.[20] By the early 13th century, when we have more texts, it appears that genitive objects were restricted to a few fixed expressions and proverbs, in the prose texts at least.[21] The absence of genitive objects and their replacement by either accusative objects or prepositional complements in inflection-rich texts such as the *V&V* cannot be due to the loss of the genitive case. It seems more likely that the causes of this change were (1) the increasingly prepositional nature of English syntax and (2) the loss of a coherent and distinctive meaning of the genitive case for objects.

5.3. Reduction of Genitive Types

We turn finally to a consideration of the loss of the dative/accusative distinction and the disappearance of certain types of genitives. In OE and to some extent in

---

[20] We only have a few original texts from the 12th century (by which I mean texts which were composed at this time, rather than ones which were copies of OE texts). My comments about the texts refer only to the original ones. Some manuscripts contain both types, and it is interesting to note that while the original homilies in Vespasian A.xxii have no genitive objects, the copied ones do have some.

[21] Poetic texts tend to be more conservative, but the replacement of genitive objects by other means is notable even in texts such as *Laȝamon's Brut* (edited by Brook & Leslie, 1963, 1978), where genitive objects are not infrequent.

EME, genitive case was used to signal many relationships that require an *of*-phrase in PDE. Space does not permit a full discussion of these types, but particularly relevant here are the use of the morphological genitive for partitives and the ability of certain adjectives to select genitive complements. These constructions both involve lexical selection of the genitive case, and it is striking that lexical selection is found with all post-head genitives in the 13th century, even in the inflection-rich texts. I will focus on partitives here. Morphological genitives are still found in partitive constructions in the inflection-poor *Ormulum*, as already illustrated in (12) and in *PCII*, as in (13):

(13) *&* *athes* *suoren* *...ðat* *her* *nouþer* *sculde*
and oaths swore that they(G) neither should
besuyken *other;*
deceive other
'and swore oaths...that neither of them (lit.'their neither') would deceive the other'
PCII: 1140.37

The existence of an example such as (13) in *PCII*, a text without any remnants of the accusative/dative distinction, seems problematic for the idea that genitive case did not survive the loss of this case distinction. The only difference between the inflection-rich texts and the inflection-poor texts with respect to partitives is that post-head partitives in the latter were restricted to pronouns due to the requirement that a post-head genitive must begin with an element which was inflected for genitive case. Naturally, since determiners were not inflected for case in the *Ormulum* or *PCII*, we do not find post-head genitives of any kind with them. While some quantifiers were inflected for the genitive case in the *Ormulum*, they were not suitable for use in partitives, so we do not find post-head genitives with them either.[22]

Nor was there a sudden decline in the use of morphological genitives in partitives in any dialect coinciding with loss of case morphology. Already in OE the *of*-genitive had become an alternative to the morphological genitive in partitives. The morphological genitive was on the decline in partitive constructions — both pre-head and post-head — in the 11th century, and by the 13th century, *of*-genitives seem to have been strongly preferred in partitive constructions even in

[22] The *Ormulum* is unusual in using post-head position for pronominal partitives; as mentioned in footnote 6, pre-head positioning for pronominal partitives was nearly 100% in most OE texts. Some partitive genitives with pronouns continue to be found into the 15th century. At some point, these must have become fixed expressions, but for EME, when adjectives could still also select for genitive complements, it makes more sense to me to assume that particular words still selected for genitive complements.

inflection-rich dialects. As we saw above, some post-head nominal partitives are to be found in *V&V*. Their existence is no surprise given the continued ability of determiners to inflect for genitive case in this dialect, but what is unexpected by the Early Reanalysis hypothesis is that there are so few of them in this text, where the genitive inflection of the determiner seems to have been very nearly obligatory. Clearly, the paucity of morphological partitives (either pre-head or post-head) in this text is not due to any disappearance of the genitive as a possible case, but rather to a strong preference for the *of*-genitive with partitives.

In sum, while the facts concerning the loss of genitive objects are not inconsistent with Early Reanalysis hypothesis but fail to give it any support, the facts concerning the partitive genitives are problematic for this hypothesis. The Early Reanalysis hypothesis not only fails to explain how morphological partitives continued to be possible after the accusative/dative distinction was lost in a given dialect but also fails to explain the similarities between the inflection-rich and case-impoverished dialects: not only were post-head genitives restricted to partitives in both types of dialects by the 13th century, but the range of the morphological genitive had become restricted regardless of the general state of case morphology.

The Ease of Processing hypothesis does not face the same difficulties. It does not predict that the changes of (6) should be simultaneous or that lexically selected genitives should be impossible once case distinctions were lost. The decline of the post-head genitives in OE is seen as a result of the increase in reliance on order-based strategies for processing grammatical relations. Then once agreement inflection was lost, post-head genitives became too difficult to process and speakers stopped using them, except when there was clear genitive marking. This was not a problem because the *of*-genitive was already available for use in situations where a post-head genitive was favoured, such as in partitives involving a noun. Once people stopped using them frequently, language learners stopped learning them, and they were no longer a part of the grammar.

6. Refinement of the Ease of Processing Hypothesis

However, there are a couple of problems with the idea that the loss of agreement morphology played a key role in the eventual loss of the post-head genitives. First, although the correlation between the frequency of post-head genitives and robust agreement morphology is good in the 12th century, the correlation disappears by the beginning of the 13th century, when we find that post-head genitives are quite uncommon even in the texts with enough agreement morphology available to make post-head genitives as unambiguously case-marked as they had been in OE.

The second problem is related to the first one: such post-head genitives, as are found in the 13th century, are restricted to nouns and quantifiers that select a genitive complement. Agreement morphology by itself does not explain why (10) is found in *V&V*, but strings like *ðare þinge* are not found in non-partitive genitives.

The answer to the first problem might be simply that by the 13th century, *of*-genitives had become so widespread that they had pretty much replaced post-head genitives even in the inflection-rich dialects. This answer, however, does not really explain why there should have been such a good correlation between loss of agreement morphology and reduction of post-head genitives in the 12th century. It also does not offer an explanation for the retention of the post-head partitives even in the inflection-impoverished dialects of the 13th century.

The restriction of post-head genitives at this time to heads which lexically selected a genitive complement suggests another explanation: let us suppose that most genitive phrases got their case marking *structurally* in OE, i.e. genitive was the case that a projection of N got when it was part of a larger nominal phrase. Structural case marking is easily replaced by word order. This cannot happen all at once. There has to be an intermediate stage between case marking being the primary signal of semantic and grammatical relations and word order being the primary signal. How do speakers and hearers cope in a language where neither word order nor case marking is an absolute guide to grammatical relations, as in ME, a language which lacked distinctive case marking but nonetheless retained a certain amount of flexibility in the ordering of objects and subjects, and direct and indirect objects, for example?[23] Let us assume that during this intermediate stage, hearers/readers relied on probabilistic processing strategies to parse sentences, such as assuming that the first NP/DP in a clause was the subject unless such an analysis was ruled out either by case marking or by pragmatics. Such an assumption would explain why the loss of case marking and the fixing of word order go hand in hand, and why word order tends to get progressively more restricted in a language once it starts down the path of relying on word order. It is reasonable to assume, for example, that the origin of SO order is the principle that more topical items are placed first, since subjects are typically more topical than objects. A language learner learns that SO order is unmarked (i.e. more probable) and makes it even more common in his or her own speech, making it even more probable that the first NP will be the subject, and so on, until this order becomes the only one which is found (except when both NPs are pre-verbal).

In the case of adnominal genitives, we have clear evidence that the majority of genitives preceded their heads, rather than followed them, especially in late OE.

[23] For a discussion of the ordering of NPs with these grammatical relations, see Allen (1995, Chap. 9).

Thus a hearer would do pretty well with a processing strategy, which assumed that an NP/DP following an N would begin a separate NP/DP unless there was some evidence to the contrary. Evidence to the contrary could be clear genitive case marking at the beginning of this NP/DP, etc. It is interesting that the non-partitive post-head genitives disappeared in the southern dialects even when clear genitive morphology was available for use. One possibility is that once agreement morphology became generally optional, agreement underwent a fundamental change of status in the grammar. Earlier, it had been important in determining the case of a noun. But we can assume that case agreement was only allowed to become optional because other means had developed to keep track of grammatical relations. This is particularly clear in the optionality of agreement markers, which were not reduced by phonological erosion, such as the masculine singular accusative *-ne* and masculine singular genitive *þæs*. Essentially, agreement morphology had become a redundant adornment to the grammar.

We can thus suggest that at the stage when agreement morphology became optional in a given dialect, case was no longer used as the primary guide to determining the grammatical relation of a noun, and the probabilistic processing strategy just outlined would become an absolute one: you can close off one NP/DP and begin another one when you hear nominal material (a determiner or N, etc.) after N. The only exception was with lexically selected complements, which did not have to be parsed as the beginning of a new NP/DP. It is not hard to understand why this should have been so; with a head such as *nan* 'none' the hearer/reader would know that there was still a genitive to come, if it hadn't come already. Thus there was not as much need to replace these morphological genitives by *of* phrases, although this process was well advanced by the early 13th century.

7. Conclusion

Old English was already undergoing a typological shift away from being a language that relied substantially on case marking. The replacement of the morphological post-head genitive is universally agreed to be part of this general shift. The post-head genitive did not disappear abruptly, however. A reduction in the overall frequency of these genitives is already apparent by late OE, with the post-head genitive increasingly restricted to certain semantic types. This initial decrease cannot be attributed to case syncretism. The *of*-genitive was introduced in OE and in EME became widely used variant even in dialects in which clear genitive marking was available.

The facts give no support for the Early Reanalysis hypothesis as an explanation for the changes of (6). This hypothesis does not explain the retention of some post-head and lexically selected genitives in all dialects in the 13th century.

The widespread acceptance of the idea that the loss of the genitive as a case category in English must have been based earlier on a dubious typological assumption: that it is impossible for a language to maintain a genitive case when other case distinctions have disappeared. A morphological system which opposes a genitive case to a single general case is an uncommon one, but apparently just this system is found in some varieties of Megleno-Romanian,[24] judging by the discussion of Atanasov (1990, pp. 195–197, 203) and Capidan (1925). In these dialects, the dative case has disappeared; instead one must use a preposition and a noun in the common case. Nouns themselves are no longer generally inflected for case. However, the genitive case still remains in the inflection of the suffixal determiner for singular feminine nouns, which is inflected for genitive versus other.[25] Also, the relative pronoun has a special genitive form which is opposed to a general form for the other cases. Another problematic language for this typological claim is Irish, which according to Duffield (1995, p. 270) is traditionally analysed as having a genitive case opposed to a single common case for nouns which are not the object of adpositions (it also has a prepositional case). Although this inflection appears on only the last of the possessor Ns in recursive genitive structures, there can be no doubt that we have an inflection, rather than a clitic, with the Irish genitive (*ibid.*, p. 320), according to the criteria outlined in Zwicky (1977, 1987), because the inflection involves lenition, rather than the addition of a suffix which could have its own syntactic position. Adjectives furthermore agree with the genitive case.

Given the existence of such systems, the systems proposed here for the inflection-poor dialects of EME (common, genitive and prepositional case in *PCII,* subject, object, genitive and prepositional case in the *Ormulum*) do not appear odd, even if they are rather unusual. Typologically unusual systems can come into existence with the right combination of circumstances, and these systems can be learned, although they are likely to disappear fairly soon because of their marked status. Our current state of knowledge about case typology gives us no justification for rejecting the clear evidence for the retention of the genitive as a

---

[24] See Acknowledgements.

[25] Atanasov, who did his fieldwork in the 1970s, gives a slightly different account from Capidan, but interestingly, the major difference is that Atanasov seems to suggest that this special genitive form can be found in plurals also. It is not at all clear that this represents an extension from an earlier situation, but certainly it does not appear that there has been any move towards abolishing the genitive/other distinction. It should be noted that Atanasov refers to this form as genitive/dative, but he indicates that it is essentially never used in a dative function. Atanasov's and Capidan's use of traditional case terminology, which is not based on the categorial distinctions which are formally distinguished, illustrates one of the difficulties which must be overcome by any typological study of possible case systems.

morphological case in English for a period after the accusative/dative distinction was lost.

On the other hand, the facts are consistent with the idea that the decline of post-head genitive originated in the increasing tendency to rely on adpositions and probabilistic processing strategies to decode grammatical relations. This tendency is apparent in OE also with sentence-level constituents, although at that stage, morphology was decisive when there was any conflict with processing strategies. There was less of a need for such a strategy with lexically selected genitives (such as partitives), and so the post-head genitive was retained the longest with these. A combination of the availability of the *of*-genitive for the situations in which a pre-head genitive was disfavoured and the reduced utility of agreement in calculating grammatical relations sealed the fate of the post-head genitive, but the loss of this construction does not reflect the loss of the genitive as a morphological case. Assuming probabilistic processing strategies, the Ease of Processing approach has the advantage of allowing for the variation that we find during the long period of typological shift.

Work on the typology of case-marking systems has focused much more on the case marking of sentence-level constituents than on case-marking within the DP/NP. Nevertheless, some influential work on the history of the genitive case in English has made crucial assumptions about how genitive case fits into case-marking systems universally. More research by typologists on NP/DP-internal case morphology and its interaction with other cases would be a boon to historical linguists. On the other hand, diachronic studies of well-documented languages increase our understanding of the interaction between morphological case systems and other means of encoding grammatical relations. In this study, I hope to have shown how a change in the status of agreement morphology in English first resulted from the increased reliance on adpositions and word order and then contributed to the further decrease in the centrality of case marking in the grammar.

Acknowledgements

I am grateful to Timothy Curnow for pointing to me the Romanian dialects as having a genitive/other system, and to Martin Maiden for kindly pinpointing Megleno-Romanian and supplying references, as well as to Laura Daniliuc for translating the Romanian of the crucial bits of Capidan for me. Thanks also to Victor Friedman for helpful comments made, concerning my interpretation of these references. Any errors I might have made in representing the Megleno-Romanian facts of course remain my own.

References

Allen, C. L. (1995). *Case marking and reanalysis.* Oxford: Oxford University Press.

Allen, C. L. (2002). Case and Middle English noun phrases. In: D. Lightfoot (Ed.), *Syntactic effects of morphological change* (pp. 57–77). Oxford: Oxford University Press.

Allen, C. L. (2003). Deflexion and the development of the genitive in English. *English Language and Linguistics, 7*, 1–28.

Allen, C. L. (to appear). Loss of the post-head genitive in English. To appear in J. Fisiak, & Kang, H. K. (Eds). *Essays on medieval English language and literature in honor of Young-Bae park: A Festschrift.* Seoul: Taehaksa Publishing Co.

Atanasov, P. (1990). *Le Mégléno-roumain de nos jours.* Hamburg: Helmut Buske Verlag.

Blake, B. J. (2001). *Case.* Cambridge and New York: Cambridge University Press.

Brook, G. L., & Leslie, R. F. (Eds). (1963, 1978). *Laȝamon: Brut*. London: Oxford University Press (Early English Text Society 250 and 270).

Calabrese, A. (1998). Some remarks on the Latin case system and its development in Romance. In: J. Lema & E. Treviño (Eds), *Theoretical analyses on Romance languages* (pp. 71–126). Amsterdam and Philadelphia: John Benjamins.

Capidan, T. (1925). *Meglenoromânii, istoria si graiul lor.* Bucharest: Cultura Nationala.

Chomsky, N. (1995). *The minimalist program.* Cambridge, MA: MIT Press.

Clark, C. (Ed.). (1970). *The Peterborough Chronicle 1070–1154.* Oxford: Clarendon Press.

Clemoes, P. (Ed.). (1997). *Ælfric's catholic homilies: The first series: Text.* Oxford: Oxford University Press (Early English Text Society SS 17).

Crisma, P. (to appear). Genitive constructions in the history of English. In: G. Banti, P. Di Giovine & P. Ramat (Eds), *Typological change in the morphosyntax of the Indo–European languages.* München: Lincom Europa.

Duffield, N. (1995). *Particles and projections in Irish syntax.* Dordrecht and Boston, MA: Kluwer Academic Publishers.

Greenberg, J. H. (1966). *Language universals. With special reference to feature hierarchics.* The Hague and Paris: Mouton and Co.

Haeberli, E. (2001). Deriving syntactic effects of morphological case by eliminating abstract case. *Lingua, 111*, 279–313.

Hawkins, J. A. (1994). *A performance theory of order and constituency.* Cambridge and New York: Cambridge University Press.

Holt, R. (Ed.). (1878). *The Ormulum: With the notes and glossary of Dr. R.M. White.* Oxford: Clarendon Press.

Jakobson, R. (1984). Contribution to the general theory of case: General meanings of the Russian cases. In: L. R. Waugh & M. Halle (Eds), *Roman Jakobson Russian and Slavic grammar: Studies, 1931–1981* (pp. 59–103). Berlin and New York: Mouton.

Janda, R. (1980). On the decline of declensional systems: the loss of OE nominal case and the ME reanalysis of *-es* as his. In: E. C. Traugott, R. Labrum & S. Shepard (Eds), *Papers from the fourth international conference on historical linguistics* (pp. 243–252). Amsterdam: Benjamins.

Janda, R. D. (1981). A case of liberation from morphology into syntax: the fate of the English genitive-marker *-(e)s*. In: B. Johns & D. R. Strong (Eds), *Syntactic change* (pp. 60–114). Ann Arbor, Michigan: Department of Linguistics, University of Michigan.

Jespersen, O. (1894). *Progress in language. With special reference to English.* London: Swan Sonnenschein and Co.

Jespersen, O. (1942). *A modern English grammar on historical principles. Morphology* (*Vol. 6*). London: Allen and Unwin.

Kellner, L. (1892). *Historical outlines of English syntax.* London and New York: Macmillan.

Kitson, P. (1992). Old English dialects and the stages of transition to Middle English. *Folia Linguistica Historica, XIII*, 27–87.

Kroch, A. (1989). Reflexes of grammar in patterns of Language change. *Language Variation and Change, 1*, 199–244.

Kroch, A. (1997). Comments on 'Syntax Shindig' papers. *Transactions of the Philological Society, 95*, 133–147.

Kroch, A., & Taylor, A. (compilers). (1999). *The Penn-Helsinki parsed corpus of Middle English 2.* Philadelphia: Department of Linguistics, University of Pennsylvania.

Lass, R. (1992). Phonology and morphology. In: N. Blake (Ed.), *The Cambridge history of the English language. Volume II: 1066–1476* (pp. 23–155). Cambridge: Cambridge University Press.

Lightfoot, D. (1999). *The development of language: Acquisition, change and evolution.* Oxford and Malden, Massachusetts: Blackwell.

McFadden, T. (2002). The rise of the to-dative in Middle English. In: D. Lightfoot (Ed.), *Syntactic effects of morphological change,* (pp. 107–123). Oxford: Oxford University Press.

McLagan, H. (2004). *The syntax of genitive constructions in Old English: The placement of genitive phrases in Ælfric's second series of catholic homilies.* Unpublished master's thesis, School of Language Studies, Australian National University.

Millar, R. M. (2000). *System collapse, system rebirth: The demonstrative pronouns of English, 900–1350 and the birth of the definite article.* Oxford and New York: Peter Lang.

Mitchell, B. (1985). *Old English syntax.* Oxford: Clarendon Press.

Morris, R. (Ed.) (1867–8). *Old English homilies of the 12th and 13th centuries (*Vol. I). London: N. Trübner and Co. [Old English Text Society 29].

Plank, F. (1985). The interpretation and development of form alternations conditioned across word boundaries. The case of *wife's, wives,* and *wives*'. In: R. Eaton, O. Fischer, W. Koopman & F. van der Leek (Eds), *Papers from the 4th international conference on English historical linguistics* (pp. 205–233). Amsterdam and Philadelphia: Benjamins.

Polo, C. (2002). Double objects and morphological triggers for syntactic case. In: D. Lightfoot (Ed.), *Syntactic effects of morphological change* (pp. 124–142). Oxford: Oxford University Press.

Randall, B. (2000). *CorpusSearch: A Java program for searching syntactically annotated corpora.* Philadelphia: Department of Linguistics, University of Pennsylvania.

Rosenbach, A. (2002). *Genitive variation in English: Conceptual factors in synchronic and diachronic studies.* Berlin and Hawthorne, NY: Mouton de Gruyter.

Saitz, R. L. (1955). *Functional word order in Old English subject-object patterns.* Unpublished doctoral dissertation, University of Wisconsin.

Thomas, R. (1931). *Statistical processes involved in the development of the adnominal periphrastic genitive in the English language.* Unpublished doctoral dissertation, University of Michigan, Ann Arbor.

Weerman, F., & de Wit, P. (1999). The decline of the genitive in Dutch. *Linguistics, 37*, 1155–1192.

Zwicky, A. M. (1977). *On clitics.* Bloomington, Indiana: Indiana University Linguistics Club.

Zwicky, A. M. (1987). Suppressing the Zs. *Journal of Linguistics, 23*, 133–148.

Appendix: Notes on the Texts Used

Most of the texts upon which I have based the factual claims made in this chapter are included in the m1 (1150–1250) and m2 (1250–1350) periods of Kroch and Taylor's (1999) *PPCME2.* I have used the CorpusSearch program (Randall, 2000) to search for pre-head and post-head genitives in these texts. Unfortunately, genitive objects cannot be specifically searched for using CorpusSearch since (with good reason, given the variable case marking systems of ME) the corpus is not annotated for case marking; however, the number of (original) inflection-rich prose texts in the 12th and 13th centuries is small enough that it was possible to read the ones used by the *PPCME2* as well as some others looking for such objects.

In addition, I have examined homilies xxv and xxvi of Morris (1867–1868). These two homilies from Cotton MS Vespasian Axxii were probably composed in the 12th century, according to Clemoes (1997, pp. 48–49). I have also read the 'Peterborough Interpolations' (edited by Clark, 1970) and I include them in any observations on *PCI.*

Chapter 10

The On-line Resolution of Subject–Object Ambiguities with and without Case-Marking in Dutch: Evidence from Event-Related Brain Potentials

Monique Lamers

1. Introduction

During the daily use of our native language, we scarcely experience any difficulty in understanding sentences, although numerous problems have to be resolved. For understanding written sentences, visual characters of each word have to be encoded so that the phonological, syntactic and semantic information associated with each word become available. While the incoming word is identified and the stored information related to this word is retrieved, integration of this new information with the already available information from the preceding words has to take place, eventually leading to the final interpretation of the written sentence. The different sorts of information that become available in this incremental process of language understanding are very diverse, varying from morphosyntactic information (e.g. case marking) to semantic/conceptual information (e.g. animacy) on a lexical and/or discourse level. To form a sentence out of the string of incoming words a syntactic relationship between the words is built up. This structure-building process is also known as parsing and forms an essential part of language understanding.

Competition and Variation in Natural Languages: The Case for Case
Edited by M. Amberber and H. de Hoop

ISBN: 0-08-044651-5

In the ideal situation a specific syntactic structure leads to one single interpretation. Natural language, however, is full of structural ambiguities. A sentence is structurally ambiguous if the sentence or a part of the sentence can be grammatically analyzed in more than one way, possibly leading to different interpretations. One of the attributes of a language to prevent these ambiguities is case marking (see also De Swart, this volume). If a sentence starts with a subject or object that is unambiguously case-marked, it is the case information that specifies whether the structure at hand is a subject initial or an object initial sentence. This is exemplified in (1) for German, a language with overtly case-marked NPs.

(1) a. *Der Lehrer* *hat* *den Jungen* *geschlagen*
the[NOM] teacher has the[ACC] boy hit
'The teacher hit the boy'

b. *Den Jungen* *hat* *der Lehrer* *geschlagen*
the[ACC] boy has the[NOM] teacher hit
'The teacher hit the boy'

Thus, unambiguous case marking provides information that guides the parser through a specific syntactic structure with only one possible final interpretation. However, even in a language such as German with a relatively rich case system, proper names, certain noun declinations and pronouns do not bear unambiguous case marking for subjects and object (e.g. die[F.NOM/ACC] Jacke, sie[F.SG.NOM/ACC]). What is more, there are also languages with a poor case system, such as Dutch and English in which personal pronouns are morphologically case-marked, but full NPs and proper names are not. However, in contrast to English, Dutch has a relatively free word order. Strictly speaking, Dutch, but also German, is full of local structural ambiguities. Nevertheless, these ambiguities hardly ever puzzle native speakers of Dutch, even if no disambiguating information becomes available. Hence, other mechanisms and sources of information must be involved in the interpretation of sentences with a (local) structural subject–object ambiguity.

This chapter addresses the question: what kind of processes are involved in the incremental processing of subject–object ambiguities in Dutch. Two experiments will be described in which the same local structural ambiguity is resolved with and without case information. Before presenting the two studies in which event-related brain potentials were registered, I discuss the structure preferences for subject initial sentences over object initial sentences (Section 2). In Section 3, the two reading studies will be presented, each followed by a discussion of the data in relation to the literature. Finally, in Section 4, an alternative interpretation of the effects found in both studies will be discussed in the general discussion.

2. Word Order Preferences in Dutch and German

In Dutch, the word order of verbs and their complements varies depending on the type of clause in which they appear. Not only does the word order differ between declarative sentences and direct questions, as in many other languages, but different declarative sentences show different word orders. As in German, many declarative main clauses exhibit the subject–verb–object (SVO) word order (1a, 2a) while most declarative subordinate clauses have subject–object–verb (SOV) order (2b).

(2) a. *Kees koopt de rode wijn*
Kees buys the red wine
'Kees buys the red wine'

b. *...dat Kees de rode wijn koopt*
...that Kees the red wine buys
'... that Kees buys the red wine'

Since the work of Bach (1962) and Bierwisch (1963) for German, and Koster (1975) for Dutch, it is assumed that the SOV word order is the underlying structure for both main clauses and subordinate clauses. This is shown for Dutch in (3a, b).

(3) a. … [$_{CP}$ dat [$_{IP}$ Kees [$_{VP}$ de wijn koop-]-t]]
... [$_{CP}$ that [$_{IP}$ Kees [$_{VP}$ the wine buy-]-s]]

b. [$_{CP}$Kees$_k$ koopt$_i$ [$_{IP}$ t$_k$ [$_{VP}$de wijn t$_i$] t$_i$]]
[$_{CP}$Kees$_k$ buys$_i$ [$_{IP}$ t$_k$ [$_{VP}$ the wine t$_i$] t$_i$]]

According to the analyses in a generative framework, the word order of a main clause is accomplished by moving the verb from the verb final position (as in a subordinate clause) to the complementizer position directly following the first constituent in Spec, CP (see 3b). In most declarative main clauses, the first constituent is the subject of the sentence that is moved from its base-generated position (Spec, IP) as in (1a) and (2a, b) to sentence initial position (Spec, CP). Such movement, however, is not restricted to subjects, direct objects can also move to the Spec, CP position preceding the verb (4a, b), as can many other sorts of constituents (topicalization). Although SVO word order is more frequently used in main clauses than the object–verb–subject (OVS) word order, the OVS word order can be used when answering a question, or for emphasis or contrast (cf. Nieuwborg, 1968; McDonald, 1987; Haegeman, 1991; Zwart, 1993).

(4) a. $[_{CP}$De wijn$_k$ koopt$_i$ $[_{IP}$ Kees $[_{VP}$ t_k $t_i]$ $t_i]]$
$[_{CP}$The wine$_k$ buys$_i$ $[_{IP}$ Kees $[_{VP}$ t_k $t_i]$ $t_i]]$

b. $[_{CP}$Den Wein$_k$ kauft$_i$ $[_{IP}$ Kees $[_{VP}$ t_k $t_i]$ $t_i]]$
$[_{CP}$The wine$_k$ buys$_i$ $[_{IP}$ Kees $[_{VP}$ t_k $t_i]$ $t_i]]$

It is a well-known fact that in both Dutch and German, object-before-subject sentences are less preferred than subject-before object sentences (Frazier, 1987, 1993; Frazier & Flores d'Arcais, 1989; Hemforth, 1993; Kaan, 1997; Konieczny, Hemforth, Scheepers, & Strube, 1997; Gorrell, 2000).

Thus, without any disambiguating information such as case marking, a main clause is (temporarily) ambiguous between a subject-initial (SI) and an object-initial (OI)[1] structure. Case information, however, is not the only source of morphological disambiguating information. In languages such as German, Dutch and English, as well as in many other languages number information might help to identify the subject of the sentence because of the number agreement between the subject and the predicate. To investigate mechanisms involved in the parsing process of Dutch main clauses with a subject–object ambiguity (Frazier & Flores d'Arcais, 1989) used number agreement in a speeded grammatical judgment study. An example of a set of sentences used in this study is given in (5).

(5) a. *De patiënt* *bezoekt* *de dokters*
the patient[SG] visits[SG] the doctors[PL]
'The patient visits the doctors'

b. *De patiënt* *bezoeken* *de dokters*
the patient[SG] visit[PL] the doctors[PL]
'The doctors visit the patient'

According to the syntactic analysis sketched above, the initial NP is moved out of IP to Spec, CP leaving a gap as is illustrated in (6) (see also (3) and (4)). A moved item, the so-called 'filler', leaves a 'gap', forming a filler–gap dependency. Frazier and Flores D'Arcais investigated their hypothesis that the first NP will be assigned to the earliest possible gap, which is the trace of the subject.

(6) a. $[_{CP}$De patiënt$_k$ bezoekt$_i$ $[_{IP}$ t_k $[_{VP}$ de dokters $t_i]$ $t_i]]$
the patient[SG] visit[SG] the doctors[PL]
'The patient visits the doctors'

[1] In this chapter, the term *subject initial* (SI) is used for all structures in which the subject precedes the object in surface structure. Similarly, the term object initial (OI) includes all sentences in which the object precedes the subject in surface structure.

b. [$_{CP}$De patiënt$_k$ bezoeken$_i$ [$_{IP}$ de dokters [$_{VP}$ t_k t_i] t_i]]
the patient[SG] visit[PL] the doctors[PL]
'The doctors visit the patient'

This so-called active filler strategy (AFS) explains the preference for SI structures over OI structures.[2] Notice that the sentence in (5a) is ambiguous until the second NP is encountered. Following the preference for subject-initial sentences in Dutch, in the incremental parsing process the initial NP will be parsed as the subject of the sentence, filling the earliest possible gap and fitting the preferred structure. The singular verb is compatible with this analysis. The second NP can thus be analyzed as the direct object. Moreover, in (5a) the second NP turns out to be plural, and thus, cannot be the subject. In (5b), on the other hand, the verb is plural and the first NP is singular. Here it is clear at the verb that the first NP cannot be the subject of the sentence because of the number features of the verb. The second NP agrees in number with the verb and is therefore suitable as the subject of the sentence. It is already apparent at the verb that the preferred structure cannot hold. Hence, reanalysis is in order, in which the first NP has to be relabelled from subject to direct object and the second NP has to be the subject, forming the less-preferred OVS structure. Results reported by Frazier and Flores d'Arcais showed faster grammaticality judgments for the SI sentences than for the OI sentences; SI sentences were also responded to more accuratcly than OI sentences. These results indicate that there is an advantage in analyzing sentences that start with a subject in comparison to those starting with a direct object.

The study of Frazier and Flores d'Arcais is one of many studies in which a preference for SVO sentences has been demonstrated. More specifically, experimental research has shown that Dutch SI structures are less difficult to process than OI structures, not only in main declarative clauses but also in relative clauses, and embedded wh-clauses (Frazier, 1987; Frazier & Flores d'Arcais, 1989; Kaan, 1997). However, the active filler strategy is not the only mechanism that can account for a preference for SI sentences over OI sentences. Among others, the Minimal Chain Principle proposed by De Vincenzi (1991), in which it is assumed that the parser always postulates the fewest number of chain members possible, and the Principle of Simplicity, a proposal based on structural parsimony put forward by Gorrell (1996, 2000) predict a preference for subject-before-object over object-before-subject word order in declarative sentences. Although the mechanisms that are held responsible for the observed preference

---

[2] Building up an initial parse as an SI structure in case there is no disambiguating information is available, is also in line with the so-called word order freezing phenomenon as described by Lee (2001) and Zeevat (2004) (see also De Swart, this volume).

for the SI structure differ between the models, they all have in common that they rely solely on syntactic information.

Recent research, however, has shown that the influence of non-syntactic information such as thematic fit, discourse information frequency, and subcategorization, all potentially contribute to the ongoing analysis of the sentence structure (MacDonald, 1993; MacDonald, Pearlmutter, & Seidenburg, 1994; Trueswell, Tanenhaus, & Garnsey, 1994; Brown, Van Berkum, & Hagoort, 2000). Part of the evidence for the use of different sorts of information, the nature and exact time course of the underlying processes in sentence processing was obtained from studies in which event-related brain potentials (ERPs) were used (for general information on ERPs: Münte, Uabach, Düzel, & Kulas, 2000; for language: Kutas & Petten, 1994; Osterhout & Holcomb, 1995; and Kutas, Federmecer, Seana, King, & Münte, 2000). ERPs are multidimensional in nature, and effects (a.k.a. components) have been described that primarily index the processing of semantic/conceptual information and syntactic information. An enhanced negative effect with a peak around 400 ms post-stimulus largest over central and right posterior electrode sites ('N400') is found for difficulties related to semantic integration processes (Kutas & Hillyard, 1980, 1983; Brown & Hagoort, 1993; Hagoort & Brown, 1994), as well as for discourse integration (St George, Kutas, Martinez, & Sereno, 1999; van Berkum, Hagoort, & Brown, 1999). In relation to syntactic processing, a late positivity with an onset of approximately 500 ms with a maximum at 600 ms after onset of the critical word at centroparietal sites has been reported ('P600'). The problems that elicit this late positivity vary from syntactic violations such as agreement violations (Hagoort, Brown, & Groothusen, 1993), phrase structure violations (McKinnon & Osterhout, 1996; Hagoort & Brown, 2000) as well as reanalysis of garden-path sentences or (sub)processes involved in computing a less-preferred structure in local structure ambiguity resolution (Osterhout & Holcomb, 1992; Osterhout, Holcomb, & Swinney, 1994; Mecklinger, Schriefers, Steinhauer, & Friederici, 1995; Rösler, Pechmann, Streb, Roder, & Hennighauser, 1998) and structure complexity (Kaan & Swaab, 2003a). In terms of its functional interpretation, the P600 has been viewed to reflect syntactic processing difficulty of different varieties, such as the inability of the parser to assign the preferred structure to the incoming words (Hagoort, 2003), syntactic reanalysis (Friederici, 1995; Gunter, Stowe, & Mulder, 1997), or syntactic integration difficulty (Kaan, Harris, Gibson, & Holcomb, 2000). It has to be noted, however, that a purely syntactic interpretation of the P600 has been challenged by findings of P600 effects to semantic anomalies (Münte, Heinze, & Mangun, 1998), and semantic reversal anomalies (Kolk, Chwilla, Herten, & Oor, 2003).

Although it became clear by making use of ERPs that different kinds of processes and different sorts of information may play an immediate role in sentence comprehension, it is still a matter of debate whether the underlying processes of subject–object ambiguity resolution differ depending on the sort of information used to disambiguate the sentence. This is because, in none of the studies mentioned so far, different sorts of information were used to resolve the same local structural ambiguity. Because of the case system in combination with the free word order in Dutch, it is possible to investigate ambiguity resolution of the same local structural ambiguity with and without case information. In the following section, two experiments will be described in which the same subject–object local structural ambiguity (NP–V–NP) in Dutch will be disambiguated either by case information of the second NP (i.e. case-marked pronoun) (Experiments 1 and 2) or because of the animacy of the first NP in combination with the selectional restriction of the verb (Experiment 2).

3. Ambiguity Resolution with and without Case Information

3.1. Introduction

In the two studies presented in this section, ERPs were registered during the online comprehension of visually presented sentences with a local structural ambiguity. The same subject–object ambiguity was used in both studies, but the kind of disambiguating information was different. Experiment 1 addresses the use of case-marking information in the resolution of a structural local ambiguity. Ambiguous NP–V–NP structures were used that were disambiguated at the second NP, a case-marked pronoun. In Experiment 2 the same NP–V–NP structure was used, but with a different manipulation. In contrast to Experiment 1, not only syntactic, but also semantic/conceptual information was available to solve the structure-building problem of the ambiguous sentences. The animacy of the first NP was manipulated in such a way that the sentences starting with an inanimate NP were disambiguated at the verb, because of the selectional restrictions of the verb. The sentences starting with an animate NP were disambiguated by morpho-syntactic information on the second NP, which was a case-marked pronoun. Using both syntactic information (i.e. morpho-syntactic information at the case-marked pronoun) and semantic/conceptual information (i.e. the animacy of the first NP and the selectional restrictions of the verb) in one experiment is thought to gain more insight into the role of the different sorts of information in resolving a local structural ambiguity. Moreover, if the use of different sorts of information leads to differences in the online process resolving the local structural ambiguity, this might be revealed by the differences in the ERP-waveforms either within Experiment 1, or in the comparison of the results between Experiments 1 and 2.

More specifically, finding an N400-like effect will be taken as evidence for making use of semantic information involving differences in lexical integration to resolve the local structural ambiguity, whereas a P600 indicates the involvement of merely syntactically driven processes possibly involving processes of reanalysis as well as differences in complexity of building the final structure.

3.2. Resolving a Subject–Object Ambiguity by Making Use of Case Information

The first experiment focuses on the use of case information in ambiguity resolution. Ambiguous NP–V–NP structures were used starting with a non-pronominal NP, as exemplified in (7). In Dutch, non-pronominal NPs are not overtly case-marked. They can be either the subject or the object of a sentence. A prepositional phrase, a transitive verb, and the second NP followed this initial NP. The second NP was a case-marked pronoun, providing disambiguating information. The accusative case-marked pronoun *hem* 'him' disambiguates the structure into an SI structure, whereas the nominative case-marked pronoun *hij* 'he' disambiguates the structure into an OI structure (7a and 7b, respectively). The sentences ended with an adverbial or prepositional phrase.

7. a. *De oude vrouw* *in de straat* *verzorgde* *hem*
   the old woman in the street took_care_of him[ACC]
   vrijwel *elke* *dag*
   almost every day
   'He took care of the old woman in the street almost every day'

   b. *De oude vrouw* *in de straat* *verzorgde* *hij*
   the old woman in the street took_care_of he[NOM]
   vrijwel *elke* *dag*
   almost every day
   'The old woman in the street took care of him almost every day'

Notice that because of the number agreement between the verb and the first NP, the point of disambiguation was set at the second noun phrase, the case-marked pronoun. To ensure that the sentences were ambiguous until the pronoun, initial NPs that were equally plausible in the subject and object function associated with the verb were selected from a pretest in which the semantic/conceptual plausibility of the first noun phrase as the subject or the object of the sentence was rated.

No differences between the two conditions are expected in the ERP-waveforms before the case-marked pronoun. As argued above, following the SI structure preference, the first NP is assumed to be parsed initially as the subject of the sentence. If the transitive verb is followed by the accusative case-marked pronoun *hem* 'him', as in (7a), no problems are expected. *Hem* fits the initial parse, and the SI structure turns out to be the final structure. In contrast, the nominative case-marked pronoun *hij* 'he', as in (7b), has to be the subject of the sentence and the initial parse has to be reanalyzed to an OI structure. Part of the reanalysis process involves the 'relabelling' of the first NP from the subject to the object of the sentence. The nominative case-marked pronoun *hij* has to be the subject, consequently the first NP can no longer be the subject of the sentence, but has to be the object. It has been suggested that a P600/SPS is elicited when a less-preferred or marked structure must be constructed after the preferred or unmarked structure from the initial parse no longer holds (Osterhout & Holcomb, 1992; Hagoort et al., 1993; Mecklinger et al., 1995). Hence, a P600 is predicted to appear in the ERP-waveform as soon as the disambiguating information of the nominative case-marked pronoun *hij* (7b) is available, compared to the ERP-waveform at the accusative case-marked pronoun *hem* (7a). Since the problem in resolving the subject–object ambiguity at the disambiguating word lies in building the appropriate structure rather than in the semantic integration of the nominative case-marked pronoun, there is no reason to expect an N400 effect at the disambiguating word due to differences in semantic processing (Brown & Hagoort, 1993; Kutas & Petten, 1994).

3.2.1. Method

3.2.1.1. Materials Sixty-three high-frequency transitive verbs, which can take animate subjects and objects, were chosen from the Celex corpus (Burnage, 1990). In order to choose a noun phrase that was equally plausible as either a subject or object, a plausibility rating was carried out. In this pretest, the verbs were presented in combination with several different noun phrases, and two types of questions were asked, one referring to the subject role as in condition (7b): "How likely is it for a ⟨noun phrase⟩ to ⟨verb⟩?" and one referring to the role of the object as in (7a): "How likely is it for a ⟨noun phrase⟩ to be ⟨verb-ed⟩?" (cf. Trueswell et al., 1994). The questions were assigned to two lists such that only the object or subject role of a verb-noun phrase combination was rated on a given list. Subjects had to rate combinations of verb and noun phrases on a scale from 1 to 20 (1 = totally unlikely, 20 = very likely). A total of 44 undergraduate students filled out the questionnaire; thus 22 ratings were obtained for each combination.

On the basis of the plausibility rating, 40 verb-noun phrase combinations were chosen. Only verb–noun phrase combinations with a rating higher than 10 for both roles and with a marginal difference between the ratings of the two roles were selected. Forty sentence pairs (see (7a) and (7b)) were constructed with the selected verb–noun phrase combinations. There was no significant difference in rating for the subject and object role (X(mean rating subject role) = 13.7, st.dev. = 1.8; X(mean rating object role) = 14.1, st.dev. = 1.8; Wilcoxon $W = 1526.5$, $Z = -0.90$; $p > 0.25$).

The experimental sentence pairs were divided across two lists, such that half of the sentence pairs appeared in SI form on one list and half on the other. The other half appeared in OI form. The 40 experimental sentences of each list and 88 filler sentences were divided over four blocks. Within each block the SI, OI and filler-sentences were randomized. Each block started with two filler sentences, ended with a filler sentence, and contained ten experimental sentences (5 SI sentences and 5 OI sentences) and 22 filler sentences with four different constructions including subject initial sentences starting with a nominative case-marked pronoun, subject–initial sentences followed with an accusative case-marked pronoun as the second argument, and some object initial sentences. The four blocks were presented to the subjects in a counter-balanced order; 24 subjects read each list.

3.2.1.2. Subjects Forty-eight healthy, right-handed with normal or corrected to normal vision, undergraduate students, all native speakers of Dutch (26 female and 22 male subjects, 18–25 years old) participated. None of them had taken part in the plausibility rating.

3.2.1.3. Procedure During the ERP registration, subjects were seated in a dimly lit, sound-attenuating room facing a colour video screen at a distance of 110 cm. Sentences were presented word by word every 500 ms (250 ms word presentation, 250 ms blank screen). After the last word of the sentence, a blank screen of 800 ms was presented, followed by a fixation asterisk (1500 ms) and a blank screen (250 ms) before the first word of the next sentence. To check whether the subjects understood the sentences, 4 experimental sentences (2 in each condition) in each block were followed by yes–no questions addressing the subject/object role of the first NP, as were 5 filler sentences. Subjects had to answer with a finger lift response (right = yes answer; left = no answer). The experiment started with a practice session during which the subjects received feedback about their performance on the questions. Instructions were given to read the sentences for content and to blink only while the asterisk was on the screen. Each block lasted 8 min; the entire experiment, including electrode application and removal, took less than 2 h.

3.2.1.4. EEG Recordings Recordings were made with tin electrodes in an electrocap. Five electrodes were positioned over the midline: FzA (AFz), Fz, Cz, Pz, Oz; and 13 over the left and 13 over the right hemisphere: F7A (AF7), F8A (AF8), F3A (AF3), F4A (AF4), F7, F8, F5, F6, F5P (FC5), F6P (FC6), F3, F4,

F3P (FC3), F4P (FC4), C3, C4, C5, C6, T3 (T7), T4 (T8), T3A (FT7), T4A (FT8), T5 (P7), T6 (P8), O1, and O2. Nomenclature within brackets follows the proposals of the Electrode Position Nomenclature Committee (1991). Bipolar EOG was recorded between electrodes at the outer left canthus and above the eyebrow of the left eye. The ground electrode was placed on the sternum. The resistance of the electrodes was kept under 5 k Ω. The signals of each electrode were amplified, bandpass filtered between 0.05 and 35 Hz, and digitized with a sampling frequency of 100 Hz.

For the two types of experimental sentences, average ERPs were computed for each subject, starting at the onset of the verb (i.e. one word before the critical word *hem/hij*) and lasting until 1000 ms after presentation of the last word of the sentence. Due to eye movements, electrode drift and amplifier blocking, approximately 4% of the trials were rejected. The 100 ms directly before the onset of the pronoun was used as a baseline.

3.2.2. Results

3.2.2.1. Comprehension question Performance on the 16 comprehension questions concerning the experimental sentences was above chance for all subjects. The percentage of correct answers varied from 56 to 100%; the average of correct answers over 48 subjects was 83%. On the basis of this result and the verbal report of the subjects that there were no problems in comprehension, it can be concluded that the subjects appropriately comprehended the sentences.

3.2.2.2. Event-related brain potentials The averaged ERPs of OI sentences show a positive shift relative to SI sentences which started approximately 150 ms after onset of the disambiguating word *hij* 'he', most prominent over the frontal and central regions (Figures 1 and 2). This positivity seems quite sustained, but statistical analyses and scalp distributions showed otherwise (Table 1). Additionally, the ERPs show a positivity for the OI sentences, starting approximately 600 ms after onset of the disambiguating word, most prominent over the right hemisphere, and continuing for approximately 400 ms (Figures 1 and 2).

3.2.3. Data analysis The data were statistically evaluated using the mean amplitudes of 40 successive windows of 30 ms each.[3] In this way it was possible to eval-

[3] The first is the interval from 0 to 30 ms after onset from the disambiguating word; the second interval is the interval from 30 to 60 ms after onset of the disambiguating word, followed by the third interval from 60 to 90 ms after onset of the disambiguating word, and so on.

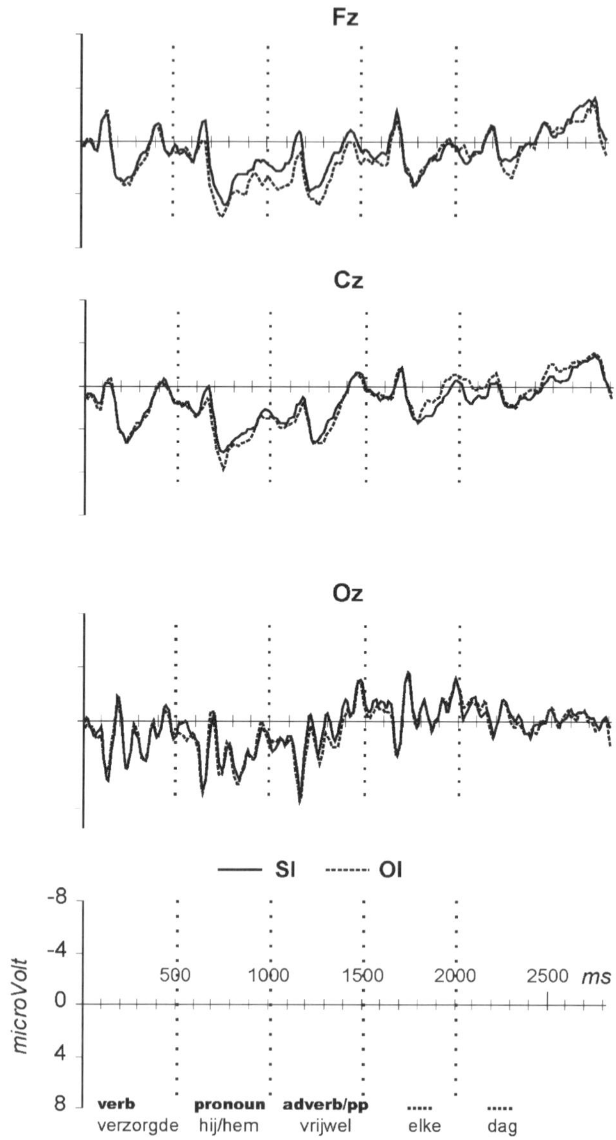

Figure 1: Average ERPs at three midline electrode sites (Fz, Cz, Oz) for SI and OI sentences with a baseline 100 ms before onset of the verb (SI = subject-initial sentences; OI = object-initial sentences; vertical dotted lines mark the onset of the words in the sentence).

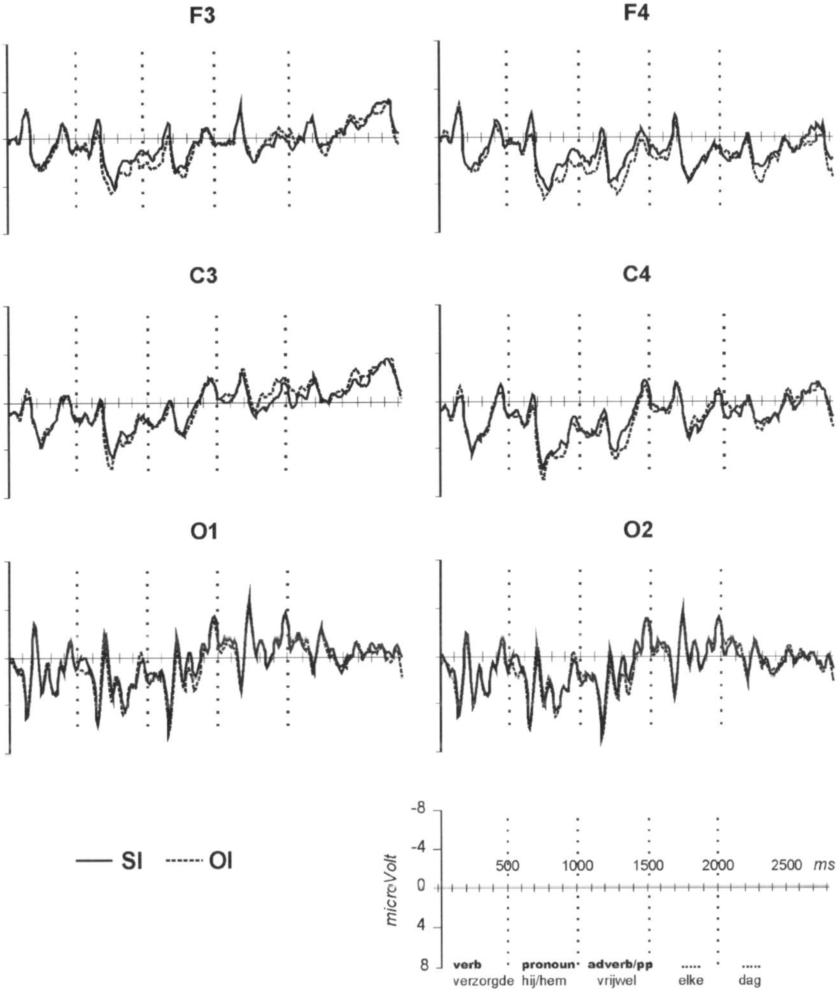

Figure 2: Average ERPs at six electrode sites (F3/F4, C3/C4, and O1/O2) for SI and OI sentences with a baseline 100 ms before the onset of the verb (SI = subject initial, OI = object initial; vertical dotted lines mark the onset of the words in the sentence).

uate the whole sentence from one word before the critical word until the word following the critical word, instead of looking at specific time windows. The results of separate MANOVAs for each electrode site with the Type of syntactic structure

Table 1: Ranges for succesive intervals with a significant effect ($p < 0.05$) according to MANOVAs (type of syntactic structure (SI, OI) as within subject factor). Significant effects were found in four time windows after onset of the disambiguating pronoun.

Midline				
	(1)	(2)	(3)	(4)
FzA	150–270		600–720	750–810
Fz	150–270	330–420	570–690	780–840
Cz	150–270			
Pz				
Oz				

Left sites					Right sites				
	(1)	(2)	(3)	(4)		(1)	(2)	(3)	(4)
F3A	150–270		510–720	750–840	F4A		540–690	750–990	
F7A					F8A		600–690		
F3	150–240		600–690		F4		600–690	780–840	
F5					F6		600–690	780–870	
F7		330–420	450–660		F8			780–870	
F3P	150–270	330–420			F4P			780–870	
F5P	180–270				F6P			780–870	
T3A	150–270				T4A			780–870	
C3	180–270				C4				
C5	180–270				C6			780–900	
T3	180–270				T4			750–930	
T5					T6				
O1					O2				

(SI or OI) as within subjects factor are shown in Table 1. Significant effects were found in four ranges of intervals:[4] from 150 to 270 ms and from 330 to 420 ms at left central frontal electrode sites, from 600 to 690 ms at the frontal electrode sites, and from 780 ms to 870 ms at the electrode sites on the right side of the scalp. In these ranges all intervals showed significant effects ($p < 0.05$). Time windows are only considered significant when three successive intervals (= 90 ms) showed these effects. This minimizes the danger of false positivities due to the large number of comparisons.

3.2.4. Discussion In this study, ERPs were registered to investigate the use of case information in the resolution of subject–object ambiguities in Dutch. The NP–V–NP main clause was disambiguated into an SI or an OI structure sentence at the second NP, which was an accusative or nominative case-marked pronoun. It was assumed that the initial parse of a subject–object ambiguity is SI. Hence, the problem of resolving the ambiguity involves reanalysis of the initial parse for the OI sentences. Therefore, a P600 was expected, elicited at the disambiguating pronoun *hij*. There was no reason to expect an N400 effect at the pronoun, since there were no semantic differences between the two structures at this point.

Four temporally distinct positivities for the OI sentences compared to the SI sentences were found after onset of the disambiguating pronoun. The first positivity (~ 150 ms after onset of the disambiguating pronoun) in the left frontal and left central regions starts very early. It seems to occur as soon as the morpho-syntactic form of the pronoun is detected. If the first noun phrase of an OI sentence is initially parsed as the subject, the positivity occurs as soon as the reader recognizes that the case-marked pronoun is not an object. So, this early positivity may be a reflection of the detection of an input to the parser, which does not fit into the most preferred structure. Alternatively, it may reflect the start of reanalysis, which will be discussed in more detail in the general discussion.

The data show a second positivity, starting approximately 330 ms after onset of the disambiguating pronoun. A positivity in approximately the same time frame is described by Mecklinger et al. (1995). Comparing ERPs of sentences with a subject-relative clause and sentences with an object-relative clause in German, they found a positivity on the disambiguating sentence-final word with a centro-parietal maximum scalp distribution, which they called the P345. Because of the similarities with the P300, they concluded that the P345 is a member of the P300 family. They argued that syntactic reanalysis occurs when the

[4] Intervals of 30 ms are designated by their midpoints. Interval 165 is thus the interval from 150 to 180 ms after onset of the disambiguating pronoun.

perceived syntactic structure (i.e. the less-preferred object-relative reading) does not map onto the initially selected structure (i.e. the subject-relative reading). The input of the disambiguating word (i.e. the sentence-final auxiliary) is an unexpected input.

The positivity that is reported here with a maximum at (left) frontal electrode sites differs in scalp distribution from the P345 with a centroparietal maximum reported by Mecklinger et al. This suggests that the effects in these two studies are distinct. Although both studies tested grammatical sentences without any violations, the structures used in both studies were very different (SI/OI main clauses in Dutch with a case-marked pronoun as the disambiguating word in the middle of the sentence vs. object/subject relative clauses with a disambiguating sentence-final auxiliary in German).[5] Nevertheless, in both studies it is possible that reanalysis is triggered by the morpho-syntactic form of the disambiguating words: the case-marked pronoun in this study and number information on the sentence-final auxiliary in the German study. It may well be that reanalyzing a syntactic structure on the basis of a case-marked pronoun is to a greater or lesser extent different from reanalysing a syntactic structure on the basis of verb-agreement information (cf. Kaan, 1997). Difference in scalp distribution, amplitude, and duration between the P345 found by Mecklinger and his colleagues and the positivity in the same range of time found in this study can possibly be ascribed to this difference in disambiguating information.

Whether the later positivities (starting at ~ 600 ms and 750 ms after onset of the disambiguating pronoun) are similar to the P600 found in the studies of Hagoort et al. (1993) and Osterhout and Holcomb (1992) is not clear. Although they partially overlap in time course and scalp distribution, there are also differences. Both of the later positivities found in this study are more prominent over the right hemisphere. The positivity starting in the range of 750 ms has a more broadly distributed maximum over the right hemisphere, while the positivity starting at approximately 600 ms is more restricted to the frontal region of the right hemisphere. The 'classic' P600 is most prominent in the centroparietal regions and starts approximately 500 ms after the onset of the disambiguating word. In this experiment, words were presented every 500 ms (250 ms word presentation, 250 ms blank screen). Therefore, the effect elicited by the presentation of the word following the disambiguating pronoun coincides with the onset of the 'classic' P600. This overlap of events complicates the comparison of the

[5] In contrast to the sentences used in this study, Mecklinger et al. (1995) used sentences with the disambiguating word at the end of the sentence. This can create a problem of overlapping components: the effects elicited by the disambiguating word and by closure effects, which occur at the end of a sentence (cf. Hagoort et al., 1993; Osterhout & Holcomb, 1992).

later positivities found in this study with the P600 reported by Hagoort et al. (1993) and Osterhout and Holcomb (1992). In the general discussion (Section 4), a functional interpretation of the positivities will be discussed in more detail.

In summary, the results of this experiment show that (1) unambiguous case information was used to resolve the subject–object ambiguity; (2) according to the early onset of the differences found in ERP waveforms, the case information was immediately used; and (3) the processes involved were merely syntactical or related to structure building processes (reanalysis) rather than semantical.

3.3. Resolving a Subject–Object Ambiguity by Making Use of Animacy Information

In the previous experiment, morpho-syntactic information from a case-marked pronoun was used to disambiguate sentences with a local structural ambiguity. In the second experiment, two different sorts of information were used to disambiguate the same subject–object ambiguity. Sentences starting with an animate NP were disambiguated by the second NP, the case-marked pronoun *hij* 'he' (8a), whereas sentences starting with an inanimate NP were disambiguated by semantic/conceptual information at the verb (8b). Notice that all experimental sentences were OI starting with either an animate or an inanimate NP.

(8) a. *De oude vrouw* *in de straat* *verzorgde* *hij*
the old woman in the street took_care_of he[NOM]
vrijwel elke dag
almost every day
'He took care of the old woman in the street almost every day'

b. *Het oude park* *in de straat* *verzorgde* *hij*
the old park in the street took_care_of he[NOM]
vrijwel elke dag
almost every day
'He took care of the old park in the street almost every day'

As for the first experiment, it is assumed that the sentences of both conditions will initially be parsed as SI sentences, which is the preferred structure. The sentence starting with an animate NP (8a) can be parsed as such without any problems until the case-marked pronoun *hij* 'he'. In contrast, for sentences starting with an inanimate NP (8b) disambiguation can already take place at the verb. Since in this experiment, verbs were chosen that can only take an animate NP, it becomes clear at the verb that the initial inanimate NP does not meet the selectional restrictions of the verb. What is more, if the problem at the verb in condition (8b) is handled as a problem of integrating the verbal semantics into the initial structure, it is expected that an N400 will occur. As explained above, the N400 is an effect known to occur when lexical integration problems arise. This prediction is in line with the results of an ERP study by Weckerly and Kutas (1999). To investigate how and when animacy influences the analysis of a complex syntactic structure, they recorded ERPs during the processing of object-relative clauses. They manipulated the animacy word order: inanimate–animate (9a), and animate–inanimate (9b).

(9) a. *The poetry that the editor recognized depressed the publisher of the struggling...*
  b. *The editor that the poetry depressed recognized the publisher of the struggling...*

Reading times showed an advantage for sentences with an inanimate main clause subject and an animate relative clause subject (9a) compared to the sentences with the inverse animacy order (9b) at and beyond the relative clause subject. Based on these findings, an N400 effect was predicted for the relative clause subject, reflecting the difficulty of integrating an inanimate NP at the relative clause subject position.[6] As soon as it is clear at the determiner following the complimentizer that an object relative clause has to be parsed, an animate NP is anticipated for to fulfil the function of subject of the relative clause.[7] Hence, no difficulties in integrating the animate NP as the subject of the relative clause were expected (9a), whereas the integration of an *inanimate* NP as the subject was thought to be problematic. Among other effects, the results showed an enlarged N400 at the inanimate second noun of the animate–inanimate relative clause (9b), compared to the animate second noun of the inanimate–animate relative clause (9a). The finding of an N400 led to the conclusion that the animacy was registered and caused lexical integration difficulties of the inanimate NP as the subject of the relative clause. As explained above, a similar lexical integration problem can be

[6] A difference in the N400 was not the only ERP effect that was expected to be elicited. Weckerly and Kutas (1999) also predicted that they would find differences in ERP-components known to be related to differences in working processes of reanalysis.

[7] Given the combination of an animate and inanimate NP as the arguments of a transitive relation it is most likely that the animate NP fulfils the function of subject, whereas the inanimate NP is the direct object. Although this is certainly not true for all transitive verbs, it holds for the largest category of transitive verbs, namely the active verbs. For this type of verb, it is the subject that performs the action expressed by the verb. Because animate NPs can be volitional and intentional, they have the right characteristics to fulfil this function, whereas inanimate NPs lack these abilities. For a more detailed description of transitive verbs and argument relation the reader is referred to Dowty (1989, 1991) and Fillmore (1968) (see also Lamers, 2001).

expected at the verb for the sentences starting with an inanimate NP in Experiment 2.

There is, however, an alternative way to handle the selectional restriction problem. Reanalyzing the sentence from an SI structure into an OI structure can also resolve the sentence processing problem. Since the verb is transitive, it needs two arguments. If the first argument does not meet the selectional restrictions as the subject, it is still possible that it suits the criteria of the other argument. Under this account, the first NP has to be relabelled from subject to object. Hence, reanalysis of an SI structure into an OI structure occurs. If reanalysis occurs, the problem lies in building the appropriate structure rather than in semantic interpretation. Therefore, a positivity like the P600 is expected at the verb for sentences starting with an inanimate NP (8b) compared to sentences starting with an animate NP (8a). Processes of reanalysis certainly do not rule out the involvement of semantic/conceptual information in resolving the local ambiguity, which may include the possible lexical integration problems of the verb. On this account an N400–P600 complex might be expected, reflecting the involvement of semantic integration processes as well as final structure building processes, whereas a single P600 indicates that the resolution is merely syntactically driven based on the syntactic information of the verb (i.e. verb transitivity).

At the case marked pronoun *hij* 'he', disambiguating information is available for sentences starting with an animate NP (2a). Although in the present experiment there is no comparison possible with an SI structure, based on the results of Experiment 1, an early positivity is predicted elicited by the pronoun for the sentences starting with an animate NP (2a). In the sentences starting with an inanimate NP (2b), disambiguating information becomes available at the verb. The case-marked pronoun *hij* 'he' following the verb fits the reanalyzed OI structure. Hence, the situation at the case-marked pronoun is different for the two conditions although reanalysis of a preferred initial parse into an OI structure occurs in both conditions. Whether this difference will be apparent in the waveform as such, is not entirely obvious. In a number of studies in which reanalysis of the first parse has been described, a sustained positivity is observed in a late time frame starting at approximately 500 ms after the critical word (Hagoort et al., 1993; Osterhout & Holcomb, 1993). If such a positivity is elicited at the verb in (2b), it could well be that the onset of the positivity elicited by the pronoun in (2a) falls within the time frame in which the late positivity elicited by the pronoun in (2b) occurs. Although a slower presentation time of the words was chosen in anticipation of a late positivity, the possible occurrence of two positivities (one in each of the two conditions) in (partly) the same time frame could result in little difference in the ERP-waveform at the case-marked pronoun between the two conditions.

3.3.1. Methods

3.3.1.1. Materials From the plausibility rating of Experiment 1, animate NPs were chosen that can either be the subject or the object of the verb, and inanimate NPs were chosen that are only as good as the object of the verb. As in the previously described experiment, only those animate NPs were selected for which the verb–noun phrase combinations were rated higher than 10 for both roles (on a scale from 1 = not plausible to 20 = very plausible), with a marginal difference between the ratings of the two roles. The inanimate NPs that were included in the experiment were rated very low (below 4) for the subject role and higher than 10 for the object role.

Forty sentence pairs (see (8a) and (8b)) were constructed with the selected verb–noun phrase combinations. There was no significant difference in rating for the animate and inanimate NPs in the role of object of the sentences (*X*(mean rating animate NP)=14.1, st.dev. = 1.8; *X*(mean rating inanimate NP)=14.4, st.dev. =1.8; Wilcoxon W = 1516.5, Z = −1.00; $p > 0.25$). Animate and inanimate NPs were matched for frequency and length ($t_{log.freq}(78) = -1.05$, $p = 0.29$; $t_{length}(78) = 0.987$, $p = 0.33$). The experimental sentence pairs were divided across two lists, such that half of the sentences of one list started with an animate NP and the other half started with an inanimate NP. There were no repetitions of verbs within a list. The experimental sentences of each list and 80 filler sentences were divided over four blocks. Within each block, the experimental sentences and filler-sentences were randomized. Each block started with two filler sentences, ended with a filler sentence, and contained ten experimental sentences (5 starting with an animate NP and 5 with an inanimate NP) and 20 filler sentences (5 OI sentences, 15 SI sentences of which 5 with an accusative case-marked pronoun as the second NP, 3 started with a nominative case-marked pronoun, and 12 with two full NPs). Each list was presented to 18 subjects with the four blocks in counterbalanced order. Each block lasted 8 min; the four experimental blocks, the practice session and the instructions took no longer than 60 min.

3.3.1.2. Subjects Thirty-six healthy, right-handed, undergraduate students, all native speakers of Dutch (24 female and 12 male subjects, 18–25 years old) participated. None of them had taken part in the plausibility rating or the previous experiment. They had normal or corrected-to-normal vision.

3.3.1.3. Procedure The experimental procedure was largely the same as described for the previous experiment. The description of the experimental procedure will be restricted to the differences.

Because a P600 effect was predicted to occur, a longer presentation time of the words was chosen compared to the presentation time used in the previous experiment: sentences were presented word by word every 657 ms (257 ms word presentation, 400 ms blank screen). After the last word of the sentence a blank screen of 800 ms was presented, followed by a fixation asterisk (1500 ms) and a blank screen (250 ms) before the first word of the next sentence.

As in Experiment 1, comprehension questions were asked in each block to check whether the subjects comprehended the sentences. Four experimental sentences (2 in each condition) and 5 filler sentences were followed by questions that could be answered by yes or no.

3.3.1.4. EEG recordings EEG was recorded over the following scalp sites: three electrodes on the midline: Fz, Cz, Pz; and 14 over the left and 14 over the right hemisphere: F3A (AF3), F4A (AF4), F7, F8, F5, F6, F3, F4, F5P (FC5), F6P (FC6), T3A (FT7), T4A (FT8), T3 (T7), T4 (T8), C3, C4, T5A (TP7), T6A (TP8), P5A (CP5), P6A(CP6), T5 (P7), T6 (P8), P5, P6, P3, P4, O1, and O2. Nomenclature within brackets follows the proposals of the Electrode Position Nomenclature Committee (1991). The ground electrode was placed on the sternum. Bipolar EOG was recorded between electrodes at the outer left canthus and above the eyebrow of the left eye. The resistance of the electrodes was kept under 5 kOhm. The signals of each electrode were amplified, bandpass filtered between 0.05 and 35 Hz, and digitized with a sampling frequency of 100 Hz.

For the two types of experimental sentences, average ERPs were computed for each subject, starting one word before the onset of the verb and lasting until 250 ms after presentation of the last word of the sentence. Due to eye movements, electrode drift, and amplifier blocking, approximately 6% of the trials were rejected. Unless otherwise described, the final 100 ms before the onset of the verb was used as a baseline.

In this experiment, the manipulation of the materials did not involve a clear morpho-syntactic difference as in Experiment 1 (i.e. nominative or accusative case marking of the pronoun), but a semantic/conceptual difference with greater variation between items within each condition and between the two conditions. To generalize the results beyond the set of materials used in this experiment, average ERPs were also computed for each item, using the same trials and baseline as for the averages per subject.

3.3.2. Results

3.3.2.1. Comprehension questions Subjects answered the 16 comprehension questions regarding the experimental sentences with high accuracy (X (Mean correct answers) =95%). Subjects did not indicate any problems with comprehending the sentences. The performance on the questions did not indicate any problems in comprehending the sentences.

3.3.2.2. Event-related brain potentials The averaged ERPs show a positivity time-locked to the verb for the sentences starting with an inanimate NP compared to the sentences starting with an animate NP (Figures 3 and 4). As in the previous experiment a positivity appears as early as approximately 200 ms after the onset of the verb. It is most prominent at the parietal electrode sites. There is also a positivity in a later time frame. This positivity appears in the same time frame in which a P600 normally occurs, starting approximately 500 ms after onset of the verb, and continues throughout the following word. This late positivity is widely distributed and more prominent over the left than the right hemisphere. There were no signs of any negativities.

3.3.2.3. Data analysis To evaluate the data statistically mean amplitude values were computed for each subject and condition for subject analyses (F1), and for each item and condition for item analyses (F2) in time windows that were suggested by earlier findings and visual inspection of the grand average waveforms. Unless otherwise described, the final 100 ms before the onset of the verb was used as a baseline. The first latency window is the window in the 200–400 ms range following the onset of the verb (cw, the critical word), the pronoun (cw+1), and the adverb (cw+2). The second latency window, in the 500–1000 ms range following the onset of the verb (cw), the pronoun (cw+1), and the adverb (cw+2), overlaps with the time range in which the P600 is commonly observed. Statistical analyses were also done on the standard time-window for N400 effects, which is the latency window in the 300–500 ms range after the onset of the verb (cw), and the pronoun (cw+1).

The overall repeated measures of analyses of variance (ANOVA) included Animacy of the first NP of the sentence (animate and inanimate), Hemisphere (left, right), Position (anterior, posterior), and a 6-level Electrode factor as the within-subject factors for the subject analyses, and the within-item factors for the item analyses. Because List (the two lists of experimental materials) as a between-subjects factor in the subject analyses and a between-items factor in the item analyses did not result in significant effects, this factor was not included in the statistical analyses reported below. Univariate F-tests with more than 1 degree of freedom in the numerator were adjusted by means of the Greenhouse-Geisser procedure.

3.3.2.4. Effects on the verb

200–400 ms after onset of the verb. In the latency window between 200–400 ms after onset of the verb the overall ANOVA showed no significant main effect of Animacy (F1(1,35) = 2.68, p = 0.11; F2(1,39) = 1.84, p = 0.182). A marginally significant Animacy × Electrode interaction was found in the item analysis F1(2.28, 79.84) = 2.22, p = 0.11; F2(2.56, 99.96) = 2.77, p = 0.054). There were no other interactions. Looking at the plots of the grand average, the

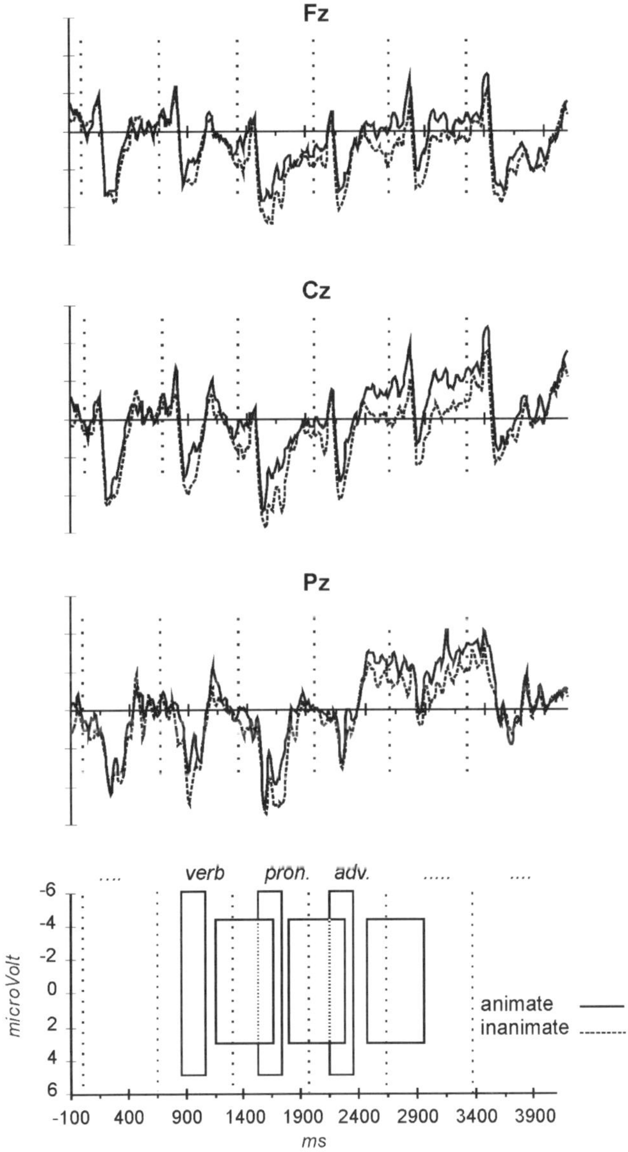

Figure 3: Average ERP waveforms of sentences starting with an animate or inanimate NP at three midline electrode sites (Fz, Cz, and Pz). The time frame starts 100 ms before the onset of the word preceding the verb, with a baseline 100 ms before the onset of this word.

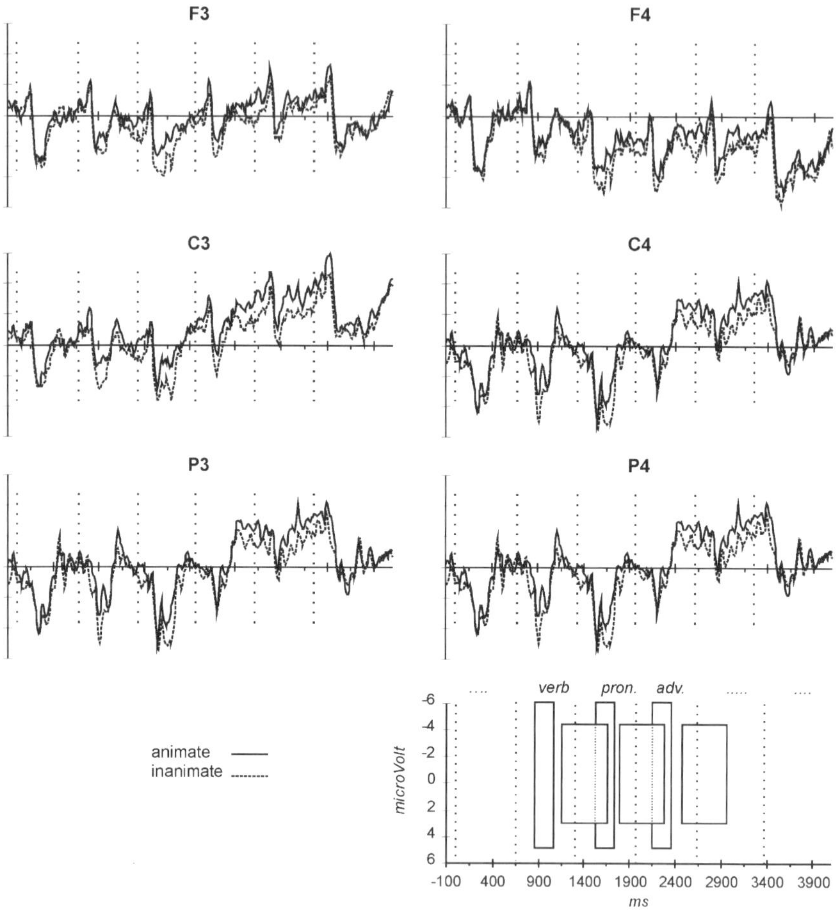

Figure 4: Average ERP waveforms of sentences starting with an animate or inanimate NP at six electrode sites (F3/4. C3/4, and P3/4). The time frame starts 100 ms before the onset of the word preceding the verb, with a baseline 100 ms before the onset of this word.

marginally significant interaction of Electrode and Animacy could indicate a focal posterior effect. To investigate this, two separate ANOVAs for the anterior and posterior electrode sites with Animacy, Hemisphere and Electrode as the within factors were computed. No significant effects involving Animacy were found in the ANOVAs with the anterior electrode sites (F3A, F4A, F3, F4, F5, F6,

F7, F8 T3A, T4A, F5P, and F6P). The ANOVAs with the posterior electrode sites (P5A, P6A, T5A, T6A, P3, P4, P5, P6, T5, T6, O1, O2) showed a significant main effect of Animacy in the subject analysis and a marginally significant effect in the item analysis ($F1(1, 35) = 4.05, p = 0.05$; $F2(1, 39) = 3.13, p = 0.09$), but no interactions. The difference between the anterior and posterior ANOVAs is indicative of a topographical difference for Animacy. The sentences starting with an inanimate NP elicited a more positive-going waveform compared to the sentences starting with the animate NP, at the posterior electrode sites in the latency window between 200–400 ms after onset of the verb, as can be seen in Figures 3 and 4.

500–1000 ms after onset of the verb. In the time frame between 500–1000 ms after onset of the critical word the overall ANOVA showed a marginally significant main effect of Animacy for subject analysis and a significant effect for item analysis ($F1(1, 35) = 3.31, p = 0.08$; $F2(1, 39) = 4.98, p = 0.03$). No significant Animacy × Electrode interaction was found for the subject analysis, but this interaction did emerge in the item analysis ($F1(2.54, 89.00) = 2.18, p = 0.11$; $F2(3.48, 135.87) = 2.99, p = 0.03$). To check whether the significant effect of Electrode and Animacy in the item analysis is indicative of differences in the distribution of the effect over the scalp, ANOVAs for each separate electrode were performed.

The ANOVAs for each electrode show significant effects on many electrode sites (Figure 5). Sentences starting with an inanimate NP show a positive shift in

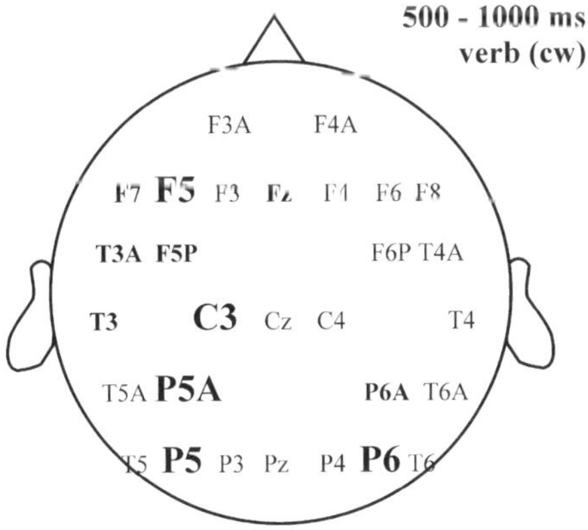

Figure 5: Significant effects in the latency window 500–1000 ms after onset of the verb (cw). Electrode sites that showed significant effects in subject- and item analysis are marked by a larger font in bold; electrode sites that showed significant effects in subject- or item analysis are marked bold.

Table 2: Subject (F1) and item (F2) effects in the latency window between 500 and 1000 ms after onset of the verb (cw). The electrode (elec.) sites at which significant effects ($p < 0.05$) occurred are marked in bold.

Left			**Midline**			**Right**		
Elec. Site	p		Elec. Site	p		Elec. Site	p	
F3	**0.07**	F1(1,35)=3.58	**Fz**	n.s		F4	n.s	
	0.00	F2(1,39)=9.23		**0.05**	F2(1,39) =4.09		**0.08**	F2(1,39)=3.26
F5	**0.03**	F1(1,35)=4.89				F6	n.s	
	0.00	F2(1,39)=7.82					**0.09**	F2(1,39)=3.13
F7	**0.07**	F1(1,35)=3.41						
	0.02	F2(1,39)=5.23						
F5P	**0.06**	F1(1,35)=3.90				F6P	**0.10**	F1(1,35)=2.88
	0.01	F2(1,39)=8.44					**0.07**	F2(1,39)=3.53
T3A	**0.10**	F1(1,35)=2.88						
	0.02	F2(1,39)=6.19						
C3	**0.08**	F1(1,35)=3.31	Cz	n.s				
	0.02	F2(1,39)=5.97		**0.07**	F2(1,39)=3.61			
T3	n.s							
	0.04	F2(1,39)=4.45						
P5A	**0.10**	F1(1,35)=2.91				**P6A**	**0.05**	F1(1,35)=4.06
	0.03	F2(1,39)=5.15					**0.06**	F2(1,39)=3.90
P3	**0.10**	F1(1,35)=2.78				P4	n.s	
	n.s						**0.10**	F2(1,39)=3.90
P5	n.s					P6	n.s	
	0.08	F2(1,39)=3.22					**0.10**	F2(1,39)=2.95

the waveform compared to the waveform of the sentences starting with the animate NP, as can be seen in Figures 3 and 4. On the left-frontal electrode sites and left- and right-parietal electrode sites significant effects were found in both subject and item analyses. Furthermore, item analyses showed significant effects on the left centroparietal and temporal electrode sites and the mid-frontal electrode site. A complete overview of significant and marginally significant effects per electrode in the time range between 500–1000 ms is shown in Table 2 (see also Figure 5).

300–500 ms after onset of the verb. The overall ANOVA in the latency window between 300 and 500 ms after the onset of the verb, showed no significant effects.

3.3.2.5. Effects on the pronoun

200–400 ms after onset of the pronoun. In the latency window between 200–400 ms after onset of the pronoun, the overall ANOVA showed a significant main effect of Animacy ($F1(1,35) = 8.51$, $p = 0.01$; $F2(1,39) = 10.10$, $p = 0.003$). There were no significant interactions. As in the earlier time windows the inanimate sentences show a more positive-going waveform than the animate sentences. Separate ANOVAs for each electrode site showed significant effects on almost every electrode site (Figure 6). A complete overview of significant and marginally significant effects in this latency window is shown in Table 3

The late positivity on the verb in the time window from 500–1000 ms for the sentences starting with an inanimate NP compared to the sentences with an

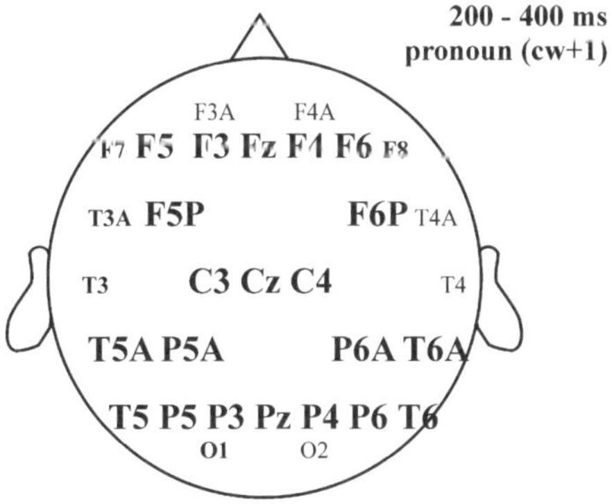

Figure 6: Significant effects in the latency window 200–400 ms after onset of the pronoun (cw+1). Electrode sites that showed significant effects in subject- and item analysis are marked by a larger font in bold; electrode sites that showed significant effects in subject- or item analysis are marked bold.

Table 3: Subject (F1) and item (F2) effects in the latency window between 200 and 400 ms after onset of the pronoun (cw+1). The electrode (elec.) sites at which significant effects ($p < 0.05$) occurred are marked in bold.

Left			Midline			Right		
Elec. Site	p		Elec. Site	P		Elec. Site	p	
F3A	n.s					F4A	**0.09**	F1(1,35)=3.05
	0.07	F2(1,39)=3.62					**0.07**	F2(1,39)=3.49
F3	**0.02**	F1(1,35)=5.97	**Fz**	n.s		**F4**	**0.07**	F1(1,35)=3.56
	0.00	F2(1,39)=9.81		**0.04**	F2(1,39)=4.86		**0.03**	F2(1,39)=4.86
F5	**0.01**	F1(1,35)=7.39				**F6**	**0.02**	F1(1,35)=5.65
	0.00	F2(1,39)=9.16					**0.03**	F2(1,39)=5.41
F7	**0.07**	F1(1,35)=3.48				**F8**	**0.07**	F1(1,35)=3.51
	0.04	F2(1,39)=4.70					**0.05**	F2(1,39)=3.97
F5P	**0.02**	F1(1,35)=5.99				**F6P**	**0.02**	F1(1,35)=6.58
	0.00	F2(1,39)=11.06					**0.02**	F2(1,39)=5.92
T3A	n.s					T4A	**0.09**	F1(1,35)=3.05
	0.02	F2(1,39)=5.67					**0.07**	F2(1,39)=3.49
C3	**0.01**	F1(1,35)=6.76	**Cz**	**0.05**	F1(1,35)=4.26	**C4**	**0.02**	F1(1,35)=5.65
	0.00	F2(1,39)=8.58		**0.01**	F2(1,39)=6.76		**0.02**	F2(1,39)=5.73
T3	n.s					T4	**0.09**	F1(1,35)=2.99
	0.05	F2(1,39)=3.94					**0.08**	F2(1,39)=3.22
P5A	**0.03**	F1(1,35)=4.97				**P6A**	**0.01**	F1(1,35)=8.65
	0.01	F2(1,39)=7.80					**0.00**	F2(1,39)=10.01

T5A	**0.05**	F1(1,35)=3.98				**T6A**	**0.06**	F1(1,35)=3.85
	0.04	F2(1,39)=4.62					**0.05**	F2(1,39)=4.27
P3	**0.01**	F1(1,35)=8.15	**Pz**	**0.08**	F1(1,35)=4.26	**P4**	**0.03**	F1(1,35)=5.34
	0.00	F2(1,39)=10.04		**0.03**	F2(1,39)=5.30		**0.00**	F2(1,39)=9.46
P5	**0.01**	F1(1,35)=6.79				**P6**	**0.03**	F1(1,35)=5.45
	0.01	F2(1,39)=8.27					**0.00**	F2(1,39)=9.30
T5	**0.03**	F1(1,35)=5.23				**T6**	**0.04**	F1(1,35)=4.97
	0.01	F2(1,39)=6.89					**0.01**	F2(1,39)=6.78
O1	**0.04**	F1(1,35)=4.77				O2	**0.10**	F1(1,35)=2.92
	n.s						n.s	

animate NP seems to continue well into the next word, the nominative pronoun. To gain more insight into this difference in the waveforms on the pronoun, statistical analyses were performed over the same time window but computed with a different baseline, were performed. Mean amplitude values were computed for each subject and condition for subject analyses (F1) and for each item and condition for item analyses (F2) with a baseline 100 ms before the onset of the pronoun in the latency window between 200–400 ms after onset of the pronoun. The overall ANOVA showed a significant main effect of Animacy ($F1(1,35) = 12.39$, $p = 0.001$; $F2(1,39) = 5.99$, $p = 0.019$), but no significant interactions. The direction of the difference between the two conditions in this time window is the same as in the time window with a baseline 100 ms before the verb: the inanimate sentences show a more positive going waveform than the animate sentences.

In Experiment 1, similar sentences starting with an animate NP followed by a transitive verb and a nominative case-marked pronoun elicited an early frontal positivity at the pronoun compared to the sentences with an accusative case-marked pronoun. The positivity found in this time window possibly overlaps an early frontal positivity for the animate sentences on the nominative case-marked pronoun. Still, finding a significant difference between the two conditions with the new baseline makes it unlikely that these differences should be completely attributed to the processes initiated by the verb in combination with the inanimate first NP. Presumably, the constraints of the verb in combination with the inanimate first NP influence the comparison at the pronoun. Separate ANOVAs for each electrode site computed on mean amplitude values with a baseline 100 ms before the onset of the pronoun support this explanation: significant effects were found on almost every electrode site, with the exception of the frontal-electrode sites. If sentences with an animate first NP result in a frontal positivity (Experiment 1), while sentences with an inanimate NP show a widespread positivity, no difference would be seen on these frontal electrode sites.

Despite significant effects over posterior- and central-electrode sites, and no effects over frontal-electrode sites, differences in scalp distribution were not reliable, which is demonstrated by the absence of significant interactions. A complete overview of significant and marginally significant effects in this latency window with a baseline 100 ms before the onset of the pronoun is shown in Table 4.

500–1000 ms after onset of the pronoun. No significant effects were found in the later time window between 500–1000 ms after onset of the pronoun.[8]

[8] To make sure that there was no effect in these later time ranges, a smaller latency window was additionally tested (500-800 ms after onset of the pronoun (cw11)). There were no significant effects in this time window.

As for the previous latency window, statistical analyses were also performed in this latency window with a baseline 100 ms before the onset of the pronoun. There was no main effect of Animacy. However, the Animacy and Hemisphere interaction was marginally significant in the subject analysis and significant in the item analysis ($F1(1,35) = 3.49$, $p = 0.07$; $F2(1,39) = 4.36$, $p = 0.04$). There were no other two-way interactions. The Animacy, Hemisphere and Electrode interaction was significant ($F1(1,35) = 4.37$, $p = 0.04$; $F2(1,39) = 6.82$, $p = 0.01$), as was the Animacy, Position, Hemisphere and Electrode interaction ($F1(2.82, 98.79) = 3.29$, $p = 0.03$; $F2(3.56, 138.69) = 3.34$, $p = 0.02$). To further investigate this interaction, ANOVAs were performed for the left and right hemisphere separately and for the anterior and posterior electrode sites. These analyses did not show any significant effects. Nor were there any effects in the analyses for each electrode site separately.

300–500 ms after onset of the pronoun. The overall ANOVA in the latency window between 300–500 ms after the onset of the pronoun showed a significant main effect for Animacy ($F1(1,35) = 6.32$, $p = 0.02$; $F2(1,39) = 7.41$, $p = 0.01$). No significant interactions were found. Taking into account that this latency window partly overlaps with the 500–1000 ms window after onset of the verb, this difference can be regarded as a continuing positive shift in the waveform for the inanimate sentences.

3.3.2.6. Effects on the adverb

200–400 ms after onset of the adverb. In the latency window between 200–400 ms after the onset of the adverb (cw+2) no significant effects were found.

500–1000 ms after onset of the adverb. In the latency window between 500–1000 ms after onset of the adverb, the overall ANOVAs showed a marginally significant main effect of Animacy in the item analysis, but not in the subject analysis ($F1(1,35) = 2.15$, $p = 0.15$; $F2(1,39) = 3.24$, $p = 0.08$). There were no interactions, barring an Animacy × Position × Electrode interaction.

On visual inspection it appears that the positivity in the inanimate sentences on the adverb (cw+2) is the ongoing positivity from the positive shift that occurred on the verb (cw) (Figures 3 and 4). Analyses with a new baseline 100 ms before the onset of the pronoun, confirmed this impression. No significant effects were found in the statistical analyses with this baseline in the latency windows between 200–400 ms and between 500–1000 ms after onset of the pronoun. These findings confirm that the effects found in these time windows with a baseline 100 ms before the onset of the verb, are elicited at the verb and not at the

Table 4: Subject (F1) and item (F2) effects in the latency window between 200 and 400 ms after onset of the pronoun. The electrode sites at which significant effects ($p < 0.05$) occurred are marked in bold.
Baseline 100ms before onset of pronoun

Left			Midline			Right		
Elec. Site	p		Elec. Site	p		Elec. Site	p	
F3	**0.05**	$F1(1,35) = 4.17$	Fz	**0.07**	$F1(1,35)=3.48$	F4	**0.06**	$F1(1,35)=3.86$
	n.s			n.s			n.s	
						F6	**0.09**	$F1(1,35)=3.02$
							0.08	$F2(1,39)=3.32$
						F8	**0.08**	$F1(1,35)=3.33$
							0.06	$F2(1,39)=3.83$
						F6P	**0.01**	$F1(1,35)=7.18$
							0.05	$F2(1,39)=3.95$
						T4A	**0.01**	$F1(1,35)=6.78$
							0.05	$F2(1,39)=4.26$
C3	**0.02**	$F1(1,35)=6.48$	**Cz**	**0.03**	$F1(1,35)=5.04$	**C4**	**0.00**	$F1(1,35)=9.95$
	0.06	$F2(1,39)=3.87$		**0.09**	$F2(1,39)=3.12$		**0.03**	$F2(1,39)=4.89$
T3	**0.03**	$F1(1,35)=4.94$				**T4**	**0.00**	$F1(1,35)=9.46$
	n.s						**0.02**	$F2(1,39)=6.15$
P5A	**0.01**	$F1(1,35)=6.72$				**P6A**	**0.00**	$F1(1,35)=11.04$
	0.03	$F2(1,39)=5.03$					**0.01**	$F2(1,39)=7.02$
T5A	**0.04**	$F1(1,35)=4.81$				**T6A**	**0.02**	$F1(1,35)=6.60$
	0.03	$F2(1,39)=4.96$					**0.02**	$F2(1,39)=6.03$

P3	**0.01**	F1(1,35)=8.88	**Pz**	**0.02**	F1(1,35)=6.55	**P4**	**0.01**	F1(1,35)=6.83
	0.01	F2(1,39)=6.66		**0.04**	F2(1,39)=4.49		**0.02**	F2(1,39)=6.04
P5	**0.01**	F1(1,35)=8.09				**P6**	**0.03**	F1(1,35)=5.17
	0.01	F2(1,39)=7.30					**0.02**	F2(1,39)=5.83
T5	**0.01**	F1(1,35)=7.63				**T6**	**0.03**	F1(1,35)=4.94
	0.01	F2(1,39)=8.46					**0.01**	F2(1,39)=7.07
O1	**0.00**	F1(1,35)=9.48				**O2**	**0.04**	F1(1,35)=4.67
	0.02	F2(1,39)=6.19					**0.05**	F2(1,39)=4.10

pronoun, indicating that we are looking at an ongoing positivity for the inanimate sentences compared to the animate sentences.

3.3.3. Discussion Expanding on the first experiment, this experiment was set up to investigate disambiguation of a local structural ambiguity by semantic/conceptual information or case information. Main declarative clauses started with an animate (8a) or an inanimate NP (8b). At the verb, constraints became apparent in which the animacy of the first NP plays a crucial role. Brain potentials showed an early and a late positivity at the verb (respectively 200–400 ms, and 500–1000 ms after onset of the verb) for the sentences starting with an inanimate NP compared to sentences starting with an animate NP. If the first NP was inanimate, the verb could not assign the subject function to this NP, since the selectional restrictions of the verb required an animate subject. The SI structure as the initial parse, could no longer be maintained. The verb, however, since it was transitive, could assign a different function to the initial inanimate NP, namely that of direct object.

Statistical analysis indicated that the early and late positivities found in Experiment 2 were related to the processes elicited by the animacy of the first NP in combination with selectional restrictions of the verb. In earlier studies in which late positivities were found and early positivities were not reported, relabelling the arguments from the subject to the object was not part of the reanalysis process (Osterhout Holcomb, 1992; Hagoort et al., 1993), whereas in studies in which a positivity with an earlier onset was found relabelling of the syntactic function was part of the reanalysis process (Mecklinger et al., 1995).

As mentioned earlier, the late positivity found in this experiment is a long-lasting positivity elicited by the constraints of the verb in combination with the inanimate NP. The time range in which this positivity was found (between 500–1000 ms after onset of the verb, between 200–400 ms, and between 500–1000 ms after onset of the pronoun) overlaps in part with the time range in which the 'classic' P600 was reported by Hagoort et al. (1993) and Osterhout and Holcomb (1992). As in the previous experiment, the scalp distribution of the late positivity found in this study differs from the scalp distribution in the two studies mentioned above. In contrast to the paradigms in which the 'classic' P600 was found, in the study reported here, no violations were used. In Experiment 2, as well as in Experiment 1, a preferred structure was changed to a less-preferred and more complex structure. Also the nature of the disambiguating information was different in this study. Where number agreement information and subcategorization information were used to resolve a local structural ambiguity in previous studies, selectional restrictions of the verb in combination with the animacy of the first NP in correct sentences were used in this study. Whether the differences in scalp distribution can be explained in terms of differences in the disambiguating information, or differences in the process of reanalysis remains unclear.

Effects at the nominative case-marked pronoun were not apparent. The nominative case-marked pronoun forced the sentences starting with an animate NP from the initially SI into an OI structure. Both an early and late positivity were predicted for this condition, based on the results of Experiment 1. In the second experiment, the process of reanalysis into an OI structure was already initiated by the verb for the inanimate sentences. Statistical analyses performed on averaged waveforms using different baselines indicated that processes at the pronoun in combination with the information of the verb and the inanimate first NP, contributed to this difference. The positivity in the inanimate sentences at the verb possibly overlaps with the effects caused by the nominative case-marked pronoun in the animate sentences, assuming that this condition resembles the materials of Experiment 1.

Additionally, it was shown that the positivity elicited by the verb was a long-lasting effect. Although the positivities found by Osterhout and Holcomb (1992), and Hagoort et al. (1993) were found in overlapping time frames, they did not last as long as the positivity reported in this study. As mentioned above, reanalysis processes in the present study involved reassignment of syntactic function to the initial NP. Whether this was initiated by case information for the sentences starting with an animate NP, or by combining the animacy of the first NP with the selectional restrictions of the verb for the sentences starting with an inanimate NP, it obviously needed more time than dealing with a subcategorization violation (Osterhout et al., 1994) or a number agreement mismatch (Hagoort et al., 1993).

No negativities were found in this study, indicating that problems that arose at the verb did not primarily reflect a lexical integration process. Although the disambiguating information in this experiment had a much more semantic nature than the morpho-syntactic information used in Experiment 1, the differences in the waveforms did not reflect ERP effects that are explained in the literature as being strongly related to semantic processes, such as the N400. This is in contrast to the findings of Weckerly and Kutas (1999), who explained the N400 they found in terms of violation of semantic expectancy and lexical integration difficulties. In the study presented in this section, animacy information became relevant because of the selectional restrictions of the verb of the main clause. That no negativity occurred in the time frame of the N400 at the verb indicates that the processes involved did not include lexical–semantic integration problems of the verb. I argue that the transitivity of the verb gives the parser the opportunity to reanalyze the structure from an SI structure into an OI structure.

In sum, disambiguating a subject–object ambiguity by making use of selectional restrictions on the verb in combination with animacy information of the first NP can force the parser to change the initial parse from a preferred SI structure into a less-preferred OI structure. The differences in the waveforms between the sentences starting with an animate and an inanimate NP indicate that resolving the ambiguity is more related to building up or changing the syntactic structure than to lexical integration processes. An early long-lasting positivity was found at the verb of the inanimate sentences. It was argued that the necessity to change the first NP from the subject in the initial parse into the object of the final structure caused the positivity. The early onset of the positivity indicates that the information that is used is available quite quickly. Furthermore, it was suggested that the duration of the positivity was influenced by the reanalysis process. In the following section, a more possible functional interpretation of reanalysis process will be presented.

4. General Discussion

In this chapter, the resolving of local structural ambiguities in Dutch was investigated. More specifically, the same subject–object ambiguity (NP–V–NP) was disambiguated by making use of the unambiguous case marking of the second NP which was a case-marked pronoun (Experiments 1 and 2), or by combining the selectional restriction of the verb with the animacy of the initial NP which was either an animate or an inanimate NP (Experiment 2). It was argued, based on parsing preferences in Dutch, that the initial parse would be the preferred SI structure. However, if it turned out that the initial parse could no longer hold at the point of disambiguation, the initial parse would be reanalyzed into the less-preferred OI structure.

In the first experiment, the disambiguating information was provided by the case marking of the second NP, which was either an accusative pronoun disambiguating the sentence into the assumed SI structure, or a nominative pronoun disambiguating the structure into the less-preferred OI sentence. In the second experiment, sentences starting with an animate NP were disambiguated into an OI sentence at the nominative case marked, whereas sentences starting with an inanimate NP were already disambiguated into an OI structure at the verb because of the selectional restriction. Differences in ERP waveforms registered during the processing of the sentences, showed early and late positivities in both experiments. No negativities were found.

It was argued that in both experiments, whether the disambiguating information was provided by the case-marked pronoun or by the selection restrictions of the verb in combination with the animacy of the initial NP, relabelling/reassigning syntactic function of the initial NP was part of the reanalysis process.

Although the difference in disambiguating information, both an early and late positivity occurred at the point of disambiguating the local structural ambiguity into the less-preferred OI structure in both experiments.

The onset of the early positivity found in the first experiment was initially explained in terms of the speed with which disambiguating information became accessible as provided by the case information of the close-class case-marked pronoun. Although it cannot be ruled out that different sorts of information can trigger early effects independently, finding an early positivity with a similar onset at the verb for sentences starting with an inanimate NP in Experiment 2 suggests that the early onset is not so much related to the sort of the disambiguating information, but rather with the characteristics of the processes of reanalysis.

Although the onset of the two early positivities was the same, there seems to be a difference in scalp distribution studies (on the left-frontal and left-central regions at the verb in Experiment 1, and on the left- and right-parietal regions at the case-marked pronoun in Experiment 2). This apparent difference might actually be explained by the difference in disambiguating information. Notice, however, that scalp distributions are known to occur if different sorts of information are involved possibly triggering different processes involving different groups of neurons, but also the complexity of the processes might result in different scalp distributions (Kaan et al., 2000; Kaan & Swaab, 2003b; see also Schmitt, Lamers, & Münte, 2002). What is more, there was not only a difference between disambiguating information and the point of disambiguation between the two studies, there was also a difference in animacy of the NPs that had to be reassigned. Recall that in Experiment 1, all initial NPs were animate, whereas in Experiment 2, the early positivity occurred at the verb of sentences starting with an inanimate NP. It cannot be ruled out that all three sorts of differences contribute to the characteristics of the ERP waveforms.

The question arises what the characteristics and the difference in characteristics of the reanalyses are. In the introduction it was argued that the initial parse was driven by mechanism that are merely syntactic in nature. However, in recent studies, Bornkessel (2002) and Bornkessel, Schlesewsky, and Friederici (2002, 2003) argue that sentence comprehension can come about following two possible pathways: the syntactic pathway and the thematic pathway. Following the syntactic pathway sentence comprehension is driven by syntactic information only. In the thematic pathway, comprehension is also based on thematic information besides the syntactic information. More specifically, Bornkessel and colleagues argue that the availability of unambiguous case-information plays a crucial role whether or not thematic information influences the comprehension process in German. What is more, testing sentences as in (10), they found an early parietal positivity for the object-experiencer verb *gefällen* 'please' in comparison to an active verb such as *folgen* 'follow', but only for sentences such as (10) where the arguments are unambiguously case-marked. This effect was independent of word order. In contrast to (10), a clear difference between the subject-initial and the object-initial word order was found for sentences such as (11) (Bornkessel, 2002). That is, a centroparietal negativity (N400) was elicited by the object-initial order in comparison to its subject-initial counterpart. There was however no effect when a different type of verb was used (such as *gefällen* 'please'). Schlesewsky and Bornkessel (to appear) claim that in case ambiguous sentences such as (11), a subject-initial reading is obtained, but without giving rise to further predictions about the thematic properties of the arguments. On the other hand, when the arguments bear unambiguous morphological case marking as in (10), the two NPs are thematically ranked. According to their case features, a (normal) thematic relation is established between the two arguments *prior to* the processing of the verb. And hence, this relation needs revision as soon as an (unexpected) object-experiencer verb is encountered resulting in an early parietal positivity.

(10) a. ...*dass* *der Mönch* *dem Bischof* *folgt/gefällt*
...that the[NOM] monk the[DAT] bishop follows/pleases
'...that the monk follows/is appealing to the bishop'
b. ...*dass Maria* *Lehrerinnen*
...that Maria[NOM/DAT/ACC] teachers[NOM/DAT/ACC]
folgt/folgen
follows/follow
'...that Maria follows teachers.'/ 'that teachers follow Maria'

If the reanalysis process described for the resolution of subject–object ambiguities into the less-preferred OI structure, as used in the present studies, entails relabelling of the syntactic function of the initial NP, it automatically imposes the reassignment of thematic roles. Bornkessel and colleagues found only an early positivity if information was available that gained access to the thematic pathway. This was either unambiguous case marking information or thematic information from the verb (see also Bornkessel et al. 2003).[9] However, in the Dutch experiments, an early positivity with a frontal distribution was elicited by sentences in which the local structural ambiguity was resolved by making use of unambiguous

[9] In contrast to the sentences used in the Dutch experiments, Bornkessel et al. (2002, 2003) used embedded sentences in which the verb followed the two NPs. As a consequence, all information about the two NPs is already available at the verb. Notice also that the case information used in the Dutch study (Experiment 1) differs from the case information used by Bornkessel and colleagues (case-marked pronoun vs. case marking of the articles of the full NPs).

case information, but a parietal distribution was obtained for sentences in which the selectional restrictions of the verb in combination with the animacy of the initial NP provided the disambiguating information. Following this least partial similarity between the German and the Dutch studies, one might suggest that the reanalysis process that occurs during the resolution of a subject–object ambiguity into a less-preferred structure involving animacy information is not only driven by purely syntactic mechanisms (i.e. late positivities), but also by mechanisms involving thematic information (i.e. early positivities). Further research in which the position of the verb relative to its arguments is manipulated, as well as the use of different sorts of verbs (active verbs in comparison to psychological verbs as well as transitive unaccusative verbs) is necessary to indicate what the exact roles of the different characteristics of the verbs are and how crucial the presence or absence of unambiguous case information is.

In summary, the experiments presented in this chapter show that case information as well as animacy information are used in resolving subject–object ambiguities. The early onset of the differences in ERP waveforms indicates that both sorts of information are used almost as soon as they become available. Both early and late positivities were found irrespective of the kind of disambiguating information for the sentences that were disambiguated into the less-preferred object-initial structure. Although the onset of the positivities were similar, there was an apparent difference in scalp distribution of the early positivity with a more frontal distribution for sentences disambiguated with a case-marked pronoun, and a more parietal distribution for sentences disambiguated making use of animacy information, but without case information. It is argued that both the early and late positivities are reflections of processes of reanalysis entailing the relabelling of the syntactic function of the initial NP. Furthermore, it was suggested that the reanalysis processes are partly driven on purely syntactically driven mechanisms (late positivity), but possibly also by mechanisms involving semantic/thematic information (parietal early positivity).

Acknowledgements

A great part of the work presented here was performed at the University of Groningen in cooperation with Laurie Stowe and Colin Brown, for whose contribution I am grateful. Furthermore, I thank the members of the PIONIER project *Case cross-linguistically* at the Radboud University Nijmegen. The Netherlands Organisation of Scientific Research (NWO) is gratefully acknowledged for financial support. Thanks also to Peter de Swart and to two anonymous reviewers, who provided useful comments on this chapter.

References

Bach, E. (1962). The order of elements in a transformational grammar of German. *Language*, *38*, 263–269.

Bierwisch, M. (1963). *The phonology of cliticization.* Dordrecht, Foris.

Bornkessel, I. (2002). The argument dependency model: A neurocognitive Approach to incremental interpretation. Unpublished doctoral dissertation. MPI-Series in Cognitive Neuroscience, Leipzig.

Bornkessel, I., Schlesewsky, M., & Friederici, A.D. (2002). Beyond syntax: Language-related positivities reflect the revision of hierarchies. *Neuroreport, 3*, 361–364.

Bornkessel, I., Schlesewsky, M., & Friederici, A.D. (2003) Eliciting thematic reanalysis effects: The role of syntax-independent information during parsing. *Language and Cognitive Processes, 18*(3), 269–298.

Brown, C., & Hagoort, P. (1993). The processing nature of the N400: Evidence from masked priming. *Journal of Cognitive Neuroscience, 5*, 34–44.

Brown, C. M., van Berkum, J. J., & Hagoort, P. (2000). Discourse before gender: An event-related brain potential study on the interplay of semantic and syntactic information during spoken language understanding. *Journal of Psycholinguistic Research, 29*(1), 53–68.

Burnage, G. (1990). *Celex centre for lexical information: A guide for users.* Nijmegen, SSn.

Committee, E. P. N. (1991). American electroencephalographic society guidelines for standard electrode position nomenclature. *Journal of Clinical Neurophysiology, 8*, 200–202.

De Swart, P. (this volume). Noun phrase resolution: The correlation between case and ambiguity.

De Vincenzi, M. (1991). *Syntactic parsing strategies in Italian.* Dordrecht: Kluwer.

Dowty, D. R. (1991). Thematic proto-roles and argument selection. *Language, 67*, 547–619.

Dowty, D. R. (1989). On the semantic content of the notion thematic role. In: B. Partee, G. Chierchia & R. Turner (Eds), *Properties, types and meanings, 2* (pp. 69–130). Dordrecht: Kluwer.

Fillmore, C. J. (1968). The case for case. In: E. Bach & R. Harms (Eds), *Universals in linguistic theory* (pp. 1–88). New York: Holt, Rhinehart and Winston.

Frazier, L. (1987). Syntactic processing: Evidence from Dutch. *Natural Language and Linguistic Theory, 5*, 519–559.

Frazier, L. (1993). Processing Dutch sentence structures. *Journal of Psycholinguistic Research, 22*, 85–108.

Frazier, L., & Flores d'Arcais, G.B. (1989). Filler driven parsing: A study of gap filling in Dutch. *Journal of Memory and Language, 28*, 331–344.

Friederici, A. D. (1995). The time course of syntactic activation during language processing: A model based on neuropsychological and neurophysiological data. *Brain and Language, 50*, 259–281.

Gorrell, P. (1996). Parsing theory and phrase-order variation in German V2 clauses. *Journal of Psycholinguistic Research, 25*, 135–156.

Gorrell, P. (2000). The subject-before-object preference in German clauses. In: B. Hemforth & L. Konieczny (Eds), *German sentence processing* (pp. 25–64) Dordrecht: Kluwer.

Gunter, T. C., Stowe, L.A., & Mulder, G. (1997). When syntax meets semantics. *Psychophysiology, 34*, 660–676.

Haegeman, L. (1991). *Introduction to government and binding theory*. Oxford: Blackwell.

Hagoort, P. (2003). Interplay between syntax and semantics during sentence comprehension: ERP effects of combining syntactic and semantic violations. *Journal of Cognitive Neuroscience, 15*(6), 883–99.

Hagoort, P., & Brown, C. (1994). Brain responses to lexical ambiguity resolution and parsing. In: C. J. C. Clifton, L. Frazier & K. Rayner (Eds), *Perspectives on sentence processing* (pp. 45–80). Hillsdale, NJ: Lawrence Erlbaum.

Hagoort, P., Brown, C., & Groothusen, J. (1993). The syntactic positive shift (SPS) as an ERP-measure of syntactic processing. *Language and Cognitive Processes, 8*, 439–483.

Hagoort, P., & Brown, C. M. (2000). ERP effects of listening to speech compared to reading: The P600/SPS to syntactic violations in spoken sentences and rapid serial visual presentation. *Neuropsychologia, 38*, 1531–1549.

Hemforth, B. (1993). *Kognitives parsing: Repräsentationen und verarbeitung grammatischen wissens*. Sact Augustin: Infix Verlag.

Kaan, E. (1997). *Processing subject–object ambiguities in Dutch*. Unpublished doctoral dissertation. University of Groningen, Groningen.

Kaan, E., Harris, A., Gibson, E., & Holcomb, P. (2000). The P600 as an index of syntactic integration difficulty. *Language and Cognitive Processes, 15*, 159–201.

Kaan, E., & Swaab T. Y. (2003a). Electrophysiological evidence for serial sentence processing: A comparison between non-preferred and ungrammatical continuations. *Cognitive Brain Research, 17*(3), 621–635.

Kaan, E., & Swaab, T. Y. (2003b). Repair, revision and complexity in syntactic analysis: An electrophysiological differentiation. *Journal of Cognitive Neuroscience, 15*(1), 98–110.

Kolk, H. H. J., Chwilla, D. J., Herten M., & Oor, P. J. W. (2003). Structure and limited capacity in verbal working memory: A study with event-related potentials. *Brain and Language, 85*, 1–36.

Konieczny, L., Hemforth, B., Scheepers, C., & Strube, G. (1997). The role of lexical heads in parsing: Evidence from German. *Language and Cognitive Processes, 12*, 307–348.

Koster, J. (1975). Dutch as an SOV language. *Linguistic Analysis, 1*, 111–136.

Kutas, M., Federmeier, K. D., Seana, C., King, J. W., & Münte, T. F. (2000). Language. In: L. G. Tassinary (Ed.), *Handbook of Psychophysiology (2nd ed.)* (pp. 576–601). New York, NY: Cambridge University Press.

Kutas, M., & Hillyard, S. A. (1980). Reading between the lines: Event-related brain potentials during natural sentence processing. *Brain and Language, 11*, 354–373.

Kutas, M., & Hillyard, S. A. (1983). Event-related brain potentials to grammatical errors and semantic anomalies. *Memory and Cognition, 11*, 539–550.

Kutas, M., & Petten, C. K. V. (1994). Psycholinguistics electrified: Event-related brain potential investigations. In: M. A. Gernsbacher (Ed), *Handbook of Psycholinguistics* (pp. 83–143). San Diego, CA: Academic Press.

Lamers, M. J. A. (2001). *Sentence processing: Using syntactic, semantic and thematic information.* Unpublished doctoral dissertation. University of Groningen, Groningen.

Lee, H. (2001). Markedness and word order freezing. In: P. Sells (Ed.), *Formal and Emperical Issues in Optimality-Theoretic Syntax* (pp. 63–128). Stanford: CSLI Publications.

MacDonald, M. C. (1993). The interaction of lexical and syntactic ambiguity. *Journal of Memory and Language, 32*, 692–715.

MacDonald, M. C., Pearlmutter, N. J., & Seidenburg, M. S. (1994). The lexical nature of syntactic ambiguity resolution. *Psychological Review, 101*, 676–703.

McDonald, J. (1987). Sentence interpretation processes: The influence of conflicting cues. *Journal of Memory and Language, 26*, 100–117.

McKinnon, R., & Ostehout, L. (1996). Constraints on movement phenomena in sentence processing: Evidence from event-related brain potentials. *Language and Cognitive Processes, 11*, 495–523.

Mecklinger, A., Schriefers, H., Steinhauer, K., & Friederici, A. D. (1995). Processing relative clauses varying on syntactic and semantic dimensions: An analysis with event-related potentials. *Memory and Cognition, 23*, 477–494.

Münte, T. F., Heinze, H. J., & Mangun G. R. (1998). Brain potentials and syntactic violations revisited: No evidence for specificity of the syntactic positive shift. *Neuropsychologia, 36*, 217–226.

Münte, T. F., Urbach, T. P., Düzel, E., & Kutas, M. (2000). Event-related brain potentials in the study of human cognition and neuropsychology. In: F. Boller, J. Grafman & G. Rizzalotti (Eds), *Handbook of neuropsychology* (pp. 139–235). Amsterdam: Elsevier Science.

Nieuwborg, E. (1968). *De distributie van het onderwerp en het lijdend voorwerp.* Antwerpen: Plantyn.

Osterhout, L., & Holcomb, P. J. (1992). Event-related brain potentials elicited by syntactic anomaly. *Journal of Memory and Language, 31*, 1–22.

Osterhout, L., Holcomb, P. J., & Swinney, D. A. (1994). Brain potentials elicited by garden-path sentences: Evidence of the application of verb information during parsing. *Journal of Experimental Psychology, 20*, 786–803.

Osterhout, L., & Holcomb, P. J. (1995). Even related potentials and language comprehension. In: M. D. Rugg & M. G. H. Coles (Eds), *Electrophysiology of Mind: Event-related Brain Potentials and Cognition* (pp. 171–215). New York: Oxford University Press.

Rösler, F., Pechmann, T., Streb, J., Röder, B., & Hennighausen, E. (1998). Parsing of sentences in a language with varying word order: Word-by-word variations of processing demands are revealed by event-related brain potentials. *Journal of Memory and Language, 38*, 150–176.

Schlesewsky, M., & Bornkessel, I. (2004). On incremental interpretation: Degrees of meaning accessed during sentence comprehension. *Lingual, 114*, 1213–1234.

Schmitt, B., Lamers, M., & Münte, T. (2002). Electrophysiological estimates of biological and syntactic gender violation during pronoun processing. *Cognitive Brain Research, 14*, 333–346.

St George, M., Kutas, M., Martinez, A., & Sereno, M. I. (1999). Semantic integration in reading: Engagement of the right hemisphere during discourse processing. *Brain, 122*, 1317–1325.

Trueswell, J. C., Tanenhaus, M. K., & Garnsey, S. M. (1994). Semantic influences on parsing: Use of thematic role information in syntactic ambiguity resolution. *Journal of Memory and Language, 33*, 285–318.

van Berkum, J. J., Hagoort, P., & Brown, C. (1999). Semantic integration in sentences and discourse: Evidence from the N400. *Journal of Cognitive Neuroscience, 11*, 657–671.

Weckerly, J., & Kutas, M. (1999). An electrophysiological analysis of animacy effects in the processing of object relative sentences. *Psychophysiology, 36*, 559–580.

Zeevat, H. (2004). Freezing and marking. *ILLC*. University of Amsterdam.

Zwart, C. J. W. (1993). *Dutch syntax: A minimalist approach.* Unpublished doctoral dissertation. University of Groningen, Groningen.

Chapter 11

Differential Subject Marking in Amharic

Mengistu Amberber

1. Introduction

The goal of this chapter is to examine the case and agreement systems of Amharic with particular reference to the phenomenon of differential subject marking (DSM). The chapter focusses on three interrelated issues: (a) the semantic factors which determine DSM, (b) the interaction between DSM and agreement on the verb, and (c) the interaction between DSM and topicality.

The main claims of the chapter are: (a) differential subject marking in Amharic is found with a semantically homogeneous class of verbs — broadly construed as "experiencer" predicates; (b) one class of experiencer predicates has an implicit Causer argument — referred to as *Ambient* causer (Pesetsky, 1995) — which occurs as the default third person masculine singular argument; (c) the Ambient causer can be morphologically suppressed through a morphological operation that is similar to passivization; (d) when the Ambient causer is not suppressed, the experiencer argument is left-dislocated and the object pronominal suffix on the verb becomes obligatory.

The chapter is organized as follows. In Section 2 a brief typological overview of the Amharic language is presented. In Section 3, the presence of the grammatical relations subject (A, S) and direct object (P) is established on the basis of language-internal formal evidence. In Section 4 the morphosyntactic and semantic properties of differential subject marking are investigated.

Competition and Variation in Natural Languages: The Case for Case
Edited by M. Amberber and H. de Hoop

ISBN: 0-08-044651-5

2. Typological Background

In this section, a brief typological overview of the Amharic language is presented in order to provide the reader with basic information about the grammar of Amharic that is relevant to the present topic. For a more detailed description of the language the reader is referred to Leslau (1995).

Amharic is a nominative/accusative language and exhibits a mixture of dependent- and head-marking. A definite object NP is obligatorily marked by the accusative suffix *-n*. The subject NP is unmarked. As can be seen in (1), the verb obligatorily takes subject pronominal affixes which cross-reference the subject NP. Object/indirect object marking on the verb is often optional (but see Section 3.2.3).

(1) a. *aster wədə bet hed - əčč*
A. to home go.PF-3F[1]
'Aster went home'

b. *aster bet-u-n ayyə -čč-(ɨw)*
A. house-DEF-ACC see.PF-3F-3MO
'Aster saw the house'

The morphological structure for a main clause verb in the perfective (past) conjugation is: [verb stem + subject + (object)] as shown in (2):

(2) *səbbər - əčč- (ɨw)*
cut.PF-3F-3MO
'She broke (it)'

The unmarked constituent order in a simple declarative clause is SOV (subject–object–verb). NP modifiers, such as possessor NPs, adjectives, demonstratives, numerals, the relative clause, precede the head noun.

Amharic has several positional prefixes (adpositions). Some of the basic ones are exemplified below (see also Hetzron, 1970):

(3) *lə-* 'to' or 'for':
a. Recipient: ***lə*** *- lɨǰ - u sət't'ə - hu - t*
to-man-DEF give.PF-1S-3MO
'I gave it to the man'

[1]The following abbreviations are used in the interlinear glossing of the Amharic data. 1: 1st person, 2: 2nd person, 3: 3rd person, ACC: Accusative, AUX: Auxiliary, CAUS: Causative, COMP: Complementizer, DEF: Definite, DT: Detransitive, F: Feminine, GER: Gerund, IMP: Imperfective, M: Masculine, O: Object, PASS: Passive, PF: Perfect, PL: Plural, POSS: Possessive, RECIP: Reciprocal, REFL: Reflexive, S: Singular, TR:Transitive.

b. Beneficiary: ***lə*** - *lɨǰ - u* *sət'tə - hu –llə- t*
for-man-DEF give.PF-1S-FOR-3MO
'I gave it (to someone) for the (benefit of) the man'

(4) *bə-* has a number of meanings including: 'with, by means of, by, through, into'
a. Instrumental: ***bə*** - *bilawa* 'with a knife'
b. Location of an event: ***bə*** - *gəbəya* 'at a market'

(5) *kə-* 'from, origin of place, out of' (source)
kə - *bet* *mət't'a*
from-house come.PF.3M
'He came from the house'

(6) *wədə* direction away from speaker (allative):
wədə *tərar -očč -u* *hed-ə*
to mountain-PL-DEF go.PF-3M
'He went to the mountains'

There are also a number of adpositions that are composed of a prepositional prefix and a postpositional element (of nominal origin) as in (7):

(7) ***bə*** *-t'ərəpp'eza-w* ***lay***
on-table-DEF **on**
'On the table'

In constructions such as (7) often the prepositional element can be omitted without affecting the adpositional meaning of the phrase. (See the Appendix for a brief comparative perspective within the context of some Ethiopian Semitic and Cushitic languages.)

3. Grammatical Relations: Coding Strategies

As already mentioned, Amharic follows the nominative–accusative case pattern. By this we mean that the subject of transitive verbs and the subject of intransitive verbs are formally similar (nominative), whereas the object of transitive verbs is distinctly marked (accusative).

3.1. Subjects (S, A) in Amharic

It is generally assumed that the subject is a grammatical relation that expresses the A and S grammatical functions (Blake, 2001; Andrews, 1985, in press).

In Amharic, A of transitive clauses and S of intransitives can be distinguished by the three coding strategies: clause-initial position, nominative case, and obligatory verb-agreement:

(8) a. *ləmma* *gəməd-u-n* *k'orrət' -ə*
L. rope-DEF-ACC cut.PF-3M
'Lemma cut the rope'

b. *ləmma* *mət't'a*
L. come.PF.3M
'Lemma came'

In both (8a) and (8b), the NP *ləmma* occurs in clause-initial position. It gets the unmarked nominative case and controls the choice of pronominal marker on the verb. When the verb is passivized, the patient NP becomes the subject and shows the same coding features as an 'agentive' subject: it occurs clause-initially, has nominative case, and agrees with the verb as in (9):

(9) *gəməd-u* *tə - k'orrət' -ə*
rope-DEF PASS.cut.PF-3M
'The rope was cut'

The coding strategies confirm that there is a grammatical relation 'subject' in Amharic that cannot be reduced to semantic factors. While the three coding strategies — word order, case and agreement – clearly show that there is a 'subject' grammatical relation in Amharic, there are some constructions where subjects exhibit 'non-canonical' or 'differential' marking, which is of course the main concern of the present chapter and is discussed in depth in Section 4.

3.2. The Direct Object (P) in Amharic

3.2.1. Transitivity and definiteness The grammatical relation of 'direct object' is defined as a relation associated with the P function (Andrews, in press). In terms of the three coding strategies – word order, case, and agreement – it can be established that there is a grammatical relation 'direct object' in Amharic.

In Amharic, the direct object of a transitive verb such as *break* normally occurs after the subject and before the verb.

(10) a. *ləmma* *and t'ərmus* *səbbər -ə*
L. one bottle break.PF-3M
'Lemma broke one bottle'

b. *ləmma* *t'ərmus-u-n* *səbbər -ə*
L. bottle-DEF-ACC break.PF-3M
'Lemma broke the bottle'

Notice that in (10a) there is no morphological distinction between the subject and the direct object as neither is overtly marked for case. However, in (10b) we see that the object receives the object marker ('accusative' case) *–n* when it is definite. The relevant parameter is 'definiteness' rather than 'animacy', as can be seen in the following example:

(11) *ləmma* *and* *lɨǰ-* *(*ɨn)* *ayy -ə*
L. one child-ACC see.PF-3M
'Lemma saw one child'

In (11) the direct object is animate but indefinite. It cannot be marked by the accusative case. Notice also that while the verb has a subject agreement marker on the verb, there is no corresponding object marker. However, the verb is optionally marked for object agreement provided that the direct object is definite:

(12) a. *ləmma* *t'ərmus-u-n* *səbbər -ə-(w)*
L. bottle-DEF-ACC break.PF-3M-3MO
'Lemma broke the bottle'

b. *ləmma* *and* *t'ərmus* *səbbər -ə-(*w)*
L. one bottle break.PF-3M-3MO
'Lemma broke one bottle'

Thus, when the direct object is not definite as in (12b) there is no object agreement marker on the verb. However, the object agreement marker on the verb becomes obligatory when the direct object is not overtly expressed:

(13) *ləmma* *səbbər -ə-*(w)*
L. break.PF-3M-3MO
'Lemma broke it'

Thus, in constructions such as (13), the object marker is not merely an expression of agreement morphology but rather itself functions as a pronominal — an argument of the verb. When the verb is passivized the direct object becomes the subject and controls subject agreement as expected:

(14) *t'ərmus-u* *tə -səbbər -ə*
bottle-DEF PASS-break.PF-3M
'The bottle broke'

In the following section we explore the formal and semantic properties that may dictate the presence or absence of the accusative suffix and the interaction between the three coding strategies — word order, case and agreement with respect to the direct object grammatical relation.

3.2.2. The accusative suffix –*n* As already mentioned, the accusative case-marking suffix –*n* (-*ɨn* after consonant-final forms) may or may not occur on the object. The traditional observation is that only 'definite' objects can and must be marked by the accusative suffix –*n*. We use the term 'definite' here as a cover term to include a number of different types of 'definite' NPs (cf. Abbott, 2004): proper names, personal pronouns, demonstrative pronouns, possessive NPs, NPs with a universal quantifier. An example of each of these categories is presented below in (15):

(15) a. *aster almaz - *(ɨn) ayy -əčč*
A. A. – ACC see.PF-3F
'Aster saw Almaz' (Proper name)

b. *aster ɨssu - *(n) ayy -əčč*
A. he – ACC see.PF-3F
'Aster saw him' (Personal pronoun)

c. *aster yɨh - *(ɨn) ayy -əčč*
A. this – ACC see.PF-3F
'Aster saw this' (Demonstrative pronoun)

d. *aster lɨj-u - *(n) ayy -əčč*
A. child-DEF – ACC see.PF-3F
'Aster saw the child' (NPs with the definite suffix –u/-wa)

e. *aster lɨǰ-e - *(n) ayy -əčč*
A. child-POSS.1S – ACC see.PF-3F
'Aster saw my child' (Possessed NPs)

f. *aster yə -ɨne- *(n) lɨǰ ayy -əčč*
A. POSS-1S-ACC child see.PF-3F
'Aster saw my child' (Possessor NPs)

g. *aster hullu – *(n) lɨǰočč ayy -əčč*
A. all children see.PF-3F
'Aster saw all the children' (Quantified NPs ('all'))

h. *aster ɨyyandandu – *(n) lɨǰ ayy -əčč*
A. each child see.PF-3F
'Aster saw each child' (Quantified NPs ('each'))

As the examples above show, in every instance the object NP must be marked by the accusative case suffix *-n* lest the construction becomes ungrammatical.

In addition to the object noun phrases exemplified above, generic noun phrases behave in a similar manner. Thus consider (16):

(16)	*ɨgziabɨher*	*səw – *(ɨn)*	*bə - məlk - u*	*fət' t'ər - ə*
	God	man-ACC	with-image-POSS.3M	create.PF-3M

'God created man in His image' (Genesis, 1:27)

Notice that the direct object *səw* 'man' is generic in the context of (16) and is marked by the accusative suffix.

3.2.3. The distribution of the object/indirect object agreement suffix The accusative suffix *–n* can also be used to mark the recipient object of a three-place predicate such as 'give'. Thus, compare the following constructions:

(17)	a.	*ləmma*	*lə - lɨǰ- u*	*məs'haf*	*sət't' - ə- (w)*
		L.	to – child – DEF	book	give.PF-3M-3MO

'Lemma gave a book to the child'

	b.	*ləmma*	*lə - lɨǰ- u*	*məs'haf-u-n*	*sət't' - ə- (w)*
		L.	to – child – DEF	book-DEF-ACC	give.PF-3M-3MO

'Lemma gave the book to the child'

(18)	*ləmma*	*lɨǰ- u- n*	*məs'haf*	*sət't' - ə -*(w)*
	L.	child – DEF-ACC	book	give.PF-3M-3MO

'Lemma gave the child a book'

Notice that in (17) the recipient is marked by the preposition *lə* 'to', whereas in (18) the same argument is marked by the suffix *–n* in a construction similar to dative-shift in English.[2] What is important for present purposes is the fact that when the recipient argument does not occur with the preposition *lə-* and is instead marked by the accusative *–n*, the object/indirect object pronominal suffix (*-w*) on the verb becomes obligatory. It is thus perhaps more appropriate in such contexts to refer to this marker as a 'pronominal suffix' rather than a mere agreement marker (see also Mullen, 1986).

It is also worthwhile to mention here that when the recipient argument is ellipsed or occurs as a pronoun (particularly with first and second person pronouns) it is

[2] The formal property of ditransitive verbs is much more complex. Unlike the verb *sət't'ə* 'to give', the closely related verb *lakə* 'to send' does not readily allow the preposition *lə-* to be substituted by the accusative suffix *–ɨn*.

agreed with even though in the latter case the recipient may occur with the preposition *lə-*, as shown in (19a):

(19) a. *ləmma* *lə- ɨne* *məs'haf* *sət't' - ə-*(ññ)*
L. to-me book give.PF-3M-1SO
'Lemma gave a book to me'

b. *ləmma* *məs'haf* *sət't' - ə -*(ññ)*
L. book-DEF-ACC give.PF-3M-3MO
'Lemma gave a book to me'

We have seen that while the object/indirect object agreement marker is often optional, there are numerous constructions where it is obligatory. Thus, whenever the direct object is ellipsed, the object pronominal suffix is obligatory. Consider (20) below, where the object agreement suffix is optional as expected:

(20) *almaz* *məs'haf – u -n* *wəssəd - əčč - (ɨw)*
A. book-DEF.M-ACC take.PF-3F-3MO
'Almaz took the book'

If the direct object ('the book') is omitted — as it would be when it is recoverable from the immediate discourse context — the object agreement marker is obligatory as shown in (21):

(21) Q. *məs'haf – u* *yət* *nəw?*
book-DEF where be.3M
'Where is the book?'

A. *almaz* *wəssəd - əčč - *(ɨw)*
A. take.PF-3F-3MO
'Almaz took it'

In response to the question *where is the book?* the direct object ('the book') is omitted but agrees with the object pronominal suffix (cf. Haile, 1970, p. 105). This suggests that, to the extent that object agreement normally occurs with definite objects, ellipsed NPs are treated as definite (see also Abbott, 2004).

Another construction in which object/indirect object agreement is normally obligatory is the causative. The embedded clauses in (22a) and (22b) below are transitive and intransitive respectively. We see that in both cases the causee can take the accusative marker *-n* and the verb occurs with an object agreement marker cross-referencing the causee.

(22) a. *aster ləmma-n məs'ɨhaf ɨnd-i-gəza*
A. L.-ACC book COMP-3M-buy.IMP
a -dərrəg -əčč- ɨw
CAUS-make.PF-3F-3MO
'Aster made Lemma buy a book'

b. *aster ləmma -n wədə bet ɨnd-i-hed*
A. L. -ACC to home COMP-3M-go.IMP
as -gəddəd -əčč - ɨw
CAUS-force.PF-3F-3MO
'Aster forced Lemma to go home'

To summarize, when a direct object is definite it is marked by the accusative case –*n*, and can be optionally agreed with by an agreement marker on the verb. The set of forms which are obligatorily marked by accusative case consists of definite NPs, including demonstrative pronouns, personal pronouns, proper names, as well as quantified and generic NPs.

While the agreement morphology on the verb which agrees with direct or indirect objects is optional, it becomes obligatory when: (a) the direct or indirect object is not overtly expressed; (b) the direct or indirect object is a pronominal; (c) the recipient argument of a ditransitive verb occurs with the accusative case marker instead of the preposition *lə*-; and (d) when there is a causee argument of a causative construction.

In the next section we investigate the roles played by overt accusative case and object/indirect object morphology on the verb in the phenomenon of differential subject marking.

4. Differential Subject Marking

In this section I show that the quirky (non-canonical) case-marking of certain subjects by and large occurs with a semantically homogeneous group of verbs — broadly construed as *experiencer* predicates. Specifically, I argue that there is a set of experiencer predicates in Amharic which take an implicit causer argument — referred to as *ambient* causer. When this argument is present, the experiencer argument receives the suffix –*n* which normally appears with direct objects of transitive clauses. When the ambient causer is not present, the experiencer argument is marked as a canonical subject. The morphosyntactic mechanism which suppresses the ambient causer is the same as other detransitivising operations such as the formation of unaccusatives, passives, and reflexives.

As already mentioned, there is enough evidence to establish the existence of the subject grammatical relation S and A in Amharic. Typically, subjects occur in clause-initial position, get the nominative case, and obligatorily agree with a pronominal marker on the verb. However, there are some constructions that are problematic in that the 'subject' appears to lack some or all of these coding properties. Consider the following examples:

(23) *aster* *tə-č'ənnək'-əčč*
A. DT-worry.PF-3F
'Aster is worried'

(24) *aster* *tə-dənnək'-əčč*
A. DT-astonish.PF-3F
'Aster is astonished'

In both (23) and (24) the subject, S, of the intransitive clause occurs in clause-initial position and agrees with the pronominal suffix on the verb as expected. Now consider the synonymous constructions in (25) and (26) below:

(25) *aster-(ɨn)* *č'ənnək'-at*
A.- (ACC) worry.PF.3M-3FO
'Aster is worried'

(26) *aster-(ɨn)* *dənnək'- at*
A.- (ACC) astonish.PF.3M-3FO
'Aster is astonished'

While (25) and (26) have the same meaning as (23) and (24) respectively, there are important formal differences between the two sets of constructions. First, in (23)–(24) the single argument of the verb — the 'subject' — is optionally marked by the suffix *–n* which is normally found with the direct object grammatical relation. Second, the single argument of the verb — semantically an Experiencer — obligatorily agrees with a pronominal suffix on the verb that normally occurs with objects. The subject pronominal suffix on the verb has the features third person, masculine and singular. Since subject agreement on the verb is obligatory in Amharic, the question becomes: what argument does the subject agree with in (25)–(26)?

Notice that the presence of the accusative case-marking on the subject noun phrase in (25) and (26) is unusual given that normally only (definite) objects of a transitive verb take the *-n* suffix which marks the accusative case. This is shown again in (27), where object noun phrase with the patient role is marked by the accusative case:

(27) *səw -ɨyyəw aster – *(ɨn) sam-(at)*
man-DEF A.-ACC kiss.PF.3M-3FO
'The man kissed Aster'

While the accusative suffix *–n* is obligatory in (27), the object pronominal suffix is not. The reader will have noticed that this scenario is the converse of what we find with the constructions in (25) – (26) where the accusative case marker on the Experiencer is optional but the object pronominal suffix on the verb is obligatory.

Constructions in which the single argument of an intransitive verb agrees with an object pronominal suffix as in (25) and (26) are not uncommon in Amharic. However, before discussing these constructions, a more detailed description of Amharic psychological (experiencer) predicates is in order.

4.1. Experiencer Predicates

Following previous work (cf. Amberber, 1996, 2000, 2001, 2002), I assume that there are two major types of experiencer predicates in Amharic: Subject Experiencer (SubjExp) predicates and Object Experiencer (ObjExp) predicates (the terms SubjExp and ObjExp are adapted from Pesetsky, 1995, and related work). These are illustrated in (28)–(29):

(28) a. *aster tə- dəssət - əčč*
A. BE-happy.PF-3FS
'Aster is happy'

b. *aster ləmma - n wəddəd- əčč - ɨw*
A. L.-ACC like.PF-3FS-3FO
'Aster liked Lemma'

(29) *sɨt'ota -w aster - ɨn as – dəssət - at*
gift-POSS.3M A-ACC CAUS-BE.happy.PF.3M-3FO
'His gift made Aster happy'

In (28a) and (28b) the Experiencer argument is in subject position (agrees with the subject pronominal suffix on the verb). The key difference between (28a) and (28b) is transitivity: the former is intransitive whereas the latter is transitive.

In (29), on the other hand, the Experiencer is in object position (is marked by the accusative suffix *–n*, and agrees with the object pronominal suffix on the verb). Note that in (29) the verb is causativized by the prefix *as-* (as opposed to the prefix *a*-) that is normally used to causativize transitive verbs or unergative intransitives. The construction in (29) becomes ungrammatical with the causative prefix *a*-:

(29′) **ləmma* *aster - ɨn* ***a*** *– dəssət - at*
A. A-ACC **caus**-BE.happy.PF.3M-3FO

Thus, it appears that the causativization pattern of the intransitive experiencer predicate is the same as the causativization pattern of unergative intransitives as shown in (30) below with the unergative intransitive verb *č'əffərə* 'dance':

(30) a. *aster* *č'əffər -əčč*
A. dance.PF – 3FS
'Aster danced'

b. *ləmma* *aster - ɨn* ***as*** *– č'əffər -at*
L. A.-ACC CAUS-dance.PF.3M-3FO
'Lemma made/let Aster dance'

c. **ləmma* *aster - ɨn* ***a*** *– č'əffər -at*
L. A.-ACC CAUS-dance.PF.3M-3FO

This contrasts with the situation of unaccusative intransitives where the verbs take the prefix *a*- in causativization. Thus, consider the following example:

(31) *a.* *aster* *mət't'a -čč*
A. come.PF – 3FS
'Aster came/arrived'

b. *ləmma* *aster - ɨn* ***a*** *– mət't'a -at*
L. A.-ACC CAUS-come.PF.3M-3FO
'Lemma brought Aster'

This raises the interesting question: why do intransitive experiencer predicates behave like unergative verbs in their causativization pattern? Is it an accident that intransitive experiencer predicates, unergatives and transitive predicates do not all take the causative prefix *a*- in causativization?

I argue that these verbs behave similarly with respect to the morphological causative because all are instances of a two-place predicate. Thus, both types of intransitive predicate (unergatives and intransitive experiencer predicates) have underlyingly two arguments. Following the suggestion made by Hale and Keyser (1993, 2002), I assume that unergative verbs appear to have a single argument because the internal argument — sometimes referred to as 'cognate' object — is not normally overtly expressed. In the case of intransitive subject experiencer predicates of the type 'be happy', 'be annoyed', etc. I assume that there is an underlying causer argument which I will refer to by the term *Ambient-causer* (A-causer) following Pesetsky (1995, p. 112). It is this argument which controls subject agreement in constructions such as (25)–(26). When this argument is 'suppressed' (by a morphological process similar to reflexivization/passivization) we get constructions such as (23)–(24) where the Experiencer occurs in the nominative.

The fact that Subject Experiencer predicates exhibit quirky Case properties cross-linguistically is by now well documented (cf. Verma & Mohanan, 1990; Belletti & Rizzi, 1988; Takezawa, 1987; Zaenen, Maling, & Thráinsson, 1985; Andrews, 2001; Aikhenvald, Dixon, & Onishi et al., 2001; Bhaskararao & Subbarao, 2004, among others). There is a body of literature both within and outside of generative linguistics that deals with quirky subjects of intransitive verbs. Often, a semantic motivation is implicitly or explicitly provided to distinguish intransitive subjects with nominative marking from intransitive subjects with accusative and dative marking. According to McCawley (1976) indirect subject constructions involve verbs that express events such as those listed in (32):

(32) a. sensory and mental experience,
b. emotional experience,
c. physical and biological experience,
d. need/duty/obligation,
e. possession/existence,
f. happenstance.

It is observed that intransitive subjects which are more 'affected', or which have less 'control' over the event, in the sense of Hopper and Thompson (1980), are likely to be marked by the accusative/dative, whereas intransitive subjects which are 'agentive', or exert more control over the event may be marked by the nominative. For a more elaborate semantic classification of verbs which exhibit differential marking ('non-canonically marked arguments') see Onishi (2001, p. 23ff); see also Malchukov (this volume) for relevant discussion.

As mentioned earlier, constructions in which the single argument of an intransitive verb agrees with an object pronominal suffix are not uncommon in Amharic. There are at least two broad classes of verbs that exhibit this behaviour. The first class involves verbs that can be characterized as experiencer predicates — broadly construed involving not only PSYCHOLOGICAL experience (as in (25)–(26)) but also *sensation* and other (physical and mental) experiences as in (33)–(35):

(33) *aster -(ɨn)* *t'əmm - at*
A.-(ACC) thirst.PF.3M-3FO
'Aster is thirsty'

(34) *səww -očč - u - (n)* *rab - aččəw*
man-PL-DEF-(ACC) hunger.PF.3M-3PLO
'The men are hungry'

(35) *aster - (ɨn)* *amməm - at*
A.-(ACC) pain. PF.3M-3FO
'Aster is sick (in pain)'

Notice that in (33)–(35) there is obligatory object agreement with the argument that undergoes some sort of sensation, just as with the argument that undergoes psychological experience in (25)–(26). Subject agreement is with a third person masculine argument.

Now if we assume that the non-canonically marked experiencers in (33)–(34) are similar to the non-canonically marked experiencers of the psychological predicates in (25)–(26), we would predict that these subjects will exhibit canonical behaviour (i.e., will be marked as nominative and control subject agreement) when the verb is inflected with the prefix *t*(ə)- similar to (23)–(24). This is exactly what we find, as shown in (36)–(38):

(36) *aster* *tə - t'əmma - čč*
A. DT-thirsty.PF-3F
'Aster is thirsty'

(37) *səww - očč -u* *tə - rab - u*
men-PL-DEF DT-hunger.PF-3PL
'The men are hungry'

(38) *aster* *t - amməm - əčč*
A. DT-sick.PF-3F
'Aster is sick'

The class of temperature verbs such as *bərrədə* 'it has become cold', *mok'k'ə* 'it has become hot' behave in a similar manner. Although these verbs are typically used to express temperature, as shown in (39), they can also be used with an experiencer argument, as shown in (40):

(39) a. *yɨ - bərd -all*
3M-IMP.be.cold-AUX.3M
'It is cold'

b. *yɨ - mok'k' -all*
3M-IMP.be.hot-AUX.3M
'It is hot'

(40) a. *aster (ɨn)* *bərrəd -at*
A.- (ACC) be.cold.PF.3M-3FO
'Aster is cold'

b. *aster (ɨn)* *mok'k'-at*
A.- (ACC) be.hot.PF.3M-3FO
'Aster is hot'

As in the psychological predicates, there is an obligatory object agreement with the argument that is experiencing the temperature.

Therefore, the presence of obligatory object agreement with an intransitive subject is reasonably widespread — it is found with predicates encoding emotion, sensation, and temperature.

Note that semantically the experiencer argument of a typical experiencer predicate exerts little or no control over the event. For example, with the verb in (41) below the event is conceptualized as something that takes place without volition, and thus the experiencer argument is agreed with by an object (indirect object) pronominal suffix:

(41) *rabə-ññ*
hunger.PF.3M-1sO
'I am hungry'
(lit. 'it hungers me')

The crucial question to address is therefore why some constructions can have arguments with two different cases and agreement properties, yet basically the same meaning. In other words, what is the proper characterization of the minimal pair repeated below as (42)?

(42) a. *aster* *tə-č'ənnək'-əčč*
A. DT-worry.PF-3F
'Aster is worried'

b. *aster-(ɨn)* *č'ənnək'-at*
A.- (ACC) worry.PF.3M-3FO
'Aster is worried'

One of the characteristic properties of the constructions we saw so far — the psychological, sensation and temperature predicates — is that they show subject agreement with a third person element. It is necessary to determine the status of this element in order to account for the quirky Case and agreement facts in (42b).

The key to the solution is to assume that the third pronominal suffix agrees with a bona fide argument of the predicate.

I follow Pesetsky (1995) in assuming that some SubjExp predicates have an additional argument which can be referred to as the *Ambient Cause*. Pesetsky (1995) argued that this argument is akin to 'weather *it*' or Ambient *it*. He articulates this notion as follows (Pesetsky, 1995, p. 111):

> Emotions like *surprise*, *annoyance*, and *amusement* are . . . like the weather in a number of respects. They are "global" (ambient), affecting one's perceptions as well as actions. They are transitory. They are somewhat unpredictable in their onset, intensity and duration (...) the proximate cause of both weather and emotions can be viewed as a force of nature, something beyond conscious control of the individual.

Thus, experiencer predicates can be conceptualized as 'experiencer weather' differing from meteorological weather in the external world, that is, in the former the natural force is within the individual. For Pesetsky, the Experiencer ambient *it* differs from the weather ambient *it* in only one way. The ambient *it* must be controlled by the Experiencer. Pesetsky (1995) speculates that this difference may in turn be a consequence of the conceptual distinction between the two different "natural forces".

Although it is generally agreed that ambient *it* is some sort of argument, the exact status of its argumenthood is a matter of some controversy. Pesetsky (1995) reports Ruwet's (1991) argument that weather predicates such as *rain* are unaccusative, suggesting that ambient *it* cannot be an external argument. However, as noted by Pesetsky (1995), there is some evidence which supports the idea that ambient *it* is an external argument. For instance, the availability of cognate objects with ambient *it* argues against an unaccusative analysis (since cognate objects are not possible with unaccusative verbs). Consider (43) (from Pesetsky, 1995, p. 110):

(43) a. *It rained a fine little rain*
b. *Il a plu une petite pluie fine*

In Amharic, weather predicates such as *zənnəbə* 'it rained' can indeed occur with cognate objects (from Leslau, 1995):

(44) *hayləñña zɨnab zənnəbə*
heavy rain rain.PF.3M
'It rained a heavy rain'

Amharic does not have a form equivalent to the English *it*. Rather, it employs an empty category (*pro*) with the grammatical features third person, masculine, singular. Thus, I assume that the Ambient-Causer is an external argument. Suppose that the Lexical Conceptual Structure (cf. Jackendoff 1990) of an experiencer verb such as *č'ənnək'ə* 'worry' is as follows:

(45) *č'ənnək'ə* 'worry'
[x A-Caus y BE *worry*]

The verb *worry* has two arguments: an A-Causer (x) and an Experiencer (y). A rough paraphrase of (45) is that "an A-Causer *x* causes an emotional state in *y*". Given fairly standard assumptions about the mapping from argument structure to syntax, one would assume that the external argument will be mapped onto subject position. This would give us the construction in (46):

(46) [e] *aster-(ɨn)* *č'ənnək'ə -at*
pro A.-(ACC) worry.PF.3M-3FO
'It worried Aster' (= 'Aster is worried')

Thus, according to (46), the subject is a third person, masculine, singular *pro* argument and the object is the Experiencer. This is consistent with the agreement facts. As already mentioned, the sentence in (46) resembles transitive predicates such as (47):

(47) *ləmma* *aster-ɨn* *ayy-(at)*
L. A.-ACC see.PF.3M-3FO
'Lemma saw Aster'

As already mentioned there are two obvious differences between (46) and (47). In (46) the accusative marking is optional, whereas the object agreement is obligatory. In (47) the opposite is true: accusative marking is obligatory but object agreement is optional. These facts are problematic if the Experiencer in (46) is simply an object on a par with the grammatical object of transitive verbs as in (47).

Therefore, the question is: how do we account for the quirky Case and agreement properties of (46)? The first thing we need to ask is: where else in the grammar is object agreement obligatory in simple transitive verbs? There are two clear cases: (a) when the object is pronominal — either null or overt as in (48) — and when the overt nominal occurs in topic position, as in (49) (see also Haile, 1970):

(48) *aster* *[e] / ɨssu-n* *mətta-čč-*(ɨw)*
A. pro / he-DEF-ACC hit.PF-3F-(3MO)
'Aster hit him'

(49) *wɨšša-w- ɨn*$_i$ *aster* *t*$_i$ *mətta- čč-*(ɨw)*
dog-DEF.M.-ACC A. hit.PF.-3FS-(3MO)
Lit. 'The dog, Aster hit him'
'Aster hit the dog'

In (48) where the direct object is either an empty pronominal or an overt pronominal, object agreement is obligatory. In (49), where the object moves from its canonical position and occurs in topic position, object agreement is also obligatory.

Now, one may argue that the obligatory object agreement in the Experiencer construction is the result of topicalization. Thus, compare (49) with the SubjExp construction in (50):

(50) *aster$_i$-(ɨn) pro [e$_i$] č'ənnək'-*(at)*
A.-(ACC) worry.PF.3MS-(3FO)
'Aster is worried'

Abstracting away from the obvious differences between topicalization in transitive verbs (49) and the construction in (50), it appears that there is some prima facie evidence for the correlation of obligatory object agreement and topicalization.

The topicalization hypothesis gains further support from the fact that in Amharic cleft questions and relative clause constructions also involve obligatory object agreement, as shown in (51b) and (52b):

(51) a. *aster bɨrtukan gəzza – čč*
A. orange buy.PF-3F
'Aster bought some oranges'

b. *mɨnd- ɨn nəw aster [e] yə- gəzza-čč-*(ɨw)*
what-ACC be.3M A. REL-buy.PF-3FS-(3MO)
'What is it that Aster bought?'

(52) a. *aster lɨǰočč ayy- əčč*
A. children see.PF.-3FS
'Aster saw (some) children'

b. *aster [e] y-ayy – əčč - * (aččəw) lɨǰočč*
A. REL-see.PF.-3FS-(3PLO) children
'The children whom Aster saw'

It is generally assumed that *wh*-questions, cleft constructions and relative clauses involve movement.[3] Thus, one may argue that the Experiencer argument is also dislocated in a similar manner.

[3] It is interesting that in *wh*-in-situ questions, the object suffix is not possible as can be seen in (i):

(i) *aster mɨn ayy- əčč -(*ɨw)*
A. what see.PF-3F-3MO
'What did Aster see?'

Alternatively, one can assume that the 'topic' Experiencer in the Experiencer construction is base-generated as an adjunct, and that the real argument of the verb is the object pronominal suffix, possibly co-indexed with an empty pronominal in argument position. This is the argument proposed for the so-called Clitic Left Dislocated constructions (CLLD) in Italian by Cinque (1990, Ch. 2). The same line of argument appealing to the Pronominal Argument Hypothesis (PAH) of Jelinek (1984) has been adopted by Baker (1996, Ch. 3; 2003, and related work) for the analysis of polysynthetic languages such as Mohawk.

Note also that the idea that NPs, which are coreferential with object pronominal suffixes, can be adjuncts is argued for by Bresnan and Mchombo (1987), based on a study of Chichewa and other related Bantu languages. Bresnan and Mchombo (1987) argue that there are two types of verbal agreement affixes: grammatical agreement affixes and anaphoric agreement affixes. An NP in grammatical agreement is an argument of the verb "...while the verbal affix expresses redundantly the person, number, and gender class of the NP". In anaphoric agreement, "...the verbal affix is an incorporated pronominal argument of the verb, and the coreferential NP has a non-argument function — either as an adjunct of the pronominal argument, or as a topic or focus of the clause or discourse structure" (Bresnan & Mchombo, 1987, p. 741). Bresnan and Mchombo argue that the coreferential NP is some kind of topic.

The important point for present purposes is that the experiencer argument appears to be outside of the minimal clause irrespective of the question of whether or not the experiencer argument is topicalized through movement or base-generation.

To summarize, constructions that exhibit differential subject properties are, to a large degree, semantically homogeneous. The argument which is optionally marked by the accusative case suffix *–n*, and agreeing with an apparent object pronominal suffix is often some kind of experiencer in different domains — including those undergoing emotional and physical sensations. I have argued that the best way to provide a unified account for such constructions is to assume that the predicates in question allow the experiencer argument to be optionally topicalized. In such topicalized contexts the agreement marker on the verb functions

---

The absence of the object pronominal suffix in the case of *wh*-in-situ questions can be accounted for naturally if we assume that *wh*-words are indefinite. Independent evidence for this assumption comes from the unavailability of the object suffix with quantified phrases (QPs), which are also indefinite (cf. Heim, 1982). Consider the following example:

(ii) *aster and nəggər ayy- əčč -(*ɨw)*
A. something see.PF-3F-3MO
'Aster saw something.'

as an anaphoric pronominal suffix. This is in line with the observation that in all constructions which involve movement — such as relativization, *wh*-questions and clefts — the object pronominal suffix is obligatory on the verb and functions as an anaphoric pronoun rather than a mere agreement marker.

5. Conclusion

The Amharic verb is obligatorily marked for subject agreement. It can often optionally be marked for object and indirect object agreement. The subject is non-overtly marked for nominative case, and the object is marked for accusative case (by the suffix *–n*) if it is definite.

Amharic, like many other languages, exhibits the phenomenon of non-canonical marking of the subject grammatical relation. First, while normally the subject is assigned nominative case and controls subject agreement, in some constructions it exhibits an optional accusative marking and agrees with an obligatory object/indirect object pronominal suffix on the verb. Predicates that exhibit such properties include verbs of emotion, physical sensation, and temperature, among others. We have argued that intransitive experiencer predicates such as 'worry' are conceptually more complex, with an abstract causer argument (Ambient Causer) which agrees with the subject pronominal suffix, whereas the experiencer argument agrees with the object/indirect object pronominal suffix on the verb. The fact that the experiencer argument obligatorily agrees with the object/indirect object pronominal suffix on the verb was accounted for by motivating a movement analysis where the experiencer argument is left-dislocated into an adjunct position. The pronominal suffix on the verb is then analyzed as an anaphoric pronoun rather than a mere agreement marker. Support for this analysis comes from clefts, relative clauses and *wh*-constructions, which all require the presence of the object/indirect object pronominal suffix on the verb.

It is hoped that this chapter will contribute to the growing body of empirical work on non-canonical subject marking. The analysis presented in the chapter shows how non-canonical behaviour is an artefact of an alternative way of encoding grammatical relations which is predictable from a semantically homogeneous class of verbs.

Acknowledgement

I would like to thank Peter Collins, Helen de Hoop, Alan Libert, Peter de Swart and an anonymous reviewer for their helpful comments and criticisms. I am responsible for remaining shortcomings. Earlier versions of the analysis of experiencer predicates incorporated into this chapter can be found in Amberber (1996, 2001, 2002).

References

Abbott, B. (2004). Definiteness and indefiniteness. In: R. Horn & G. Ward (Eds), *Handbook of pragmatics* (pp. 122–149). Oxford: Blackwell.

Aikhenvald, A., Dixon, R. M. W, & Onishi, M. (2001). *Non-canonical marking of subjects and objects*. Amsterdam: John Benjamins.

Amberber, M. (1996). *Transitivity alternations, event-types and light verbs*. Unpublished doctoral dissertation, McGill University, Montreal.

Amberber, M. (2000). Valency-changing and valency-encoding devices in Amharic. In: R. M. W. Dixon & A. Y. Aikhenvald (Eds), *Changing valency: Case studies in transitivity,* (pp. 312–332). Cambridge: Cambridge University Press.

Amberber, M. (2001). Testing emotional universals in Amharic. In: J. Harkins & A. Wierzbicka (Eds), *Emotions in crosslinguistic perspective* (pp. 39–72). Berlin and New York: Mouton de Gruyter.

Amberber, M. (2002). *Verb classes and transitivity in Amharic*. Lincom studies in Afro-Asiatic Linguistics, 6. Lincom, Munich.

Andrews, A. D. (1985). The major functions of the noun phrase. In. T. Shopen (Ed.), *Language typology and syntactic description: Vol 1, clause structure* (pp. 62–154). Cambridge: Cambridge University Press.

Andrews, A. D. (2001). Non-canonical A/S marking in Icelandic. In: A. Y. Aikhenvald, R. M. W. Dixon & M. Onishi (Eds), *Non-canonical marking of subjects and objects* (pp. 85 111). Amsterdam: John Benjamins.

Andrews, A. D. (In press). The major functions of the noun phrase. In: T. Shopen (Ed.), *Language typology and syntactic description*: Vol. 1 (2nd ed.). Cambridge: Cambridge University Press

Baker, M. (1996). *The polysynthesis parameter*. Oxford: Oxford University Press.

Baker, M. (2003). Agreement, dislocation, and partial configurationality. In: A. Carnie, H. Harley & M. Willie (Eds), *Formal approaches to function in grammar: In honor of Eloise Jelinek* (pp. 107–132). Amsterdam: John Benjamins.

Belletti, A., & Rizzi, L. (1988). Psych-verbs and θ-theory. *Natural Language & Linguistic Theory, 6*, 291–352.

Bhaskararao, P., & Subbarao, K. V. (2004). *Non-nominative subjects*. Amsterdam: John Benjamins.

Blake, B. (2001). *Case* (2nd ed.). Cambridge: Cambridge University Press.

Bliese, L. F. (1981). *A generative grammar of Afar*. The Summer Institute of Linguistics, The University of Texas, Arlington.

Bresnan, J., & Mchombo, S. (1987). Topic, pronoun, and agreement in Chichewa. *Language, 63*, 741–782.

Cinque, G. (1990). *Types of A'-Dependencies*. Cambridge: The MIT Press.
Dixon, R. M. W. (1994). *Ergativity*. Cambridge: Cambridge University Press.
Gragg, G. (1997). Ge'ez (Ethiopic). In: R. Hetzron (Ed.), *The semitic languages* (pp. 242–260). London: Routledge.
Gutt, E. (1997). The Silte group (East Gurage). In R. Hetzron (Ed.), *The semitic languages* (pp. 509–534). London: Routledge.
Haile, G. (1970). The suffix pronouns in Amharic. In: Chin-Wu Kim & H. Stahlke (Eds), *Papers in African linguistics* (pp. 101–111). Edmonton: Linguistic Research, Inc.
Hale, K., & Keyser, S. J. (1993). On argument structure and the lexical expression of syntactic relations. In: K. Hale & S. J. Keyser (Eds), *The view from building 20. Essays in linguistics in honor of Sylvian Bromberger* (pp. 53–109). Cambridge: The MIT Press.
Hale, K., & Keyser, S. J. (2002). *Prolegomenon to a theory of argument structure*. Cambridge: MIT Press.
Heim, I. (1982). *The semantics of definite and indefinite noun phrases*. Unpublished doctoral dissertation, University of Massachusetts, Amherst.
Hetzron, R. (1970). Toward an Amharic case-grammar. *Studies in African Linguistics, 1*, 301–354.
Hopper, P., & Thompson, S. (1980). Transitivity in grammar and discourse. *Language, 56*, 251–299.
Jackendoff, R. (1990). *Semantic structure*. Cambridge, MA: The MIT Press.
Jelinek, E. (1984). Empty categories, case, and cofigurationality. *Natural Language and Linguistic Theory, 2*, 39–76.
Kogan, L. E. (1997). Tigrinya. In: R. Hetzron (Ed.), *The semitic languages* (pp. 424–445). Oxford: Routledge.
Leslau, W. (1995). *Reference grammar of Amharic*. Wiesbaden: Harrassowitz.
Malchukov, A. (this volume). Case pattern splits, verb types, and construction competition.
McCawley, N. (1976). From OE/ME "impersonal" to "personal" constructions: What is a "subject-less" S? *Papers from the Parasession on Diachronic Syntax* (pp. 192–204). Chicago: Chicago Linguistic Society.
Mullen, D. S. (1986). *Issues in the morphology and phonology of Amharic: the lexical generation of pronominal clitics*. Unpublished doctoral dissertation, University of Ottawa, Ottawa.
Onishi, M. (2001). Introduction: Non-canonically marked subjects and objects: parameters and properties. In: A. Y. Aikhenvald, R. M. W. Dixon & M. Onishi (Eds), *Non-canonical marking of subjects and objects* (pp. 1–51). Amsterdam: John Benjamins.
Owens, J. (1985). *A grammar of Harar Oromo (Northeastern Ethiopia): Including a text and a glossary*. Hamburg: H. Buske.
Pesetsky, D. (1995). *Zero syntax: Experiencers and cascades*. Cambridge: The MIT Press.
Ruwet, N. (1991). *Language and human experience*. Chicago: University of Chicago Press.
Takezawa, K. (1987). *A configurational approach to case-marking in Japanese*. Unpublished doctoral dissertation, University of Washington, Washington.
Verma, M., & Mohanan, K.P. (1990). *Experiencer subjects in south Asian languages*. Stanford: CSLI Publications.

Wagner, E. (1997). Harari. In: R. Hetzron (Ed.), *The semitic languages* (pp. 486–508). Oxford: Routledge.
Zaenen, A., Maling, J., & Thráinsson, H. (1985). Case and grammatical functions: The Icelandic passive. *Natural Language and Linguistic Theory, 3*, 441–483.

Appendix: Case-marking in some Ethiopian Semitic and Cushitic languages

Ge'ez (Ethiopic)

In Geez (Ethiopian Semitic) the nominative and accusative are contrasted as follows (Gragg, 1997, pp. 248–249):

(1) a. *bet* 'house' (nominative)
b. *betə* 'house' (accusative)

(2) *sərhə* *nɨgus* *bet- ə*
built king[NOM] house-ACC
'the/a king built the/a house'

Tigrinya

In Tigrinya (Ethiopian Semitic), case relations are expressed mainly by the use of prepositions (Kogan, 1997, pp. 432–433). Both the direct and indirect object are formally marked by *nɨ*-:

(3) ***nə****ta* *səbbəyti* *ʔaytɨnkɨʔəwwa* 'Do not touch this woman'

Interestingly, it appears that the use of *nɨ*- for marking direct objects is optional — particularly if the object is inanimate.

Silte

In Silte (Ethiopian Semitic), the following markers are used to indicate various case relations (Gutt, 1997, pp. 513–514):

(4) *la*- 'to' dative: experiencer, goal, location, benefactive, theme
-a 'accusative': theme

Interestingly, similar to Amharic, "the accusative is used for definite objects, indefinite objects have no overt case marker…" (Gutt, 1997, p. 514). However, Gutt further states that some dialects may use the accusative marker even with indefinite objects.

Harari

Harari (Ethiopian Semitic) is also a nominative–accusative language where the nominative is unmarked, whereas the accusative can be indicated by the suffix –*w* (Wagner, 1997, pp. 486–508). Wagner (1997, p. 492) points out that "the *w* could have been taken over from the verbal object suffix (*sabare*:***w***) 'he broke him')"

The use of the accusative suffix is not obligatory in Harari. "Definiteness…does not play a part in it" (Wagner, 1997, p. 492). This is perhaps due to the fact that the marking of definiteness is not obligatory. In Harari, other case relations "between the verb and its objects must be expressed by post-positions, e.g. the dative –*le* (< la-), the locative instrumental – *be* < ba)" (Wagner, 1997, p. 492).

Oromo

Oromo (Cushitic) exhibits case relations markedly different from the Ethiopian Semitic languages. Thus, for example in the Eastern dialect of Oromo (Owens, 1985) both transitive and intransitive subjects (A, S) are marked by the nomi-native case. Object NPs occur in the unmarked accusative (traditionally known as 'absolutive') case. It appears that the accusative is both formally and functionally unmarked: it has no overt marking and is the form used in citations.

(5) *innîi* ***ibidda*** *sún* *bobees - e*
he.NOM **fire** that lit-3M.PAST
'He lit that fire'

Oromo is one of the few languages of the world which have marked nominative case and unmarked accusative case.[4]

[4] See Dixon (1994, p. 64ff) for discussion. Dixon (1994, p. 64) notes "languages showing the 'marked nominative' include some members of the Cushitic family, from north-east Africa."

Afar

Afar (Cushitic) also exhibits case relations similar to Oromo: marked nominative, unmarked accusative. The nominative is marked by the suffix *–i* (Bliese, 1981):

(6) *bad – i* *xado* *yakme*
son-NOM meat eats

"Those nouns which are not nominative-genitive may be grouped together as accusative (direct objects, postpositional objects, and predicate complements). They all take the base form with no marker." (Bliese, 1981, p. 173)

(7) *usuk* *xan* *yo-h* *yexe*
he milk me-to gave
'He gave me milk'

Chapter 12

Differential Case-Marking in Hindi

Helen de Hoop and Bhuvana Narasimhan

1. Patterns of Case-Marking

There are cross-linguistically two major patterns of case-marking (Blake, 2001). *Nominative-accusative* refers to a grammatical pattern in which the subject of an intransitive verb receives the same morphological treatment as the subject of a transitive verb, while the object of the transitive verb receives a different treatment. *Ergative-absolutive* refers to a grammatical pattern in which the subject of an intransitive verb receives the same morphological treatment as the object of a transitive verb, while the subject of the transitive verb receives a different treatment (Dixon, 1979). Some examples from a (nominative-) accusative language (spoken Tamil) (examples (1) and (2)) and an ergative (-absolutive) language (West Greenlandic) (Manning, 1996, p. 3) (examples (3) and (4)) follow:[1]

spoken Tamil

(1) *avan-Ø* *puLLay-E* *paattAA*
he-NOM boy-ACC see.PST.3SG.M.
'He saw the boy'

[1] Glossing conventions follow the Leipzig Glossing Rules (http://www.eva.mpg.de/lingua/files/morpheme.html):
1:first person; 2:second person; 3:third person; ABL:ablative; ABS:absolutive; ACC:accusative; AUX:auxiliary; DAT:dative; ERG:ergative; F:feminine; FUT:future; GEN:genitive; IMP:imperative; INF:infinitive; INS:instrumental; INTR:intransitive; IPFV:imperfective; LOC:locative; M:masculine; NOM:nominative; PFV:perfective; PL:plural; PRS:present; PROG: progressive; PST:past; SG:singular; TR:transitive

Competition and Variation in Natural Languages: The Case for Case
Edited by M. Amberber and H. de Hoop

ISBN: 0-08-044651-5

(2) *avan-Ø* *ooDinAA*
He-NOM run.PST.3SG.M.
'He ran'

West Greenlandic

(3) *Oli-p* *neq* *neri-vaa*
Oli-ERG meat.ABS eat-IND.TR.3SG.3SG.
'Oli eats meat'

(4) *Oli* *sinippoq*
Oli.ABS sleep:IND.INTR.3SG
'Oli sleeps'

In many languages, nominative or absolutive case is the unmarked (morphologically zero) case and functions as the 'elsewhere' case for either the subject or the object of a transitive clause. We assume that nominative (or absolutive) case is in fact a label for 'no case': that is, we assume that the absence of special morphological marking indicates the absence of case.[2] Thus, whereas morphologically zero cases are, in some accounts, treated as 'unmarked' (Marantz, 1991) or 'obligatory' cases (Bobaljik, 1993), we treat them as 'no case', which is also the approach advocated by Aissen (2003). In both types of case systems, the 'primary' (unmarked) argument in transitive constructions as well as the single argument of intransitive predicates remains null-marked, hence case-less.

2. Two Functions of Case-Marking

In the functional-typological literature, two main functions of case-marking are distinguished (cf. a.o., Mallinson & Blake, 1981; Comrie, 1989; Kibrik, 1985; Song, 2001; Malchukov, this volume).

One motivation for case-marking the subject or object of a transitive clause "is *to disambiguate arguments* in terms of their function(s) in the clause, and this is only relevant in clauses in which there is more than one argument" [emphasis ours] (Van Valin, 1992, p.19). In order to differentiate the subject from the object it is not necessary to mark them both; a case marker on one of them serves to distinguish them. Hence, if the marker is given to the object, the subject is left in the citation form, with the result that the subject of both a transitive as well as an

[2] We will, however, continue to gloss null-marking as 'nominative case' in our Hindi examples in this chapter.

intransitive clause is morphologically unmarked and the object is marked, yielding an accusative pattern. Conversely, if the marker is applied to the subject of a transitive clause, the object remains unmarked, resulting in an ergative pattern in which the subject of an intransitive clause and the object of a transitive clause are morphologically unmarked and the subject of a transitive clause is marked (Van Valin, 1992). This function of case marking is generally referred to as the *distinguishing* or *discriminating* function.

Another widely attested function of case has to do with the expression of specific semantic information in the case morphology. Morphological cases are generally considered to express some kind of specific (e.g. thematic) information of the argument under consideration. That does not only hold for lexical (or semantic) cases such as, e.g. locative cases, but to a certain degree for structural cases as well. For example, dative case is associated with goal and experiencer semantics, ergative case is usually associated with agentivity, or more specifically with properties such as volitionality or control, and accusative case is associated with patienthood. This function of case marking is usually referred to as the *identifying* or *indexing* function of case.

We assume that in the case-patterns we find cross-linguistically both functions play a role. For instance, ergative case marking may serve either of two purposes: it may mark and identify agentivity or it may mark the first argument of a transitive clause in order to distinguish the two arguments. In principle, these two functions of ergative case marking can go hand in hand, but below we will encounter a specific problem with that view.

In this chapter, we propose that a model of case-marking based on the two functions of case should be augmented with the notion of *argument strength*. We show how this extended approach derives split case-marking in subject and object position in Hindi. The identification of *argument strength* is needed to distinguish between strong arguments of a certain type and their weak counterparts. We suggest that in languages with differential case-marking, subjects and objects that are 'strong' are likely to be overtly marked as such. The question is how we measure the 'strength' of arguments, since languages seem to vary in this respect. Notoriously, the degree of transitivity varies on the basis of a number of factors, including animacy, specificity or referentiality, volitionality, perfectivity, etc. in different languages (Hopper & Thompson, 1980). These can be represented in terms of features that play a role in some languages but not others, or as semantic scales.

One possibility of measuring the 'strength' of arguments is by using the notion 'discourse prominence'. Legendre et al. (1993) recognize the importance of argument prominence and translate the notion into their Optimality Theoretic constraints. They use constraints such as "high-prominence arguments receive C_1" and "low-prominence arguments are not case-marked C_1 and C_2", where C_1 in their framework refers to both nominative and ergative, and C_2 to both accusative and absolutive. According to Legendre et al., intransitives always come with a high-prominent argument, while transitives always come with two high-prominent arguments. They write high-prominent arguments with capital letters. Thus, the subject and object of a transitive clause are written as *A* and *P*, while low-prominent subjects and objects are written as *a* and *p*. As a consequence, the two arguments of a transitive predicate are universally marked A_1P_2 (where the subscript indicates the type of case). If one of the arguments of a transitive predicate is low-prominent, the construction becomes 'passive' or 'antipassive', i.e. with either an implicit or an obliquely case-marked argument. Obviously, high-prominent arguments are typically animate and referential or specific. However, one cannot maintain that in ordinary transitive clauses the arguments are always 'high-prominent' in the discourse, nor that a 'low-prominent' argument of a transitive predicate always gives rise to a passive or antipassive construction (cf. De Hoop, 1999). Legendre et al. (1993) are not able to make a distinction between a high-transitive clause such as *Jane hit Jacky* and a low-transitive clause such as *Jane was drinking wine* (both are *AP* in their model), although intuitively the wine is definitely less prominent in the discourse than Jacky. A related problem for Legendre et al. is that their system does not account for phenomena of differential case-marking of subjects and objects in a straightforward way. However, we think that the basic insight that the 'discourse prominence' of arguments plays a crucial role in case-marking patterns can still be maintained. In our view 'prominence' is one of the factors that can be used to measure the 'strength' of arguments.

Another way of measuring the 'strength' of an argument is by looking at its prototypicality, i.e. the degree to which it possesses certain features that characterize the argument's role in the expressed event (Dowty, 1991). On the one hand, a feature such as volitionality (being in control) determines a 'prototypical' agent. Note that volitionality usually implies animacy as well. On the other hand, a 'prototypical' patient is usually characterized in terms of its 'affectedness'. Strikingly, animate and referential or specific arguments are intuitively perceived as 'more affected' by an action than inanimate non-specific arguments. 'Prototypicality' is another factor that can be used to measure the 'strength' of arguments.

We conclude that in fact the different ways of measuring the 'strength' of arguments point in the same direction: arguments which are high-prominent in the discourse are usually animate and specific, whereas arguments which are prototypical, that is, either in control of the action (agents) or affected by the action (patients) are usually animate and specific as well. Therefore, we have chosen to use 'strength' as a comprehensive term for the different factors. We will write strong arguments with capital letters *A* and *P*, following Legendre et al. (1993) in this respect.

At this point, let us return to the two functions of case-marking. Which arguments of a transitive clause are case-marked, the strong or the weak ones? We may illustrate case-marking in order to *identify* arguments as in figure 1 below.

Clearly, in its identification function, case-marking identifies *strong* arguments, as argued above, the prototypical agents (the high-prominent ones, which are in control) and the prototypical patients (the high-prominent, affected ones). So, the identifying function of case predicts *A* and *P* to be case-marked in a transitive clause.

However, as we argued above, differential case-marking can also be employed to distinguish between arguments which are potentially ambiguous with respect to their role (that is, case-marking is used to distinguish between agents and patients). Since subjects are usually stronger than objects (or, agents stronger than patients), obviously, differential case-marking on the basis of distinguishability marks weak agents rather than strong agents because the weak ones are 'closer' to the patients in strength. On the other hand, strong patients get case-marked and not the weak ones, because in the case of patients, the strong ones are 'closer' to the agents. Hence, the weak agents and the strong patients represent the borderline where the danger of ambiguity is maximal and are therefore potential candidates for case-marking on the basis of distinguishability. This is illustrated as in figure 2.

Comparing Figures 1 and 2, an essential difference between the two functions of case-marking is revealed. While the identifying function explains case-marking of the strong agent and the strong patient (the *A* and the *P*), the distinguishing function explains case-marking of the weak agent and the strong

```
|------- |------- |--------|--------|-→ (Argument strength)
A        a        P        p
|                 |
case              case
```

Figure 1: The identifying function of case.

```
|------- |------- |--------|--------|- → (Argument strength)
A        a        P        p
         |        |
         case     case
```

Figure 2: The distinguishing function of case.

patient (the *a* and the *P*). Another difference becomes obvious as well. When case-marking is merely used to distinguish two arguments in a transitive clause, it is of course sufficient to mark only one of the two arguments. Case-marking applies to the *relation* between the subject and the object then, rather than to both arguments independently. By contrast, when case-marking is used to identify the strength of an argument, it may apply to each argument independently, both the subject and the object of a transitive clause, as well as to the sole argument of an intransitive construction. That is, in the identifying view, we do not necessarily expect a difference in case-marking between intransitive and transitive sentences. There is a possibility to get a difference indirectly, when we say that an agent is only 'strong' in the presence of a patient and vice versa. That would be possible when we measure strength in terms of 'prototypicality' (a prototypical agent needs a patient and vice versa), but not when we measure it in terms of discourse prominence, nor when we measure it in terms of intrinsic notions such as volitionality or affectedness or features like animacy and specificity.

Thus, in the distinguishing approach of case, ergative case should be assigned to agents that are somehow 'marked' (in the grammatical role of subject) or 'weak' (compared to the object) and hence confusable with patients. Overtly marking a weak agent with 'agentive case-marking' or a strong patient with 'patientive case-marking' helps to tag the role of one argument with respect to the other. Obviously, this appears to be the case in a language such as Dyirbal, where both the weak agent and the strong patient receive overt case-marking.

On the one hand, the transitive subjects that get overt marking are the 3rd person pronouns, proper names and common nouns, i.e. the weak agents. The 1st and 2nd person pronouns, however, are prototypical (strong) subjects that do not need any morphological marking. On the other hand, 1st and 2nd person pronouns are non-canonical objects (strong patients), hence they get morphological marking, while canonical objects (that is, 3rd person pronouns, proper names and common nouns) lack overt marking. This would support Aissen's idea that weak agents as well as strong patients receive case-marking as a function of distinguishing the two arguments of a transitive clause.

A	–∅	*–ŋgu*	*–ŋgu*	*–ŋgu*
S	–∅	–∅	–∅	–∅
O	*–na*	–∅	–∅	–∅
	1st & 2nd person pronouns	**3rd person pronouns**	**proper names**	**common nouns**

Figure 3: Case-marking in Dyirbal (from Table 2 in Dixon, 1979, p. 87).

Note that with regard to *object marking* in a transitive construction, an explanation on the basis of distinguishability (Aissen, 1999, 2003) points into the same direction as an explanation on the basis of identification: in both approaches it is the strong patient that gets case-marked. However, with respect to the *subject* of a transitive clause, the two approaches are diametrically opposite: the identifying function marks the strong agent, while the distinguishing function marks the weak one.

We recognize that both types of agent marking are indeed exemplified in languages across the world, but we claim that in Hindi it is the identification function that is dominant with respect to the case-marking of the subject in a transitive clause. Distinguishability, on the other hand, plays a crucial role in the case-marking of the object of a transitive clause. We will analyze the Hindi case-marking patterns in transitive clauses within a bidirectional Optimality Theoretic approach.

3. Differential Object Marking in Hindi

Consider the sentences in (5) and (6). The Hindi–Urdu case system allows for two case-alternations in ordinary transitive clauses. That is, transitive subjects can occur in nominative (null-marked) or ergative case, and objects in nominative or accusative case. Apparently, the two case alternations occur independently of each other (see (5),(6)).

Hindi

(5)	*wo-Ø*	*ek bakraa-Ø /ek bakre=ko*	*bec-taa*	*hae*
	He-NOM	one goat-NOM/one goat=ACC	sell IPFV.SG M.	be.PRS.3SG.
	'He sells a goat / the goat'			

(6)	*us=ne*	*ek*	*bakraa-Ø / ek bakre=ko*	*bec-aa*
	He=ERG	one	goat-NOM / one goat=ACC	sell-PFV.SG.M.
	'He sold a goat / the goat'			

The *subject* case alternation in (5) and (6) is dependent on the perfectivity of the transitive sentence. Ergative case is assigned in the context of a perfective verb phrase. The *object* case alternation in (5) and (6) correlates with the specificity of the object: accusative case marks the specific object. The latter phenomenon is referred to in the literature as *differential object marking* (cf. Aissen, 2003). Similarly, we may refer to the ergative-nominative alternation in subject position as *differential subject marking*.

As Aissen (1999, 2003) argues, sometimes the function of case-marking is not to mark all subjects or all objects of a transitive clause, but rather to mark those that are 'less prototypical' subjects or objects. Citing the work in Silverstein

(1976) and Comrie (1989), Aissen argues that differential case-marking on objects occurs whenever the object is *non*-canonical, or 'marked': "The higher in prominence a direct object, the more likely it is to be overtly case-marked" since "it is those direct objects which are most in need of being distinguished from subjects that get overtly case-marked"(Aissen, 2003). Since agents tend to be more prominent, or in our terms are stronger than patients, the ones that should get overtly case-marked under this disambiguating perspective are the 'weak' subjects and the 'strong' objects (as illustrated in Figure 2 above).

De Swart (2003) partly rejects Aissen's analysis since, as he points out, it is not the (un)markedness of either the subject or the object in a transitive clause that is crucial, but rather the *relation between the two*. That is, we might argue that as long as a subject outranks an object in a hierarchy of strength, there is no danger of ambiguity, and no case-marking is necessary to distinguish the two. No case-marking has the advantage of satisfying an *economy* constraint which penalizes morphological case marking. However, as De Swart (2003) points out, when the arguments do not minimally differ semantically (in terms of animacy or specificity), the object is case-marked to *avoid ambiguity*.

Differential object marking in Hindi can also be accounted for in this way, i.e. as the result of an interaction between *economy* and *distinguishability* (cf. Aissen, 2003, pp. 447–448).[3] Indeed, those patients which are most agent-like in the sense that they are animate and/or specific are most likely to be confused with true agents, and hence receive overt accusative *ko* marking. However, as we observed in the previous section, differential object marking cannot make a principled distinction between the identifying and the distinguishing function of case, since both functions predict the same type of objects, *viz.* the strong ones, to be case-marked. In other words, the case alternation in object position in Hindi could just as well be an alternation based on the *identifying* function of case. But when does differential object marking in Hindi occur?

Differential object marking, between *ko* and null-marking, shows up with a variety of transitive and ditransitive predicates.[4] Strikingly, these are the same predicates that allow case alternations on the subject as well, between *ne* and null-marking. Examples are predicates like *bec* 'sell' (see examples (5),(6)

[3] As we discuss further in later sections of the chapter, distinguishability in Hindi is not determined on the basis of nominal semantics alone. For example, an animate noun (e.g. *bakraa* 'goat') is more likely to receive object case-marking with the verb *maar* 'hit' than with a verb such as *bec* 'sell'. Thus, the possibility of differential object-marking depends not only on nominal properties (e.g. animacy, volitionality) but also on verb semantics, among other factors.

[4] When (specific) objects are marked with *ko* in ditransitives, they undergo obligatory scrambling (Bhatt & Anagnostopoulou 1996)

above), *maar* 'hit', and *de* 'give'. When there is no *possible* case alternation in subject position, there is no *possible* case alternation in object position either (Davison, 1999a, b).[5] So, the subjects of predicates such as *mil* 'receive' or *pasand ho* 'like' are not real agents and get obligatorily dative marked irrespective of the (im)perfectivity of the predicate. The object of these predicates does not receive case marking, that is, irrespective of its animacy and/or specificity.

(7) *raam=ko* *halwaa-∅* *pasand* *hae*
Raam=DAT halwaa-NOM liking be.PRS.3SG.
'Raam likes halwaa'

(8) *raam=ko* *wo* *laDkii-∅* *pasand* *hae*
Raam-DAT that girl-NOM liking be.PRS.3SG.
'Raam likes that girl'

We thus observe a *partial dependency* between subject and object case-marking in (5)–(6) above (Davison, 1999a, b) because both case alternations only occur with a certain class of predicates. This dependency is of course different from the dependency pointed out by De Swart (2003). That is, De Swart refers to the relation between subject and object in individual cases. For example, he pointed out that in some languages that have differential object marking on the basis of animacy, surprisingly an *inanimate* object is case-marked when the subject is inanimate as well. In Hindi, however, the case-alternations seem to be more restricted to certain grammatical contexts. Yet, the case-alternations of subjects and objects of a transitive clause are dependent on each other *in a certain way* because both alternations are only possible with a verb that allows the other alternation as well. That suggests that we cannot explain the complete pattern of case-alternations in transitive clauses in terms of the identifying function of case only. Indeed, the identifying function of case operates independently of the grammatical context and we would expect the case-alternations to show up on the basis of certain properties of the arguments alone, not necessarily in relation to the type of verb. In the remainder of this section, we will analyse the pattern of differential object marking in Hindi as an interaction between the identifying function of case, distinguishability and economy, resulting in the case-marking of an object only when the predicate is of a certain type, and the object is animate or specific.

First of all, we will call the class of predicates that *allow for* the two types of case alternations (differential subject marking as well as differential object marking)

[5] Davison's (1999b) formulation is stated in the other direction: "if the direct object can be marked with the dative *–koo* (specific/animate referent), then the subject can be marked with the ergative *–nee* (in perfective, finite clauses)."

highly transitive predicates. Transitivity in Hindi has been described as constituting a continuum (Bhatia, 1981; Kachru, 1981; Pandharipande, 1981; Hock, 1985). For the purposes of this chapter, we designate a set of predicates at the 'high' end of the transitivity continuum which allow passivization of the object, (and case-alternations between *ne* and null marking on the subject, and *ko* and null-marking on the object), as being 'highly transitive'.

As we witnessed above, it is exactly with this class of 'highly transitive' predicates that we find the object case-alternation on the basis of argument strength. That is, strong objects of highly transitive verbs are case-marked, while neither weak objects of highly transitive verbs nor strong objects of less transitive verbs are. We claim that it is the identifying function of case that is responsible here for the identification of strong patients. However, we appeal to the distinguishing function of case to explain why objects of other predicates do not receive case-marking, independently of whether or not they are strong.

We assume that there is a general principle of preserving distinguishability of the subject and the object in a transitive clause (basically following Aissen, 2003; De Swart, 2003; De Hoop & Lamers, to appear). This principle is captured by the following universal constraint (De Hoop & Lamers, to appear):

(9) DISTINGUISHABILITY
The two arguments of a transitive relation should be distinguishable.

The principle of DISTINGUISHABILITY specifies that, in a transitive relation, the two arguments should be distinct from each other (variants of this constraint and elaborate discussions are also found in other recent work, e.g. Bornkessel, 2002; Næss, 2004).

As De Hoop and Lamers (to appear) point out, DISTINGUISHABILITY can be satisfied in many different ways, such as by case morphology, argument strength, and agreement. As we have seen above, two arguments can be distinguished when only one bears unambiguous morphological case marking, e.g. when the object bears accusative case or the subject ergative case. We also pointed out that arguments may be distinguished on the basis of argument strength, measured along different scales, such as animacy or specificity. Obviously, the verb itself may provide the crucial information that helps to distinguish the two arguments. In English, for example, *please* takes an experiencer (animate) object, while *like* takes an experiencer (animate) subject. Therefore, when there is one animate and one inanimate noun phrase available, it is clear which one is the subject and which one is the object in the context of one of these verbs. Last but not least, agreement with the verb may determine the subject in languages such as German and English, and both subject and object in languages with subject and object agreement. In Hindi, the verb agrees with the highest nominative argument, which can be either the subject or the object (the latter when the subject bears non-nominative case, e.g. ergative dative). Obviously, there may be other ways to distinguish between the two arguments of a transitive relation, such as word order, prosody or surrounding context, but in this chapter we will not be concerned with all of these factors.

In Hindi, the two arguments of less transitive verbs are distinguishable on the basis of case-marking. To see this, reconsider the examples in (7)–(8) above, repeated below as (10)–(11).

(10) *raam=ko* *halwaa-∅* *pasand* *hae*
Raam=DAT halwaa-NOM liking be.PRS.3SG.
'Raam likes *halwaa*'

(11) *raam=ko* *wo* *laDkii-∅* *pasand* *hae*
Raam=DAT that girl-NOM liking be.PRS.3SG.
'Raam likes that girl'

With these predicates, the subject is invariably marked with dative case. So, DISTINGUISHABILITY is already satisfied whether or not we case-mark the object and for reasons of economy we do not expect the object to be case-marked if not necessary. In fact, the dative morphology happens to be identical to what has been analysed as accusative morphology in Hindi. Thus, marking the object with accusative case would lead to a blunt violation of DISTINGUISHABILITY. Hence, with less transitive predicates, there is a fixed case-pattern that already adequately deals with distinguishing the subject from the object. In other words, when there is no case alternation in subject position there should be no case-alternation in object position either.

Highly transitive verbs, on the other hand, run the danger or ambiguity because both the subject and the object can be unmarked (i.e. both can end up in the nominative case). With those predicates, DISTINGUISHABILITY is saved by making use of the identifying function of case, which enables us to case-mark strong objects. With less transitive verbs, however, DISTINGUISHABILITY is already satisfied by the case-marking of the subject, and therefore, case-marking of strong objects becomes superfluous. Thus, differential object marking in Hindi can be straightforwardly accounted for in terms of the *distinguishing* function of case, as was also argued by Aissen (2003) and De Swart (2003).

The constraints DISTINGUISHABILITY (De Hoop & Lamers, to appear) and IDENTIFICATION-P in combination with the economy constraint PAIP (Malchukov, to appear) give the right results for differential object marking in Hindi. The relevant constraints are given below:

(12) DISTINGUISHABILITY
The two arguments of a transitive relation should be distinguishable.

(13) IDENTIFICATION-P
Strong patients receive accusative marking.

(14) PAIP
"Primary Argument Immunity Principle": don't case-mark the unmarked — nominative — argument.

DISTINGUISHABILITY was extensively discussed above. At this point we make a split between two subconstraints of DISTINGUISHABILITY, viz. a 'form'-constraint DIST-CASE and a 'meaning'-constraint DIST-STRENGTH. The form-constraint refers to distinguishability of the two arguments on the basis of case-marking, while the meaning-constraint deals with the distinguishability of the two arguments on the basis of argument strength. The latter one can be formulated as "Subjects must outrank objects in strength" and is thus equivalent to De Swart's (2003) constraint MINIMAL SEMANTIC DISTINCTNESS. Our constraint IDENTIFICATION-P is used to identify and case-mark a 'strong' patient. Malchukov's PAIP is a constraint that penalizes morphological case-marking of an argument that could in principle remain unmarked. In Hindi, PAIP is satisfied when an object or subject is in the nominative (unmarked) case and it is violated whenever an object or subject bears an overt (ergative, accusative/dative, etc.) case-marker. We prefer to use PAIP rather than Aissen's more general constraint $*Struc_C$, since the latter assumes a special notion of 'structural case,' which our account lacks. PAIP penalizes morphological case-marking on the argument that could (should) remain unmarked (the nominative argument), i.e. the obligatory case bearing argument in terms of Bobaljik (1993).

In the following bidirectional OT tableau, the different form-meaning pairs are evaluated against the set of constraints. Nominative (i.e. the absence of) case is written as ∅, accusative case is represented by the marker *ko*, a pair such as *[ko,* ***p****]* is to be read as "[form *ko*; meaning: weak patient]".

Note that the object of a highly transitive verb in Hindi fulfills DIST-CASE when it bears the accusative case-marker, while a null-marked object violates this constraint. This is evaluated independently of the case of the subject. So, even when the subject is assigned ergative case in a perfective highly transitive clause in Hindi, we say that DIST-CASE is violated when the object is in the nominative. This is so because the case-marking pattern is grammaticalized in Hindi such that we cannot compare individual cases to see whether the subject is actually ergative or nominative (cf. Davison, 1999a, b).[6] In the context of a highly transitive verb, it

[6] Davison (1999b) notes that a small class of verbs (e.g. pahcaan 'recognize', bhuul 'forget') are optionally ergative and also allow alternation of ko and null marking on the object, providing further support for the partial independence of object case alternations from subject case alternations in the same (perfective) context.

Table 1: Bidirectional OT table: Form and meaning of the object of a highly transitive verb.

Object of highly transitive verb	**PAIP**	**DIST-CASE**	**DIST-STRENGTH**	**IDENTIFICATION-P**
✌ [∅, p]		*		
[*ko*, p]	*			
[∅, P]		*	*	*
✌ [*ko*, P]	*		*	

is the accusative case marker that in fact guarantees *distinguishability* if the object is strong instead of weak. However, as we pointed out above, we assume that in sentences with less transitive verbs, one of the arguments is obligatorily case-marked and then DIST-CASE is straightforwardly satisfied already.

The table reads as follows. The null-form is better than the form *–ko* because the economy constraint PAIP outranks the distinguishability constraint DIST-CASE as well as the identification constraint IDENTIFICATION-P. The meaning *p* wins over the meaning *P* because the latter violates meaning constraint DIST-STRENGTH. Thus, the first candidate form-meaning pair combines an optimal form with an optimal meaning, hence is super-optimal (Blutner, 2000). The second candidate pair is not super-optimal, because it combines an optimal meaning with a sub-optimal form (hence, it is blocked by the first super-optimal pair). The third candidate pair is not super-optimal either, because it combines an optimal form with a sub-optimal meaning (hence, it is blocked by the first super-optimal pair as well). Finally, the fourth candidate pair combines a sub-optimal form with a sub-optimal meaning, and this is super-optimal again (it cannot be blocked by the first pair, because both its form and its meaning are different from that one).

Whenever we encounter a pattern in language where in the same context two forms are available as well as two meanings, we may find a one-to-one mapping of forms and meanings that is in accordance with a principle of markedness that states that the (more) marked form is used to express the (more) marked meaning (Horn, 1984). When this is the case, the pattern can be analysed in terms of bidirectional Optimality Theory (Blutner, 2000). Blutner (2000) derives the markedness principle by means of bidirectional optimization (from form to meaning and from meaning to form).

In Blutner's (2000) framework a form-meaning pair $\langle f, m\rangle$ is called *super-optimal* if and only if there is no other super-optimal pair $\langle f', m\rangle$ such that $\langle f', m\rangle$ is more harmonic than $\langle f, m\rangle$ and there is no other super-optimal pair

⟨*f*, *m*′⟩ such that ⟨*f*, *m*′⟩ is more harmonic than ⟨*f*, *m*⟩. The reader may verify that according to this definition, there are two super-optimal pairs in the diagram in the table above, namely ⟨∅, p⟩ and ⟨*ko*, P⟩. Indeed, although *ko* is not an optimal form itself and *P* is not an optimal meaning, the pair ⟨*ko*, P⟩ is super-optimal, because there is no super-optimal pair that blocks it (i.e. the two candidates ⟨*ko*, p⟩ and ⟨∅, P⟩ are not super-optimal, because they are both blocked by the super-optimal pair ⟨∅, p⟩).

Thus, bidirectional OT provides us with an analysis of differential object marking in Hindi that is in accordance with the markedness principle (our analysis is similar to De Swart's 2003 analysis). Note, by the way, that the constraint IDENTIFICATION-P does not play a role at all in the analysis because it is harmonically bound by DIST-STRENGTH. Let us now turn to differential subject marking.

4. Differential Subject Marking in Hindi

When we consider differential *subject* marking, Hindi presents a serious problem for an analysis along the lines of the distinguishing function of case. First of all, whenever the subject and the object in a transitive clause are not minimally distinct, then in principle, the case marking of the specific object should already be sufficient to do the job. In both Aissen's (1999, 2003) and De Swart's (2003) approaches, there would be no need to mark additionally the subject of a transitive clause in the perfective with ergative case.

Moreover, the ergative case-marking of the subject of a perfective transitive clause in Hindi is clearly not triggered by a *decrease in strength* of that subject. That is, subjects of perfective clauses are not weaker nor in any other way more similar to objects than subjects of imperfective clauses. Following Aissen's account, we would expect non-canonical subjects (i.e. those agents which are weak, e.g. non-volitional, inanimate) to receive overt case-marking. By contrast, the subject of a perfective transitive clause, in other words the agent of an action, which (successfully) brings about an event to completion, is usually conceived of as being typically agentive (Dowty, 1991), corresponding to a high degree of transitivity (Hopper & Thompson, 1980), hence strong in our account, and not weak as predicted by Aissen.

Further evidence that the ergative case-marker *ne* marks properties associated with strong agents can be seen in constructions where *ne* is used optionally. Even though there is no second argument in intransitive predicates, case-marking is used to resolve a certain ambiguity allowed by the predicate semantics. In Hindi, the single argument of a small set of primarily 'bodily emission' verbs is optionally ergative. In such cases, ergative case-marking associates with an interpretation of the agent as being volitional, i.e. in *control* of the event the predicate refers to and as such more like a prototypical agent (Butt & King, 1991, 2003; Mohanan, 1994; Lee, 2003).

(15) *Raam=ne* *chiikh-aa*
Raam=ERG scream-PFV.SG.M.
'Raam screamed (purposefully)'

(16) *raam-∅* *chiikh-aa*
Raam-NOM scream-PFV.SG.M.
'Raam screamed'

As we have argued in Section 2, only the identifying function of case is in principle compatible with a differential case-marking pattern in the subject position of an intransitive clause. To sum up, it appears that both ergative subjects of perfective transitive sentences as well as ergative subjects of intransitive verbs such as *to scream* mark strong agents rather than weak ones. These subjects definitely do not need the ergative case marking to avoid ambiguity with respect to the object. This all strongly suggests that ergative case-marking in Hindi is a reflection of the identifying function of case.

In other words, it seems that we cannot account for the case marking patterns in Hindi only by appealing to the distinguishing role of case, but we have to take into account the identifying role of case as well. That is, ergative case-marking seems to fulfill an identifying role in Hindi, marking strong agents, hence coding prototypical agent properties such as volitionality. But as we will argue below, we cannot account for differential subject marking in Hindi by (directly) relating ergative case to volitionality. Rather, as we will argue, ergative case is reserved for strong agents in Hindi, but volitionality is not among the factors that make an agent strong in Hindi. Elsewhere we provided an alternative explanation for the relation between ergative case and volitionality, namely that it is an indirect relation that comes about as the result of bidirectional optimization (De Hoop & Narasimhan, 2004).

Let us begin our examination of the ergative case with the assumption that ergative case identifies (true, prototypical, strong) 'agentivity'. Agentivity may capture different types of properties, such as successful completion of the event, the presence of a patient, and volitionality or control. In transitive predicates, we find that a decrease in 'agentivity' in one way or another may result in the loss of ergative case marking, which is grammaticalized in specific construction types (an overview of the main predicate classes in Hindi is provided in the appendix). For instance, in 'inability passive' constructions (Mohanan, 1994) where the subject is unable to accomplish the action encoded by the verb, it is marked with the instrumental case *se* (17).

(17)	*Raam=se*	*kaam-∅*	*nahII*	*ki-yaa*	*ga-yaa*
	Raam=INS	work-NOM	not	do-PFV.SG.M.	go-PFV.SG.M.
	'Raam couldn't do the work'				

The ergative case-marking on the subject is also blocked with transitive predicates (e.g. *khaa* 'eat', *pii* 'drink') when the intransitive light verb *jaa* 'go' is added (18). This light verb is typically used with intransitive main verbs with a theme argument which undergoes change of location or state (19, 20). In (18), we may say that the cumulative effect of the highly transitive verb in combination with the intransitive verb is a decrease in agentivity, again blocking the use of ergative case in favour of the subject to be null-marked, even in perfective contexts.

(18)	*siitaa-∅*	*saaraa*	*khaanaa-∅*	*khaa*	*ga-yii*
	Siitaa-NOM	all	food-NOM	eat	go-PFV.SG.F.
	'Siitaa ate up all the food'				

(19)	*siitaa-∅*	*kamre=se*	*nikal*	*ga-yii*
	Siitaa-NOM	room=ABL	emerge	go-PFV.SG.F.
	'Siitaa left the room'			

(20)	*siitaa-∅*	*so*	*ga-yii*
	Siitaa-NOM	sleep	go-PFV.SG.F.
	'Siitaa went to sleep'		

Note that the objects in (17) and (18) can still alternate between null and *ko* marking, that is, both in the inability passives and in the *jaa* light verb constructions, reinforcing our point regarding the principled independence of the object case-alternation with a highly transitive verb.

(21)	*raam=se*	*mohan=ko*	*maar-aa*	*nahII*	*ga-yaa*
	Raam=INS	Mohan=ACC	beat-PFV.SG.M.	not	go-PFV.SG.M.
	'Raam could not beat Mohan.'				

(22)	*siitaa-∅*	*us*	*baDe*	*seb=ko*	*puuraa*	*khaa*	*ga-yii.*
	Siitaa-NOM	that	big	apple=ACC	all	eat	go.PFV.SG.F.
	'Siitaa ate up the whole big apple'						

To sum up the discussion so far, the examples in (17)–(18) and (21)–(22) suggest that a reduced agentivity blocks ergative case-marking in Hindi, independent of the case-marking of the object, thus in accordance with our initial assumption that ergative case-marking in Hindi cannot be accounted for in terms of the distinguishing function of case.

However, if we wish to argue that ergative case in Hindi is indeed used to identify (strong) agentivity, then we should be able to characterize the exact agentive property or set of agentive properties that is associated with ergative case. In the examples (5)–(6) above we saw that it was the perfectivity of a highly transitive predicate that triggered ergative case marking on the subject. In (18) and (22), however, the presence of a perfective highly transitive predicate is overruled by the influence of a certain type of light verb, such that ergative case-marking gets blocked. In (17) and (21), we might say that the agent has less control, and that this blocks the ergative case. Another factor which might influence ergative case-marking in Hindi, is volitionality. Note that the agent of an inability passive as in (17) and (21) may have less control, but must still have 'internally determined capability' (Pandharipande, 1981) to carry out an action; an independent measure of this property is the compatibility of a predicate which occurs in the inability passive construction to also occur felicitiously in imperatives (e.g. the verb *maar* 'hit' can occur in sentences such as *us = ko maaro* 'hit him'; however, a verb such as *dikh* 'glimpse', which does not entail an 'internally capable' subject, can occur in neither construction).

As we pointed out above, in Hindi, the single argument of a small set of primarily 'bodily emission' verbs (e.g. *khAAsnaa* 'cough', *bhAUknaa* 'bark', *chiikhnaa* 'scream', *muutnaa* 'urinate', *thuuknaa* 'spit', *muskuraanaa* 'smile') is optionally ergative (see examples (15)–(16) above). The case alternation in (15)–(16) above clearly performs an identifying role, coding a prototypical agent property such as volitionality. In some dialects of Hindi, the subject argument of 'desire' constructions can optionally receive ergative case-marking as well to imply volitionality (Butt & King 2002):

(22) *raam=ne* *ghar* *jaa naa* *hae*
Raam=ERG home go-INF be.PRS.3SG.
'Raam wants to go home'

When the subject is marked with *ko* (dative case), the sentence is ambiguous between a volition and obligation reading:

(23) *raam=ko* *ghar* *jaa-naa* *hae*
Raam=DAT home go-INF be.PRS.3SG.
'Raam wants to/has to go home'

The obligation reading suggests that the subject is non-volitional, or less in control, i.e. less of a prototypical agent; this might be argued to make the ergative case marking less adequate, and that is why this reading is only available with the dative case marker. Note, however, that the unmarked form (dative case) can be associated with both readings. Ergative case thus expresses the volitionality of the agent argument in the intransitive examples above, *but volitionality is not necessarily realized as ergative case*. This is also obvious when we consider other types of intransitive verbs, which do not optionally allow for ergative case irrespective of whether their subjects are volitional agents as with *bhaag* 'run' or not, as with *gir* 'fall':

(24) *mohan-∅* *ghar* *bhaag-aa*
mohan-NOM home run-PFV.SG.M.
'Mohan ran home'

(25) *aruNaa-∅* *zamiin=par* *gir-ii*
aruNaa-NOM ground=LOC fall-PFV.SG.F.
'Aruna fell on the ground'

So, in Hindi the role of the verb class in determining the case-marking patterns is quite strong (as also argued by Davison, 1999a, b). Similar to what we have observed with respect to differential object marking, differential subject marking in transitive clauses is restricted by the verb class and aspect, in the sense that agents of *perfective highly transitive* predicates are assigned ergative case. In a small class of intransitive predicates, ergative case is used to expresses the volitionality of the agent argument, but in general, volitionality of the agent is not realized as ergative case in Hindi. That is, volitional agents of imperfective transitive clauses do not bear ergative case, nor usually do volitional agents of perfective intransitive clauses.

Moreover, subjects of perfective predicates which are lexically characterized as highly transitive receive ergative case-marking, even if the subject is clearly non-volitional, in fact a patient rather than an agent (Davison, 1999b). This is shown in example (28), where the use of a highly transitive predicate, *khaa* 'eat', results in ergative case on the subject even when the subject in this idiomatic psych-predicate based on the verb *khaa* 'eat' has all the properties of a prototypical patient. That is, the verb *khaa* can occur as the light verb in a compound verb such as *maar khaa* 'suffer beatings' (lit. 'eat beating') or *dhokhaa khaa* 'suffer deception'. But even in such cases, the patient subject receives ergative case:

(26) *raam=ne* *apne* *shaitaanii=ke* *kaaraN* *bahut* *maar* *khaa-yaa*
Raam=ERG self bad.behaviour=GEN reason much beating eat-PFV.SG.M.
'Raam suffered much beating for his bad behaviour'

Examples such as (26) as well as the examples in (18) and (22) above might suggest that ergative case-marking on the subject is merely lexically determined by the (light) verb and that the semantic notion of agentivity of the subject is irrelevant. However, the lexical and semantic contributions of the light verb are not easily teased apart. For instance, the compound verb construction in (26) differs from other (psych–verb) constructions ((30)–(32)) in being able to occur in the progressive, compatibility with an adverb of duration, and (marginally) passivization ((27)–(29)):

(27) *raam-∅ bahut maar khaa rahaa hae*
Raam-NOM much beating eat PROG be.PRS.3SG.
'Raam is getting a lot of beating'

(28) *raam=ne ek ghaNTe=ke liye bahut maar*
Raam=ERG one hour=GEN for much beating
khaa-yaa
eat-PFV.SG.M.
'Raam got a lot of beating for an hour'

(29) *?raam=se aaj bahut maar khaa-yaa ga-yaa*
Raam=INS today much beating eat-PFV.SG.M. go-PFV.SG.M.
'Raam was beaten a lot today'

(30) **raam=ko seb-∅ pasand ho rahaa hae*
Raam=DAT apple-NOM liking be PROG be.PRS.3SG.
*'Raam is liking the apple'

(31) **raam=ko ek ghaNTe=ke liye maamlaa-∅ pataa hu-aa*
Raam=DAT one hour=GEN for matter-NOM know become-PFV.SG.M.
*'Raam got to know the matter for an hour'

(32) **raam=se seb-∅ pasand hu-aa ga-yaa*
Raam=INS apple-NOM liking become-PFV.SG.M. go.PFV.SG.M.
'The apple was liked by Raam'

Since not all psych–predicate constructions necessarily pattern in a similar way with respect to these tests, the evidence is not conclusive that *maar khaa* involves different agentive and aspectual properties than other psych–predicates. Nevertheless, it is suggestive of the fact that the light verb may not be mechanically assigning case to the subject owing to arbitrarily specified lexical information, but it is the interaction of (grammaticalized) predicate semantics and other semantic and aspectual features such as control and perfectivity, which eventually determines the degree of agentivity (and thereby the case-marking) of the subject.

In our analysis of differential subject marking, we assume that languages vary in which features determine argument strength (De Hoop, 1999). De Hoop (1999) argues that the mapping from semantic input to case morphology takes place in two optimization cycles, such that strength is determined at a separate (intermediate or hidden) level. As such, argument strength can be viewed as an abstract notion that is neither part of the actual meaning nor of the actual form. Nevertheless, in this chapter we represent argument strength as a feature of the abstract meaning, by using *A* for a strong agent and *a* for a weak one. In addition to factors such as perfectivity, verb semantics, and volitionality, the strength of the agent can be determined by additional sources, including, as discussed above, the strict dominance of the light verb in compound verb constructions, and reduced control (as in the inability passives).

We claim that in Hindi, it is perfectivity which determines the strength of the agent argument, while the factor of volitionality in Hindi is too weak to influence the assignment of ergative case (contra Lee, 2003, among others). Elsewhere, however, we accounted for the indirect influence of volitionality on ergative case assignment in Hindi (De Hoop & Narasimhan, 2004).

In the remainder of this section, we present our bidirectional Optimality Theoretic analysis of the differential subject marking patterns we find in Hindi transitive clauses. We assume, as we pointed out in the previous section, that argument strength is determined at an intermediate level between input (predicate and argument properties) and output (case-marking). As we mentioned above, we assume that in Hindi a perfective highly transitive predicate makes an agent strong.

The three relevant constraints are listed below:

(12) DISTINGUISHABILITY
The two arguments of a transitive relation should be distinguishable.

(33) IDENTIFICATION-A
Strong agents receive ergative marking.

(14) PAIP
"Primary Argument Immunity Principle": don't case-mark the unmarked — nominative — argument.

With the help of the three constraints formulated above, we can account for the nominative–ergative case alternation we find in the subject of transitive clauses in Hindi.

(34)	*Raam=ne*	*Shyaam=ko*	*maar-aa*			
	Raam=ERG	Shyaam=ACC	hit-PFV.SG.M.			
	'Raam hit Shyaam'					

(35)	*Ravii-∅*	*(ek)*	*gaay=ko*	*khariid-naa*	*caah-taa*	*hae*
	Ravi-NOM	one	cow=ACC	buy-INF	wish-IPFV. SG.M.	be.PRS.3SG.
	'Ravi wishes to buy the cow'					

Table 2: Bidirectional OT table: Form and meaning of the subject of a highly transitive verb.

Subject of highly transitive verb	IDENTIFICATION-A	PAIP	DISTINGUISHABILITY
☞ [∅, a]			
[*ne*, a]		*	
[∅, A]	*		
☞ [*ne*, A]		*	

The analysis is based on the assumption that in Hindi, as we pointed out above, in the context of a highly transitive predicate, perfectivity determines the strength of the agent argument in the input. We assume that DISTINGUISHABILITY is completely taken care of by the pattern of differential object marking (as pointed out in the previous section), so that this constraint is vacuously satisfied and does not play a role in differential subject marking (Table 2). Unlike its counterpart IDENTIFICATION-P, IDENTIFICATION-A is an influential constraint and it is even ranked higher than PAIP.

The first candidate pair combines an optimal form with the meaning *a*. Because this pair does not violate any constraints, it is super-optimal. The second candidate pair violates the economy constraint as it has a marked form. Therefore, this pair is blocked by the first one. The third candidate pair violates the constraint that requires a strong agent to be case-marked. Because the latter constraint is ranked higher than the economy constraint, this pair is blocked as well. The fourth candidate violates PAIP in order to satisfy IDENTIFICATION-A. Hence, this pair combines an optimal form with the meaning *A*. Because both the form and the meaning of the fourth pair differ from the form and meaning of the first super-optimal pair, this pair is super-optimal as well, i.e. a winner.

5. Conclusion

The class of highly transitive predicates in Hindi is distinguished by case-alternations on the object, between *ko* and null-marking. These are also the predicates that allow case-alternations on the subject, among *ne*, *se*, and null-marking. We can thus posit a partial dependency between subject and object case-marking in these instances. Case alternations on the subject and the object do not depend on predicate semantics alone, but also on grammatical aspect (for the subject) and nominal properties (for the object).

The main motivation for the overt case-marking of the object in Hindi is to *distinguish* argument roles (agents from patients) in transitive predicates. For the class of highly transitive predicates, there seems to be a general principle of preserving the relative distance between the subject and the object (in a hierarchy of argument strength) such that case-marking appears when the object becomes more agent-like (in particular, when it is animate and specific).

On the other hand, the motivation for overt case-marking of the subject cannot be to distinguish agents from patients. If it were, we would expect patient-like or 'weak' agents to be case-marked. By contrast, ergative case in Hindi transitive clauses is assigned to 'strong' agents. Moreover, we find ergative case-marking on some subjects of intransitive clauses as well. Therefore, we claim that the main motivation for the overt case-marking of the subject in Hindi is to identify (strong) agentivity of a subject.

We have proposed a model of case-marking based on two functions of case (the identifying and the distinguishing function of case) augmented with the notion of *argument strength*. We assume that languages can vary in which features determine argument strength. In Hindi, perfectivity of a highly transitive verb makes an agent strong. We argue that the interaction of the language-specific determination of argument strength, the two main functions of case, and an economy constraint that penalizes morphological case adequately characterizes the complex case-system that shows differential object marking as well differential subject marking, which is found in transitive clauses in Hindi.

Acknowledgements

We greatly benefited from discussions with and comments of Miriam Butt, Andrej Malchukov, Peter de Swart, Mengistu Amberber, Hanjung Lee, an anonymous reviewer, and various audiences at occasions where we presented a version of this chapter. Helen de Hoop gratefully acknowledges the Netherlands Organisation of Scientific Research (NWO) for financial support, grants 220-70-003 (PIONIER-project "Case cross-linguistically") and 051-02-070 (NWO-Cognition project "Conflicts in interpretation").

References

Aissen, J. (1999). Markedness and subject choice in optimality theory. *Natural Language and Linguistic Theory*, *17*, 673–711.

Aissen, J. (2003). Differential object marking: Iconicity vs. economy. *Natural Language and Linguistic Theory*, *21*, 435–483.

Bhatia, T. (1981). The treatment of transitivity in the Indian grammatical tradition. *Studies in the Linguistic Sciences*, *11*(2), 125–208.

Bhatt, R., & Anagnostopoulou, E. (1996). Object shift and specificity: Evidence from ko-phrases in Hindi. Paper presented at the 32th Regional Meeting of the Chicago Linguistic Society. Chicago.

Blake, B. J. (2001). *Case*. (2nd ed.) Cambridge: Cambridge University Press.

Blutner, R. (2000). Some aspects of optimality in natural language interpretation. *Journal of Semantics*, *17*, 189–216.

Bobaljik, J. D. (1993). Nominally absolutive is not absolutely nominative. In: J. Mead (Ed.), *Proceedings of the Eleventh West Coast Conference on Formal Linguistics*, Stanford, CSLI: 44–60.

Bornkessel, I. (2002). *The argument dependency model: A neurocognitive approach to incremental interpretation*. Leipzig: MPI-Series in Cognitive Neuroscience.

Butt, M., & King, T. H. (1991). Semantic case in Urdu. Papers from the 27th Regional Meeting of the Chicago Linguistic Society. Chicago (31–45).

Butt, M., & King, T. H. (2002). *The status of case*. Unpublished manuscript, University of Konstanz.

Butt, M., & King, T. H. (2003). Case systems: Beyond structural distinctions. In: E. Brandner, & H. Zinsmeister, (Eds), *New perspectives on Case Theory*. Stanford: CSLI Publications.

Comrie, B. (1989). *Language universals and linguistic typology*. Chicago, University of Chicago Press.

Davison, A. (1999a). Ergativity. Functional and formal issues. In: M. Darnell, E. Moravcsik, F. Newymeyer, M. Noonan, & K. Wheatley (Eds), *Functionalism and formalism in linguistics*. Amsterdam, John Benjamins pp. (177–207).

Davison, A. (1999b). Dependent structural case as a consequence of VP structure. Paper presented at the Perspectives on argument structure:1999 Texas Linguistics Society Conference.

Dixon, R. M. W. (1979). Ergativity. *Language*, *55*, 59–138.

Dowty, D. (1991). Thematic proto-roles and argument selection. *Language*, *67*(3), 547–619.

Hock, H. H. (1985). Transitivity as a gradient feature? Evidence from Indo-Aryan, especially Sanskrit and Hindi. *Proceedings of the Conference on Participant Roles: South Asian and adjacent areas, an ancillary meeting to the CLS regional meeting*. Chicago: Indiana University Linguistics Club.

de Hoop, H. (1999). Optimal case assignment. *Linguistics in the Netherlands 1999*. In: R. van Bezooijen en R. Kager (Ed.). Amsterdam, John Benjamins pp. (97–109).

de Hoop, H., & Lamers, M. (to appear). Incremental distinguishability of subject and object. *Case, valency, and transitivity*. In: L. Kulikov, A. Malchukov & P. de Swart (Eds), Amsterdam, John Benjamins.

de Hoop, H., & Narasimhan, B. (2004). Differential subject marking in Hindi. Paper presented at the PIONIER Workshop on Differential Subject Marking. Nijmegen.

Hopper, P. J., & Thompson, S. A. (1980). Transitivity in grammar and discourse. *Language*, *56*(2), 251–299.

Horn, L. (1984). Towards a new taxonomy of pragmatic inference: Q-based and R-based implicature. In: D. Schiffrin (Ed.), *Meaning, form, and use in context: Linguistic applications* (pp. 11–42). Washington: Georgetown University Press.

Kachru, Y. (1981). Transitivity and volitionality in Hindi. *Studies in the Linguistic Sciences, 11*(2), 181–193.

Kibrik, A. E. (1985). Toward a typology of ergativity. In: J. Nichols & A. Woodbury (Eds), *Grammar inside and outside the clause* (pp. 268–324). Cambridge: Cambridge University Press.

Lee, H. (2003). Parallel optimization in case systems. In: M. Butt & T. King (Eds), *Nominals: Inside and out*. Stanford, CSLI.

Legendre, G., Raymond, W., & Smolensky, P. (1993). An optimality-theoretic typology of case and grammatical voice systems, *Proceedings of the 19th Meeting of the Berkeley Linguistics Society*. Berkeley, Berkeley Linguistics Society.

Malchukov, A.L. (this volume). Case pattern splits, verb types, and construction competititon.

Malchukov, A.L. (to appear). Transitivity parameters and transitivity alternations: Constraining co-variation. In: L. Kulikov, A. Malchukov & P. de Swart (Eds), *Case, valency, and transitivity*. Amsterdam: John Benjamins.

Mallinson, G., & Blake, B. (1981). *Language typology: cross-cultural studies in syntax.* Amsterdam: North Holland Publishing.

Manning, C. (1996). *Ergativity: Argument structure and grammatical relations*. Stanford: CSLI.

Marantz, A. (1991). Case and licensing. *Proceedings of the Eastern states Conference on Linguistics'* 91, Ohio State University (pp. 234–253).

Mohanan, T. (1994). *Argument Structure in Hindi.* Stanford: Stanford University.

Næss, Å. (2004). What markedness marks: The markedness problem with direct objects. *Lingua, 114*, 1186–1212.

Pandharipande, R. (1981). Transitivity in Hindi. *Studies in the Linguistic Sciences, 11*(2), 161–179.

Silverstein, M. (1976). Hierarchy of features and ergativity. In: R. M. W. Dixon (Ed.), *Grammatical Categories in Australian Languages* (pp. 112–171). Canberra: Australian Institute of Aboriginal Studies.

Song, J. J. (2001). *Linguistic typology: morphology and syntax*. London: Longman.

de Swart, P. (2003). *The case mirror*. MA. thesis, University of Nijmegen.

Van Valin, R. (1992). An overview of ergative phenomena and their implications for language acquisition. In: D. I. Slobin (Ed.), *The crosslinguistic study of language acquisition*. (Vol. 3, pp. 15–39) Hillsdale, NJ: Lawrence Erlbaum.

Appendix

The main predicate classes in Hindi are listed in Table A1.

Table A.1

A	P	P properties	A properties	Grammatical aspect	Class	Examples of predicates
ne	***ko***	animate/specific		perfective	HT	*likh* 'write'
ne	***0***	elsewhere		perfective	HT	*dekh* 'see'
0	***ko***	animate/specific		imperfective	HT	*maar* 'hit'
0	***0***	elsewhere		imperfective	HT	*de* 'give'
se	***ko***	animate/specific		passive + negative	HT	*dikhaa* 'show'
se	***0***	elsewhere		passive + negative	HT	
ko	*0*			(im)perfective	MT	*mil* 'receive' *bhukh+lag* 'feel hungry'
0				(im)perfective	IV	*gir* 'fall'
0				(im)perfective	IV	*ciikh* 'shout' *khAAs* 'cough'
ne			+ volitional	perfective	IV	

Note: A and P are syntactically 'core' arguments; A controls binding of reflexives; is obviated by pronouns; controls PRO of participial clauses; gaps subjects in coordinate constructions; P sometimes passivizes, and determines verb agreement when it is the highest nominative argument in the clause (Mohanan, 1994); HT: high transitive; MT: mediotransitive; IV: intransitive.

Author Index

Subject Index

Language Index